NONDESTRUCTIVE TESTING

NONDESTRUCTIVE TESTING

SECOND EDITION

WARREN J. McGONNAGLE

Consultant in
NONDESTRUCTIVE TESTING

GORDON AND BREACH
Science Publishers
NEW YORK • LONDON • PARIS

First edition copyright © 1961 by The McGraw-Hill Book Company,
assigned to Warren J. McGonnagle. Second edition © Warren J. McGonnagle

Published by GORDON AND BREACH, Science Publishers, Inc.,
One Park Avenue, New York, New York 10016

Library of Congress Catalog Card Number: 60-9850
ISBN 0-677-00500-8

Gordon and Breach Science Publishers Ltd.
42 William IV Street
London WC2N 4DE

Gordon & Breach
58, rue Lhomond
75005 Paris

Printed in the United States of America

Second Printing 1969   (0050)
Third Printing 1971
Fourth Printing 1975
Fifth Printing 1977
Sixth Printing 1982

To my sons, James and Robert

PREFACE

The application of physical principles for detecting inhomogeneities in materials without impairing the usefulness of the material has brought into being a technique known as "nondestructive testing." The rapid growth in the use of nondestructive testing methods and techniques in the past few years has resulted from demands by industry for improved quality and especially by the critical requirements in the fields of jet aircraft, missiles, and nuclear energy. Every day more and more is being demanded of materials because materials are being subjected to service and environmental conditions never before encountered. To continue to make engineering advances, to meet the requirements of a rapidly advancing technology and the critical applications of tomorrow, nondestructive testing must and will be used to a still greater extent. Thus, the need and demand for the use of these methods and techniques will continue to grow at an even more accelerated rate. If this field of technology is to fulfill and achieve its full potential, not only more but also better nondestructive testing must be done.

There is a need for more and better trained personnel in the area of nondestructive testing. There is need for personnel who have received formal training in nondestructive testing fundamentals, who understand the physical principles on which the various test methods are based, and who in the application of the tests will not lose sight of the physical properties involved. It is essential that these personnel know the advantages, disadvantages, and limitations of the various methods and techniques.

One of the primary requirements in expanding the technology demanded for the solution of tomorrow's problems is education in the field of nondestructive testing. As yet few institutions of higher learning are offering formal courses in this field. This book was written in an attempt to fulfill the need for a textbook and to help orient the beginning student and technician in the field.

Since nondestructive testing is a technology based on applied physics, physical principles have been stressed. Nondestructive testing is a science developing in a rapidly changing technical society; consequently, the emphasis is on more and better physical measurements, with quan-

vii

titative information rather than qualitative information. It is hoped that this text will stimulate improvement in existing methods and techniques, inspire the development of new methods and techniques, and foster the evolution of new and more adequate instrumentation.

The term nondestructive testing is often considered to be concerned only with the detection and location of flaws. Actually, the methods and techniques used in nondestructive testing measure physical properties or nonuniformity in physical properties of materials. Variations or nonuniformities in physical properties may or may not affect the usefulness of a material, depending upon the particular application under consideration. These methods and techniques can be used to determine what variations or nonuniformities in properties can be tolerated in the anticipated service.

The author is indebted to many persons for advice and guidance, for technical information, and for illustrative material used in preparing this text. The author is particularly indebted to Dr. Stuart McLain for his continued encouragement and assistance. He is also indebted to the ASTM and its staff, especially Mr. Fred Van Atta, for their cooperation in supplying illustrative materials. Many stimulating and interesting discussions with C. J. Renken, R. G. Myers, R. H. Selner, R. A. diNovi, and R. G. Peterson of the Nondestructive Testing Group at Argonne National Laboratory, and W. G. Marburger, Professor of Physics at Western Michigan University, Kalamazoo, Michigan, are gratefully acknowledged. Finally, he thanks his wife, June D. McGonnagle, for help in the preparation and critical reading of the completed manuscript.

Warren J. McGonnagle

CONTENTS

ix

CHAPTER 1

INTRODUCTION

During the past few years the industrial world has witnessed an amazing growth in the use of nondestructive testing methods and techniques. The demand for and use of nondestructive testing will continue to grow at an even more accelerated rate. For example, jet airplanes are operating at higher and higher speeds, jet and conventional airplane engines are developing greater horsepower per pound, and pressure vessels and pipes are operating at higher pressures. As our technology advances further, there will be a demand and need for new performance from engineering materials. To continue to make engineering advances and to meet the needs of the world's advancing technology, nondestructive testing methods and techniques must find wider application and new methods and techniques must be developed. Our expanding technology is demanding not only more but better nondestructive testing.

The purpose of this book is to acquaint engineers, metallurgists, and other technical personnel both practicing and "to be" with the (1) basic principles of nondestructive testing, (2) kind of results which can be expected when using these tests, and (3) typical applications.

The term "nondestructive testing" is used in this book as a general name given to all test methods which permit testing or inspection of material without impairing its future usefulness. From an industrial viewpoint, the purpose of nondestructive testing is to determine whether a material or part will satisfactorily perform its intended function.

It must be realized that the desired properties or qualities must be built into a product; they cannot be inspected into it. The primary purpose of a nondestructive inspection is to determine the existing state or quality of a material, with a view to acceptance or rejection. By use of nondestructive testing methods and techniques it has been possible to decrease the factor of ignorance about material without decreasing the factor of safety in the finished product. Absolute, perfect, and sound industrial material does not exist. Any correctly applied nondestructive test can tell only whether the relative soundness of a specimen lies within specified tolerances. The use of nondestructive testing has been and is being more fully recognized by management as a means of meeting consumer demands for better products, reduced cost, and increased production.

1

The art and science of nondestructive testing are very old. Probably one of the most famous and well-known examples is that of Archimedes and Hiero's crown. In performing a test to determine if the king had been defrauded by the silversmiths, Archimedes discovered the principle that now bears his name. The art of nondestructive testing is used in many fields of endeavor without even being considered in the realm of nondestructive testing. For example, the fruit vendor who can tell if a watermelon is ripe by "thumping" or if a canteloupe is ripe by shaking and listening for the "rattle" of the seeds is using nondestructive testing.

Test Methods. Nearly every form of energy has been utilized in nondestructive testing. Likewise nearly every property of the materials to be inspected has been made the basis for some method or technique of nondestructive testing. For the purposes of this book all these test methods have been divided into 10 groups as follows:

Visual   Acoustic (sonic and ultrasonic)
Pressure and leak   Magnetic
Penetrant   Electrical and electrostatic
Thermal   Electromagnetic induction
Radiography (X ray and gamma ray)   Miscellaneous

In the visual method, the test specimen is illuminated with light, usually in the visible region. The specimen is then examined with the eye, aided or unaided, or with other light sensitive devices such as a photocell. Visual examination is sometimes not considered as a nondestructive testing method. However, the author considers it as one of the important tests. Not only is it the oldest and simplest test, but it sometimes reveals defects that are not readily detected by other tests. Pressure and leak tests are characterized by the flow of a liquid or gas into or through defects. Penetrant testing can be used to find defects only open to the surface of the specimen. A liquid having low viscosity and surface tension enters into defects by capillary action and is then partially removed from the defect by the blotting action of the developer. In thermal tests the test specimen is heated and the resulting temperature distribution revealed by heat sensitive chemicals, phosphors, thermocouples, thermometers of various types, and infrared sensitive cells.

In radiography X-ray and gamma radiation the so-called "penetrating radiation" is used. Variations in thickness and density modify the passage of radiation through the test specimen. This variation in the intensity of the transmitted radiation can be detected in a variety of ways, by use of film, semiconductors, photoconductors, and scintillation crystals. The information on radiography has been divided into two chapters, one dealing primarily with X-ray radiography and the other dealing with gamma-ray radiography. In the latter chapter a section on personnel protection has been included. Likewise the information and material on acoustic testing has been divided into two chapters. One chapter deals primarily with acoustic energy above the audible range—ultrasonics. The other chapter deals primarily with frequencies in the audible range and is entitled Dynamic Testing. In magnetic testing the specimen is magnetized, and a distortion of the magnetic field is produced by defects. The magnetic field distortion can be measured in a number of ways; the most widely used employs "magnetic particles" such as iron oxide. Electrical methods of nondestructive testing depend primarily on the differences in electrical resistivity. In the electrostatic method, variations in the static electric field are detected by use of charged particles. When a coil carrying high-frequency alternating current is brought into the neighborhood of a conductor, eddy currents are induced in the conductor. There is a magnetic field associated with the induced currents. Flaws as well as physical, chemical, and other structural variations cause resistivity changes in the specimen and affect the induced eddy currents and consequently the induced magnetic field. These variations in the induced magnetic field can be detected by their effect on the coil inducing the currents or by a search coil in the neighborhood of the specimen.

A chapter has been included to cover gaging the thickness of materials and the thickness of coatings or claddings. Both contact and noncontact gages are discussed here. Radiation gages are probably the best known and most widely used. There are a number of nondestructive tests which are of interest and use, but which do not warrant such a detailed discussion to require a separate chapter each. Consequently, such tests as spot tests, spark tests, and stress and such instruments and techniques as Stresscoat, strain gages, and surface analysis equipment are included in a chapter entitled Other Useful Testing Techniques.

Selected bibliographies of pertinent literature are given at the end of each chapter. These comprise General References, Specific References, and Additional References, all of which, however, may not be cited for every chapter. Specific References are indicated in the text by superscript numbers.

Components of Tests. The essential parts of any nondestructive test are (1) *application* of a testing or inspection medium, (2) *modification* of the testing or inspection medium by defects or variations in the structure or properties of the material, (3) *detection* of this change by a suitable detector, (4) *conversion* of this change into a form suitable for interpretation, (5) *interpretation* of the information obtained. In some cases, there may be a sixth step to consider, that is, convincing the supplier, fabricator, or engineer that the interpretation of the test results is correct. Interpretation is the most important step in the procedure. In succeeding chapters it will be shown that most nondestructive testing results are indirect measurements. For example, these include deflections or readings on a meter, deviations in a recorder trace, dark or light areas on an X-ray film, the appearance of a pip or a change in the pattern on an oscilloscope screen, or a change in the color of a chemical or phosphor. Consequently, it is essential that the interpretation be made by an experienced person. The person interpreting the results sometimes determines the success or failure of a test method or technique.

Although most of the tests are simple in principle, the choice of method, the techniques used, and the interpretation of the results require skill and experience. The method of application of a nondestructive testing method or technique can and will produce varied results. Incorrect test methods or techniques will lead only to wasted effort, rejection of perhaps acceptable material or acceptance of defective material, and condemnation of the test method or technique and perhaps the whole field of nondestructive testing. It is the author's opinion that a test wrongly applied is worse than no test because it may give a false sense of security. Instead of decreasing the factor of ignorance, the factor of ignorance has been increased and the factor of safety decreased. Just to acquire the necessary instruments or the tools to make a test is not sufficient; they must be intelligently applied. Likewise a serious mistake is made if an instrument which works satisfactorily and adequately for one testing problem is applied "across the board" in an attempt to solve all the problems which are presented.

The five essential parts of a nondestructive test listed above are reviewed in the two following examples. If a piece of paper is inspected by holding it between a source of light and the eye, then (1) visible light is the testing or inspection medium, (2) the intensity of the transmitted light is modified by any flaws in the paper or by any markings on the paper, (3) the human eye is the detector and can determine variations in the intensity of the transmitted light, (4) the nervous system converts the intensity variations observed by the eye into impulses, and (5) the brain interprets these impulses. In the case of X-ray film radiography, (1) the X rays are the testing or inspection medium, (2) any defects in the material being radiographed will modify the intensity of the radiation reaching the film on the opposite side of the specimen, (3) certain silver bromide emulsions are sensitive to X rays and can be used as a detector, (4) the emulsions are capable of recording variations in X-ray intensity and by the proper developing procedures can be made to give a permanent record, and (5) interpretation is then a process of explaining variations in density of the radiograph.

Verification of Results. In developing nondestructive testing techniques, a destructive test must be performed to verify and confirm the results obtained. During routine testing, periodic destructive tests should and must be made to ascertain that the test and the associated instrumentation are operating properly and at the desired sensitivity and resolution. It should be pointed out that valuable information can be obtained from visual examination made during destructive testing or from a broken or damaged part.

It is also important to make periodic checks on the test operators, especially if a visual type of indication is used. For example, during World War II it was found in a plant using the magnetic particle technique that a certain inspector early in a shift was observing all the indications given by defective materials, even finding some defects not considered detrimental. Later in the day because of fatigue and boredom, this same operator was not finding the magnetic particle indications, and thus permitted defective material to be installed in critical assemblies. In visual inspection it is absolutely essential that there be adequate illumination, and that it be glare-free and free from dazzle and specular reflections. Even then, fatigue and boredom are very serious problems.

The fact that workmen such as welders know that their work is to be subjected to a nondestructive test helps produce improved quality. For example, during World War II all the welds on a completed naval vessel were radiographed, and approximately 40 per cent of the welds were found to be defective. Because of this experience, it was decided that all welds on this type of naval vessel must be radiographed. One vessel was partly completed when this order was issued. In this vessel 15 per cent of the welds were found to be defective. In welds made after this order was issued only 2 or 3 per cent were found to be defective.

Nuclear Energy Requirements. In the nuclear energy field, great emphasis has been put on the use of nondestructive tests. Nondestructive testing of fuel elements for heterogeneous reactors and of the components of the reactor system, such as the heat-exchanger tubes, differs from ordinary industrial nondestructive testing in that the specifications for reliability in the nuclear field are much more rigid than those which have been used until recently in other industries. The sensitivity and resolution required of the tests are in excess of those which can be justified in most other industrial applications. The consequences of fuel element failures in heterogeneous reactors or leaks in any reactor coolant system are great enough to justify relatively high expenditures in time, money, and manpower for nondestructive testing. Consequently, considerable effort has been expended in the development of various types of test procedures. The application of the nondestructive testing techniques has greatly reduced component failures of the fuel elements and other components in the various reactors in operation. The methods and techniques developed have many industrial applications.

Radiography used in nuclear energy basically involves only the application of industrial radiography to different materials and with increased sensitivity. Industrial inspection codes generally have been based on the specification of the size of the defect that must be observed radiographically. Such codes are developed by relating destructive tests, performance, and strength of materials calculations. For nuclear materials, however, industry-wide codes do not yet exist generally because of lack of operating experience in reactor components and insufficient experience with the new developments in reactor materials. Also, the basis for inspection differs from that of other industries. In the metals industries, radiography is used mainly to find internal defects that affect the strength of the specimen inspected. In the nuclear industries, radiography is used in addition to find conditions that would affect nuclear reactions, heat transfer, and corrosion. This, generally, is not compatible with a code based on a per cent thickness specification, but rather on the best radiography attainable. The conditions thus observed often are the basis for specifications. Also, radiographic inspection reports state the smallest defect of interest that the inspection technique detects, rather than a per cent sensitivity.

Benefits from Nondestructive Testing. The contribution which non-destructive tests can make to industry may be divided into four categories: increased productivity, increased serviceability, safety, and identification of materials. The benefits which can be derived from nondestructive tests include the following:

1. Increased productivity and profits
   a. Prevent wastage of material; less scrap; better use of raw material
   b. Prevent wastage of manpower; less rework needed
   c. Prevent wastage of shop time—time which can never be replaced
   d. Prevent divergence from standard of quality; better quality level and better uniformity of quality; customer satisfaction
   e. Lowered servicing costs
   f. Lowered operating and production costs
   g. Process control and improvement—monitor manufacturing processes, pinpoint cause of trouble, and find cure
   h. Salvage of material; some defective material may be usable
   i. Guide in engineering design
   j. Nature and location of abnormalities
   k. More efficient use of equipment
   l. Find unacceptable material
   m. Proper processing
2. Increased serviceability
   a. Locating regions of mechanical stress
   b. Locating fatigue failures
   c. Prevent malfunctioning of vital equipment
   d. Eliminating breakdown of equipment
   e. Lowered servicing and operating costs

3. Safety
   a. Preventing accidents
   b. Preventing loss of life
   c. Preventing loss of property
4. Identification
   a. Sorting
   b. Differences in chemical composition
   c. Differences in heat-treating
   d. Differences in physical properties
   e. Differences in metallurgical properties

By detecting faulty material and thus preventing loss of material, manpower, and shop time nondestructive tests will increase productivity, and with the increased productivity will come economic gains. Nondestructive tests can be used as an aid in new process and manufacturing techniques. Preventative maintenance tells if parts are still satisfactory for use; it pays off in dependable predictable production, fewer repairs, less accidents, and lower over-all operating costs. Increased serviceability of equipment and material will result through the application of nondestructive testing methods and techniques, by finding and locating defects which may cause malfunctioning or breakdown of equipment. In the field of safety proper use of nondestructive tests will aid in the prevention of accidents, with their possible loss of life, property, and vital equipment. The identification of materials differing in metallurgical, physical, or chemical properties can often be done by using nondestructive testing methods. The correlation of various kinds of defects and their effect on serviceability of a part or component has not been investigated as fully as needs be. Some work is being done in this area, but much more is needed.

Types of Defects. The type of defects that nondestructive testing is called upon to find can be classed into three groups:

1. Inherent defects—introduced during the initial production of the base or raw material
2. Processing defects—introduced during processing of the material or part
3. Service defects—introduced during the operating cycle of the material or part

Some of the kinds of defects or structural variations which may exist in these three groups are: cracks, surface and subsurface, arising from a large number of cases; porosity; tears; machining, rolling, and plating defects; laminations; lack of bond; inclusions; segregation; lack of penetration in welds; pipe; fatigue defects; seams; blowholes, dross shrinkage, flakes; pitting; and laps.   There may also be variations in grain size, differences in heat-treatment, and variations in chemical composition.

A number of defects are considered of special significance in nuclear reactor design, construction, and operation and in the atomic energy field in general.   Among these are the following: lack of bonding (liquid-metal or metal-metal bonds) between cladding and core; thick or thin spots in the clad; cracks and holes in the cladding; porosity, inclusions and lack of penetration in welds; porosity, pipe, seams, and laps in the core material. Metallurgical properties such as grain size, grain orientation, and heat-treating, chemical properties such as variation in carbon content, and certain physical properties are often of special significance in reactor design, construction, and operation.   Many of these properties or variations in these properties can be found by use of nondestructive tests.

It must be realized that all commercial materials are not perfectly free from certain common defects or are not homogeneous even under the best conditions of manufacture.   The degree of purity and the absence of surface and subsurface defects are only relative, because of the fact that the modern processes of metal reduction from ores and the mechanical and thermal treatment are not perfect.   The quality of material required depends upon the application; for example, it is obvious that the quality of steel for tools or for rapidly rotating parts of machinery should be higher than that of structural steel.   Exact knowledge of actual conditions of service should always be a guide to the quality of material required or selected.   To draw a line between permissible deviations from the normal and those which are undesirable for various applications, the reader should consult the standard specifications of the American Society for Testing Materials, the American Standards Association, and the specifications issued by the United States military establishments.

Segregation.   The distribution of the chemical elements in alloys is not always uniform.   For example, in ingots and castings some regions in an alloy may be enriched in certain elements while other regions are improverished in these same elements.   These local deviations originate from conditions during solidification of the alloy.   While the whole melt is liquid, all components are more or less uniformly distributed, but when solidification begins, the melt starts to precipitate first the most refractory portion of the solute followed by the less refractory portions; consequently the composition of both phases, the solid as well as the liquid, changes every moment and remains different.   Diffusion tends to equalize the compositional differences set up in the solid, but does not take place at the speed of differential freezing so that variations in composition remain in the solid solution form.   Desirable properties in commercial alloys can be attained only with uniform distribution of all chemical elements, because the physical properties in these areas will differ considerably from the normal properties of material having an average composition or will vary from point to point along the casting or ingot. Certain elements such as sulfur and phosphorus in steel are especially harmful. Corrosion may be accelerated by local differences in chemical composition.

Blowholes. Blowholes are cavities in metals caused by gas entrapped during the solidification process. In the molten state, metals are capable of dissolving large volumes of gases. When the temperature is lowered to the freezing point, a rapid decrease in solubility occurs and the gases are liberated from the liquid metal. If liberation of the entrapped gases occurs at a time when the metal is almost completely solidified, all the gas cannot escape and some of it is held in cavities which form during solidification. In castings, the origin of gases may also be traced to the moisture in the sand. The most common shape of blowholes is oval, but when they occur in the outside zone of ingots their shape is extended perpendicular to the walls of the mold and they assume an oblong shape. The size of blowholes may vary considerably. The location of blowholes in ingots is of great practical importance. When the holes are on the surface, the cavity walls will be exposed to the oxidizing action of the air, and once oxidized they cannot be welded by rolling and may cause serious surface defects. Deep-seated holes are comparatively harmless. An advantage of rimming steel is that a solid steel shell free from holes protects the metal from the harmful effects of superficial porosity.

Another type of porosity found in metals and alloys is pinholes. These are very small holes visible only under the microscope in polished and etched sections of the metal. Like blowholes, they are produced by evolution of gas. It is probable that pinholes originated in minute gas bubbles during the last stages of solidification and had no chance to accumulate into larger bubbles.

The term "pinholes" is also used to designate small blowholes, Fig. 1-1, found in large numbers on the surface of metals and not visible to the unaided eye. Since under the microscope they have exactly the same appearance as blowholes of larger size and are of the same origin, there is no necessity for a special term. Pinholes have their own distinctive characteristics: they are extremely small in size, have regular

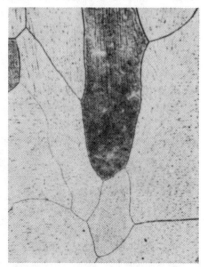

FIG. 1-1. Pinholes in cast copper. (*Reproduced from E. P. Polushkin, "Defects and Failures of Metal, Their Origin and Elimination," Elsevier Publishing Company, Amsterdam, 1956.*)

arrangements within the grains, and are visible only after etching. The term "microporosity" is occasionally used to cover large accumulations of small blowholes on the metal surface. The harm caused by porosity depends chiefly on the position of blowholes in the metal and of course on their size and number. Blowholes reduce the useful section of the cast piece of metal and lower its strength in a disproportionately greater degree because of the notch effect. Since blowholes are concealed from direct observation in castings, castings are less reliable in important applications than rolled raw metals unless thoroughly examined for porosity. Blowholes inside steel ingots may be tolerated or even desired, for they reduce the size of the pipe.

Pipe. The majority of metals and alloys contract during solidification and in cooling from melting point to room temperature. The amount of linear contraction in steel, for example, is about 2 per cent. When a liquid metal is poured into a metallic mold, the heat is taken away rapidly by the walls of the mold and a thin shell of solid material is formed almost immediately. The solid shell grows thicker through further deposits of the solid, and when the liquid metal in the center of the mold begins to freeze it tries to pull the shell, but the latter being already rigid resists deformation, and a rupture occurs throughout the last portion solidified, forming a cavity. At the top of the ingot the pipe is wider and has a funnellike shape, but generally it has an irregular shape. The region around the pipe is always porous. This porosity differs from the porosity caused by blowholes both in origin and appearance. The porous dendritic structure, shown in Fig. 1-2, is characteristic for the metal surrounding the pipe. The detrimental effects of piping are self-evident, for the surface of the pipe is oxidized by air and cannot be welded and rolled. Deep-seated pipe with clean walls will be closed up under the pressure of rolls. The presence of slag in the pipe may interfere with satisfactory welding. Pipe entails a great waste of metal, and a considerable portion

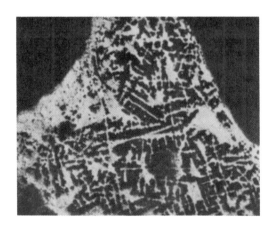

Fig. 1-2. Structure of metal surrounding pipe. (*Reproduced from E. P. Polushkin, "Defects and Failures of Metal, Their Origin and Elimination," Elsevier Publishing Company, Amsterdam, 1956.*)

of an ingot may have to be discarded. Cracks may start from the remote branches of the pipe if they are not properly welded.

Inclusions. Industrial materials always contain impurities which affect the properties of the material. Impurities may be present in the metals either in solution with the metal or as separate mechanically held particles, called "inclusions." Colloidal or emulsified inclusions represent an intermediate type. Inclusions of ferric alloys are divided into three groups: oxides, sulfides, and silicates. There are also complex intermediate compositions. The characteristic inclusions such as oxides and slag can be used to identify commercial metals if the chemical composition of these inclusions is known. Distribution of inclusions in metals is seldom uniform, being mostly concentrated in the central region of ingots or castings because of segregation. In rolled metals, inclusions are strung out in the direction of mechanical work, and if they form continuous lines the transverse strength of the metal is lowered. Inclusions may decrease ductility, impart brittleness, and decrease the strength of the metal. When two contacting surfaces are in rapid motion such as ball bearings, inclusions are readily spalled; for this reason steel for balls and raceways of ball bearings should be exceptionally clean and free from nonmetallic inclusions. In wrought metals where inclusions are aligned in more or less continuous rows, this happens not only in rolling, but also in other kinds of mechanical work. Such an arrangement of inclusions may lead to the formation of cracks. In heat-treated steel, inclusions are objectionable because they may start quenching cracks. Inclusions in metal should be considered as internal notches which localize stresses and thus lead to premature fatigue failure. Inclusions may interfere with uniformity of case-hardening, affect the strength of welds, and cause machining and polishing difficulties. Steels containing a large number of inclusions are not suitable for rails because the working surface of a rail soon becomes pitted by the action of heavy pressure and atmospheric corrosion. In some commercial metals, nonmetallic inclusions produced beneficial effects. For example, they may facilitate machining, and when greater speed of machining is desired inclusion-forming elements are purposely introduced in the alloy, as in the case of sulfur added to low carbon steel or lead added to brass or steel.

Scale. The layer of oxide formed on the surface of a metal heated to high temperatures in air or in some other oxidizing atmosphere is called "scale." The thickness of the oxide layer depends on the nature of the scale, its structure, chemical composition, and melting point. The scale may consist of several oxides, giving a composition gradient.

Stresses. Stresses remaining in metals after cold-working or rapid cooling are called "residual" or "internal" stresses. Such stresses are referred to as "macrostresses" when they extend through a relatively large portion of the metal and "microstresses" if confined to the individual grains.   Each stress is identified by its magnitude and direction.   The nature of stresses and the fact that they shift with any change in loading or with temperature make them especially difficult to analyze and measure.   The designers of machines and structural members often consider only the magnitude of stresses that are possible in service and overlook the residual stresses which may be incurred during the fabrication of the metal, although the latter may exceed the service stress.   It should be mentioned that the presence of residual stresses in metals is not always detrimental.   For example, there is a large group of commercial products in which the stresses are induced intentionally to improve their mechanical properties.   Cold-working in general increases the tensile strength and yield point of metals and alloys and if carried out in moderate degree will not reduce ductility to such an extent as to impair toughness.   Cold-drawn bars and cold-rolled shafts and axles are examples of products in which residual stresses are desired.   The surface of structural materials is compressed by shot peening.   This produces the beneficial effect of counteracting destructive tensional stresses and thus preventing rupture.   In all useful applications the intensity of residual stresses must be below the dangerous range leading to rupture.   In welded structures such as ships and buildings, residual stresses are harmful and their presence does not serve any useful purpose, but decreases strength. They are especially undesirable in metals subjected to corrosion.   Some of the conditions which give rise to residual stresses include quenching stresses and forging of steel with too small a hammer since the extension of the metal in the inside and the outside portions of the ingot will not be at the same rate.   Other conditions include too low or nonuniform heating of the forging, cold-working, machining if a blunt tool is used, notches, chaffing, and fretting.   Cold-working-induced stresses can be of two kinds: elastic through a large portion of the material, and plastic microstresses within the deformed grains.   Residual stresses produced by cold-working change the physical properties of metals, increasing their tensile strength, yield point, and hardness and will lower ductility. Unequal heating of hot pieces of air-hardened steel will develop residual stresses which can produce warping.

The presence of residual stresses in metals can be revealed by several methods.   However, accurate measurement of their magnitude is extremely difficult and possible only in exceptional cases.   The chief cause of difficulty in accurate measurements is due to the triaxial nature of stresses, since in experimental procedures the stress can be determined in one direction only.   The mechanical operations involved in the process of determining the stress may introduce additional stresses.

Fatigue.   Failure of metals under repeated stresses is known as "fatigue."   The maximum stress that a material can withstand failure for a specific number of cycles of stress is called the "fatigue limit" or "endurance limit." Fatigue limits of carbon and alloy steels usually amount to 35 to 55 per cent of the tensile strength. Examination of many steel parts which have failed in fatigue has shown that the cause of failure can be attributed to the following conditions: an excessive number of inclusions; internal fissures; improper design; local concentration of stress at holes, grooves, and sharp changes in contour; surface scratches; and overstraining during service by high vibration stresses.

Flakes. The 1948 "Metals Handbook" defines flakes as internal fissures in ferrous metals. In a fractured surface these fissures may appear as sizable areas of silver whiteness in coarse texture; in wrought products this fissure may appear as short discontinuities on an etched section. Figure 1-3 show typical flake.

Cracks. Cracks, both surface and subsurface, constitute a major source of defects. Two types of cracks are considered here, actual and potential. The latter are not visible even under the microscope but may be suspected from indirect evidence. Either they are too small to be seen under the highest magnifications of the optical microscope, or they are not entirely open. Microscopic cracks are sometimes referred to as hairline cracks or microcracks. Investigation into the cause of cracking is important not only from the scientific point of view but also for practical purposes where it is necessary to place responsibility for failures of metals in service and also for failures revealed during inspection of metal products. Numerous factors may be responsible for cracks, and a detailed examination of any evidence related to the cause is necessary in order to determine the actual cause.

A general description of the appearance and the origin of the cracks may be helpful. Cracks may be transcrystalline or intergranular. Quenching cracks usually belong to the latter type. In some cases the path of rupture partly intersects the grain and partly follows grain boundaries. Cracks may propagate in a great variety of directions. Likewise, cracks may occur in a great variety of locations. The interior space of cracks may be empty, filled with oxidation products, or filled with extraneous material. Some of the common types of cracks and their

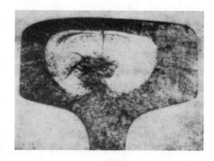

Fig. 1-3. Flake.
(*From E. P. Polushkin, "Defects and Failures of Metal, Their Origin and Elimination," Elsevier Publishing Company, Amsterdam, 1956.*)

causes are listed as follows: quenching or hardening cracks caused by rapid volume changes, tempering cracks caused by rapid heating, shrinkage cracks caused by too rapid cooling, hot tears caused by improper design of mold and faulty method of pouring, and grinding cracks caused by local heating by friction of an emery wheel. Cracks can also be caused by residual stresses, large reduction during cold-working, improper forging, laps, low melting inclusions, segregation, faulty design, improper rolling, collapsed blowholes, sharp edges on die stamp, and etching. Among the surface-type defects are cold shuts, laps, folds, seam, hairline cracks, and scratches.

Standardization. Improvement is needed in the present techniques for evaluating discontinuities and flaws in raw material and in fabricated parts. In order to do this it is necessary to standardize the approach to nondestructive testing, the test procedures, the interpretation of the results, and standards of acceptance and rejection. Having standardized the above it is most important that the training of persons engaged in shop and field inspection be standardized. The complaint is sometimes heard that nondestructive tests are "operator sensitive." The kind of standardization mentioned above should help to eliminate this complaint. Nondestructive testing specifications must state exactly what is desired and required. Then manufacturers and fabricators can carry out the necessary procedures to produce the material or part to the desired specifications.

Even after many years of using radiography, magnetic particle and fluorescent or dye penetrant inspection, there is no universal "go-no-go" standard because it is difficult in many cases for the various groups concerned to reach a common agreement. Therefore, many disagreements are to be expected, particularly in the newer fields of nondestructive testing and inspection in which at present there are no references as to acceptability. It is well to point out that finding of defects and setting of standards are two distinct processes.

Summary. One of the most difficult problems that the nondestructive testing engineer often encounters is that the problems to be solved often are not too well defined. So it becomes necessary to evaluate carefully the testing problem presented by ascertaining what quality or characteristics are to be found. In order to obtain this information, it is often necessary to reverse the process and ask questions of those who proposed the problem. Then and only then can a decision be made as to what test method and technique will furnish the necessary answers most expediently and most reliably. Selection of methods and techniques can be decided only by experience and an understanding of the principles involved.

Nondestructive testing is still in its infancy. If it is going to fulfill the needs of our advancing technology, old methods and techniques must be improved, new methods and techniques developed, and adequate instrumentation evolved. In order to do this, basic research and development must be carried out and accelerated. In addition, the inspection process must be speeded up and made less operator dependent. This means that the inspection processes must be automated and have some judgment or intelligence built into them. In general, nondestructive tests are qualitative. There is a great need for quantitative information and the correlation of nondestructive test results with performance, service, mechanical test results, and destructive test results.

GENERAL REFERENCES

American Society of Mechanical Engineers: ASME Handbook, "Metals Engineering —Design," McGraw-Hill Book Company, Inc., New York, 1953.

American Society for Metals: "Metals Handbook," American Society for Metals, Cleveland, Ohio, 1948.

American Society for Metals: "Metals Handbook 1954 Supplement," American Society for Metals, Cleveland, Ohio, 1954.

American Society for Testing Materials: Symposium on Nondestructive Tests in the Field of Nuclear Energy, *ASTM, Spec. Tech. Pub.* 223, 1958.

Davis, Harmer E., George Earl Toxell, and Clement T. Wiskocil: "The Testing and Inspection of Engineering Material," McGraw-Hill Book Company, Inc., New York, 1941.

Hanstock, R. F.: "The Nondestructive Testing of Metals," The Institute of Metals, London, 1951.

Lewis, D. M.: "Magnetic and Electrical Methods of Nondestructive Testing," George Allen & Unwin, Ltd., London, 1951.

McGonnagle, W. J.: Nondestructive Testing of Reactor Fuel Elements, *Nuclear Sci. and Eng.*, vol. 2, no. 5, pp. 602–612, September, 1957.

McGonnagle, W. J., Stuart McLain, and E. C. Wood: Applications of Nondestructive Testing to Fuel Elements for Nuclear Reactors, *J. Soc. Nondestructive Testing*, vol. 15, no. 2, pp. 86–90, March–April, 1957.

Polushkin, E. P.: "Defects and Failures of Metals," Elsevier Publishing Company, Amsterdam, 1956.

Society for Nondestructive Testing: "Handbook of Nondestructive Testing," The Ronald Press Company, New York, 1959.

CHAPTER 2

VISUAL TESTING

Visual inspection is probably the most widely used of all the non-destructive tests. It is simple, easy to apply, quickly carried out, and usually low in cost. However, because of its simplicity, visual examination should never be ignored. Even though a specimen is to be inspected using other nondestructive testing methods, it should be given a good visual examination. For example, visual examination of a completed weld by an experienced inspector can reveal the following information about the quality of the weld: the presence or absence of cracks, orientation and position of cracks relative to the various zones in the weld, surface porosity, unfilled craters, contour of the weld beads, and the probable orientation of the interface between the fused weld head and the adjoining parent metal. Other observations include the extent of penetration, the presence or absence of oxide film inclusions near the surface, undercutting, and potential sources of mechanical weakness such as sharp notches or misalignment. In addition, the results of visual examination may be of great assistance as a guide in other tests. For example, visual examination of the weld bead may aid in deciding upon the angle of incidence of an X-ray beam when examining the weld for cracks not visible at the surface. As in the case of all nondestructive tests, proper application and correct interpretation of the results are essential to their usefulness and success.

The basic principle used in visual nondestructive tests is to illuminate the test specimen with light, usually in the visible region. The specimen is then examined with the eye or by light sensitive devices such as photocells or phototubes. The equipment required for visual inspection is extremely simple, but adequate illumination is absolutely essential. The surfaces of the specimen should be adequately cleaned before being inspected. Sandblasting or shot blasting may be required for properly preparing the surface. Surface preparation is especially needed when inspecting heavy plate as adhering mill scale often hides defects such as scale pits, rough areas, and surface laps.

The Eye. As a registering device, the eye is "notorious for lack of accuracy." Vision is a variable thing considered from the viewpoint of a single individual and is an even more variable quantity when con-

16

sidered from the viewpoint of a number of individuals. This is due to variations in the eye itself as well as to variations in the brain and nervous system. It is well known that in certain so-called "optical illusions" the surroundings of an object lead to some misleading visual conclusions. The eye is especially unreliable when it comes to recognizing variations in light intensities. The relative brightness of two light sources can be judged only approximately, and such an approximation is possible only when the light sources have the same order of brightness.

When observed directly with the eye, an object seems large when the retinal image is large. The angle subtended at the eye by the object, called the "visual angle," is a measure of the apparent size of the object. In Fig. 2-1, θ is the visual angle. In order to examine an object in detail, it is brought as near to the eye as possible in order to obtain a large visual angle. Because the eye cannot focus sharply on objects closer than approximately 10 in., the maximum visual angle obtainable by the unaided eye is limited by the power of accommodation. When a converging lens is placed in front of the eye, the visual angle is increased and the eye looks at an enlarged virtual image, as shown in Fig. 2-2. The diameter of the pupil of the eye is 0.1 in. for light of 5,500 A wavelength. The minimum angular separation of two point objects resolvable by the eye is about one minute of arc.

The minimum size of a defect that can be seen depends on the surface being examined, the brightness level, and the brightness contrast between the specimen and the immediate background. The brightness of the image on the retina is of more importance than the brightness of the object being viewed. The brightness of the retinal image is determined by the area of the pupil. The size of the pupil is variable from 1 to 6 mm, so that the pupil varies in area by a factor of 36.

The sensitivity of the human eye varies greatly for different wavelengths. Under ordinary conditions the eye is most sensitive to yellow-green light, which has a wavelength of 5,560 A. The human eye will give satisfactory vision over a wide range of conditions. For this reason it is not a good judge of differences of brightness or intensity except under the most restricted conditions. The sensitivity of the normal eye as a function of wavelength is shown in Fig. 2-3. The human eye has excellent visual perception. However, adequate lighting is of prime

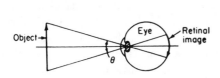

FIG. 2-1. Retinal image.

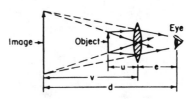

FIG. 2-2. Simple magnifier.

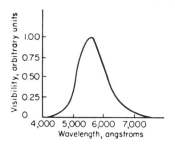

FIG. 2-3. Relative sensitivity of the eye.

importance. The period of time during which a human inspector is permitted to work should be limited to avoid errors due to decrease in visual reliability and discrimination.

It is important that the amount of light reaching the eye should be sufficient for the best definition obtainable either with the unaided eye or with an optical instrument. Also the relation between visual acuity and brightness must be considered. According to reliable experimental evidence, the ratio of the least perceptible brightness difference to the brightness at which it is measured ($\Delta B/B$) is fairly constant over a large range. This range is from about 1 to 100,000 candles/m² and is approximately the range from ordinary interior illumination to the brightest daylight. On the other hand, the visual acuity of the eye varies quite sharply over the lower and middle portions of the same range. For the normal eye, with no lens defects, visual acuity is generally considered to be dependent on the threshold response of the cones in the retina. Unfortunately, insufficient anatomical data exist to give a completely satisfactory explanation of these factors. Figure 2-4 shows the relationship between visual acuity and brightness.

Optical Aids to Vision. Optical aids such as mirrors, lenses, microscopes, periscopes, and telescopes provide a means of compensating for the limits of visual acuity of the human eye by enlarging small discontinuities. Enlarging projectors and comparators provide a means for improving viewing conditions for rapid inspection of small precision parts. Borescopes permit direct visual inspection of the interior of hollow tubes, chambers, and other internal surfaces. Photoelectric and other light sensitive systems can often be used to replace direct visual inspection and compensate for errors due to operator fatigue.

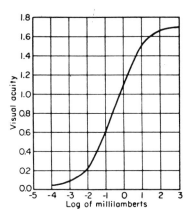

One of the simplest ways of looking into an inaccessible or a forbidden area is by means of large mirrors. The principal objection to their use is the difficulty of getting large mirrors sufficiently free from surface irregularities. Mirrors have an advantage in that only a portion of the mirror is used for a

FIG. 2-4. Relationship between visual acuity and brightness. [*From George S. Monk and W. H. McCorkle (eds.), "Optical Instrumentation," McGraw-Hill Book Company, Inc., New York, 1954.*]

given object point since the cone of rays is limited by the pupil of the eye. If, however, an auxiliary instrument such as a telescope is used, the entrance pupil is no longer the pupil of the eye but the much larger telescope objective. All mirror surfaces must be extremely flat if definition is to remain good. Mirrors must be kept clean and free from dust. If the right number of mirrors is not used, the image will not be erect or right side to. Moreover, the reflecting power of a single mirror is rarely above 75 per cent after a period of use, and the reflecting power of two or more mirrors is much less. However, the reflecting power can be improved by the use of special coating.

A microscope is used to obtain a greatly magnified image of a small object. The specimen is placed close to the objective in order to make the magnification of the latter as large as possible. The simplest microscope is a single converging lens, sometimes referred to as a simple magnifier. The magnification of a single lens is given by

$$\text{Magnification} = \frac{10}{f} \tag{2-1}$$

In this equation, f is the focal length of the lens in inches and 10 represents the average minimum distance of an object from the "normal" unaided eye. This value is based on the practical experience that, nearer than 10 in., objects cannot be seen distinctly by the unaided eye. Using the formula, a lens with a focal length of 5 in. has a magnification of 2 or is said to be a two-power lens (sometimes written 2X). The focal length of a simple magnifier and its working distance are approximately the same. For example, suppose that it is desired to examine a part being made on a machine, without removing it from the machine. Further, suppose that the part is so mounted within the machine that the magnifier cannot be brought any closer than 3 in. This means using a lens with a working distance of at least 3 in., which from Eq. (2-1) requires a three-power lens.

The field of view is the area seen through the magnifier. The diameter of the field of view of a simple magnifier is less than its focal length. Suppose that one is examining the surface of a large specimen for defects. It would take hours to cover the entire surface using a 20-power glass whose field of view is just a little more than ⅜ in. The proper procedure would be first to use a low-power glass, marking the questionable areas, then to inspect the questioned areas with a high-power glass. Depth of field is the name given to the distance the magnifier can be moved toward or away from a specimen and still have the specimen in good focus. Beyond these limits the specimen is said to be "out of focus," and the image is not sharply defined. Depth of field varies with the power of the lens. It is comparatively great in lower-power magnifiers, but decreases along with working distance as the power increases.

To get the best possible performance from a magnifier, it should always be held as close to the eye as possible. This permits the greatest number of rays from the specimen to enter the eye. It also cuts down the reflected light striking the surface of the lens, thus minimizing reflection and glare. In using a magnifier, the plane of the lens should be parallel to the plane of the object. For example, in examining a horizontal surface, the lens should be in a horizontal position; otherwise part of the image will be out of focus. Obviously, this is less critical in low-power lenses and more critical in higher-power lenses, where the focal lengths are shorter and depth of focus is smaller. It should be remembered that, as an area is magnified, detail is lost. Consequently, for examining radiographs a low-power magnifier should be used.

The exception to the above rules is the use of a big lens-type magnifier such as a reading glass. The reading glass has been purposely designed to be used with both eyes and should be held far enough away from the eyes to permit use in this manner. It is a low-power glass for use over extended periods of time, and although it has many important uses in science and industry, it cannot be classified by the rules governing most magnifiers.

Optical microscopes have been used to detect and study fatigue cracks. In these studies the optical microscope has been used for three purposes: to determine the location of the first macroscopic crack in a specimen where cracks are equally likely to occur at many places in the specimen, to follow the course of the fatigue crack and determine the manner in which it is affected by grain boundaries, inclusions, etc., and to study the development of microscopic cracks. In the first case a low-power microscope having a magnification of 2 to 20X is used, in the second case a magnification of 100 to 500X is used, and in the latter case a magnification of 1,500 to 2,000X is needed. Minute surface cracks can be detected by making plastic replicas of the surface of the test specimen and examining the replicas with an optical microscope. Peterson, Jones, and Allen[1,*] used a process called Faxfilm.[2] At 500X magnification they are able to detect cracks in the surface of enamelware of 1 to 1.5 μ. Hunter and Fricke[3] used this technique for detecting the beginning of cracking in fatigue specimens of aluminum and aluminum alloys.

The optical microscope is limited by the ability of the lens to resolve detail, but the use of the electron microscope results in a manyfold increase in resolving power. Cracks of the order of 2 to 3 μ in length have been observed using this technique. Electron micrographs are made from very thin plastic "castings" or replicas of the surface of the metal. The electron beam in the microscope is caused to traverse the thickness of the replica; variations in thickness cause variations in the intensity of the beam recorded on a photographic film. Other techniques

* Superscript numbers indicate Specific References listed at the end of the chapter.

such as "shadow casting," described by Craig,[4] are used for the purpose of increasing contrast.

One optical instrument of importance and usefulness to the nondestructive testing engineer is the metallograph. As the name implies, this instrument is for recording the structure of metals. However, it can be used to study virtually all materials of an opaque nature. It is really a duplex compound microscope, that is, one microscope for viewing the structure and a second microscope employing the same objective for projecting the image on a screen for further viewing or photographing. In a metallograph, the illuminating rays are introduced between the ocular and objective and are reflected to the specimen by a transparent mirror, the objective serving as a condenser. Light in varying degrees of intensity is reflected through the objective to form a primary image which is remagnified by the ocular, Fig. 2-5a,b. In dark field illumination applied to opaque microscopy, the illumination, after passing through the lamp condenser, as shown in Fig. 2-5c, strikes a plane mirror surface as a cylinder of light, the central portion of the cylinder having been already restricted by the dark field center stop. The plane mirror reflects this cylinder of light to the surface of the annular mirror and in turn reflects the rays to the specimen. (In many cases the annular mirror and objective are an integral component, the focal planes having been computed for coincidence.) From this it may be seen that, if an illuminating ray strikes a highly polished surface which is normal to the optical axis, it will be reflected away from the optical system and cause the surface to appear black or "dark field." If, however, there are any surface irregularities, these will become luminous and reflect in varying degrees. It is with this feature that the use of the metallograph has been further extended into many other unconventional fields of application.

In the conventional application of the metallograph many preparation techniques and illuminating features employed for metallography may be used directly or slightly modified. Some of the harder nonmetallic materials, such as certain plastics and ceramics, may be prepared by polishing and examined under bright field illumination or polarized light at low, medium, and high magnifications. Materials having irregular surfaces may be examined and photographed at low magnifications to great advantage by means of dark field illumination. This is especially true of fibrous-type substances or specimens having painted or plated surfaces. With the decentering iris, a shadow-casting technique may be used to emphasize the relief of some surfaces. Figures 2-6 and 2-7 show typical micrographs.

The periscope is an instrument in which the general direction of the rays is not in a straight line but is deflected one or more times for the purpose of giving the observer a view from a position in which he cannot put his head. The simplest form of periscope obviously would be

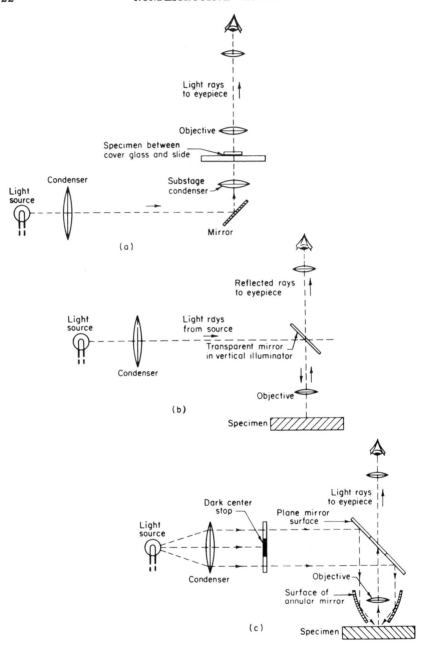

Light rays
to eyepiece

Objective

Specimen between
cover glass and slide

Condenser

Substage
condenser

Light
source

Mirror

(a)

Reflected rays
to eyepiece

Light
source

Light rays
from source

Transparent mirror
in vertical illuminator

Condenser

Objective

(b)

Specimen

Dark center
stop

Plane mirror
surface

Light rays
to eyepiece

Light
source

Objective

Surface of
annular mirror

Condenser

(c)

Specimen

Fig. 2-5. Metallograph. *(F. Gordon Foster, courtesy of the Bell Telephone Laboratories.)*

FIG. 2-6. Photomicrographs, by oblique illumination, of (*left*) cleavage plane in calcite (175×) and (*right*) quartz crystal (100×). (*F. Gordon Foster, courtesy of the Bell Telephone Laboratories.*)

one or more mirrors.   The requirements of brightness, image orientation, size of field, and maneuverability have made it necessary for a periscope to contain image-forming optical parts.   With modern low-reflection coating methods and wide-aperture optical design, almost any periscope can be expected to have a transmission of approximately 40 per cent or better.

The telescope is used to obtain a magnified image of a distant object or to gather more light than reaches the unaided eye.   If the magnifying power of the telescope is great, the field is correspondingly limited.   The telescope gives an inverted image, but reinversion of the image can be achieved by an erecting eyepiece.   Another limitation of the telescope is the excessively large range of motion of the eyepiece required for focusing on nearby objects.   This latter difficulty can be avoided in the periscope.

FIG.   2-7.   Photomicrograph of relay contact by replication (dark field illumination; 50×).
(*F. Gordon Foster, courtesy of the Bell Telephone Laboratories.*)

As the name implies, a borescope is an instrument designed to enable an observer to inspect the inside of a narrow tube, bore, or chamber.   Borescopes are precision-built optical systems having a complex arrangement of prisms, achromatic and plain lenses through which light is passed to the observer with maximum efficiency.   The light source located in front or ahead of the object lens provides illumination for the part being examined.   The optical train for a borescope is shown in Fig. 2-8.   The brightest images are obtained with borescopes of large diameter and short length.   As the length of the borescope is increased, the image becomes less brilliant because of light losses.   In most borescopes the observed visual area is approximately 1 in. in diameter at 1 in. distance from the objective.   The size of the visual field usually varies with the diameter for a given magnification system.

The design of the objective determines the angle of view, size of visual field, and amount of light gathered by the system.   The middle lenses conserve the light entering the system and pass it through the telescope tube to the eye with a minimum loss in transmission.   Design of the middle lenses has an important influence on the character of the image obtained.   Most middle lenses are achromatic for the purpose of preserving sharpness of the image and true color values.   Depending on the length of the telescope, the image requires reversal, inversion, or both at the eyepiece.   This is done by means of a correcting prism within the eyepiece for borescopes of small bore and by the use of erecting lenses for larger systems.   The prisms and lenses are precision polished to the highest optical standard.   The systems are carefully assembled and accurately adjusted to provide clear, brilliant, distortion-free images.   At the same time, they must be ruggedly constructed.   Figure 2-9 shows a Tuboscope for inspecting pipe.

Borescopes can be designed and constructed to provide any desired angle of vision.   Figure 2-10 shows four angles of view.   Figure 2-10a shows a right-angle borescope, with the integral lamp positioned ahead of the objective lens.   The right-angle system provides vision at right angles to the axis of the telescope and a visual working area of about 1 in. at 1 in. from the objective lens.   It permits inspections of the most inaccessible corners and internal surfaces.   It can be supplied for insertion into apertures as small as 0.020 in. in diameter and in a wide range of lengths in the larger sizes.   It is the perfect instrument for the examination of such objects as rifle and pistol barrels and the walls of cylin-

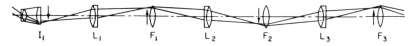

FIG. 2-8. Optical train of a borescope.   [From George S. Monk and W. H. McCorkle (eds.), "Optical Instrumentation," McGraw-Hill Book Company, Inc., New York, 1954.]

Fig. 2-9. (*Top*) Tuboscope being used to examine the inside surface of an oil-field casing. (*Bottom*) Tuboscope in its carrying case with the necessary accessories. (*Courtesy Tuboscope Company.*)

drical or recessed holes. The foroblique system, Fig. 2-10b, provides vision forward and at an angle of about 55° from axis of the telescope. A unique feature of this optical system is that by rotating the borescope the working area of the visual field is greatly enlarged. This design permits the mounting of the lamp at the end of the instrument and still allows forward and oblique vision. The lamp does not appear in the field of vision. The direct system, Fig. 2-10c, provides vision directly forward with a visual area of approximately 1 in. at 1 in. from the objective lens. The retrospective borescope, Fig. 2-10d, has an integral light carrier, the lamp being mounted at the side of the objective lens. The retrospective borescope provides the only method of accurately visually inspecting a bore with an internal shoulder.

The ability of an optical instrument to resolve two objects close together depends not only on the perfection of design and construction of the instrument but also on the laws of physical optics. Since the image of a point object is not a point but a diffraction pattern of finite size, it becomes necessary to specify the conditions under which two such images

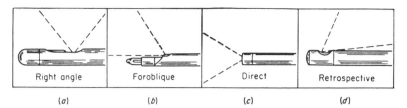

FIG. 2-10. Angles of view.   (*Courtesy American Cystoscope Makers, Inc.*)

of a close pair of points can be distinguished from each other. Obviously, this depends on the quality of the image and the eye of the observer. There also must be considered the ability of the observer, possibly based on familiarity with optical observations or mental adaptability, to draw reasonable inferences from borderline cases. The formula of Rayleigh[5] is generally accepted. If d is the diameter in the image plane of the first dark ring of the diffraction pattern, λ the wavelength, and u' the half angle of the cone of rays contributing to the image point, by Rayleigh's criterion,

$$d = \frac{1.22\lambda}{\sin u'} \tag{2-2}$$

It is generally considered that in actual practice the eye can do better than the limit set by Eq. (2-2), and it may be accepted that with sufficiently good optical design two point images may be separated if their distance $D = (d/2)$ apart is given by

$$D = \frac{0.5\lambda}{\sin u'} \tag{2-3}$$

In an optical comparator, a magnified image of the specimen being inspected is projected on a screen. The accurate projection of profiles necessitates that the specimen be illuminated with parallel light rays. In addition, the distance between the screen and condenser lens must be held constant, which assures exact magnification on the screen. The optical comparator, Fig. 2-11, uses a parallel beam of light which is directed on the part to be inspected or measured. A projection lens system and an optically flat mirror throw a magnified shadow of the part on the receiving screen, where it can be inspected and measured by comparison with a master chart or outline. Inaccuracies can be measured by using a micrometer for moving the work-holding table laterally and a graduated handwheel for moving it vertically. For measuring angles and angular deviations, the graduated chart ring can be rotated and measurements taken on a vernier. To assure the highest degree of perfection in the optical system of an optical comparator, and to assure a dense, distortion-free shadow, precision lenses and mirrors must be used. A high-intensity illuminating unit should be used to assure maximum illumination and the sharpest shadow outlines. The illumination units consist of a low-voltage bulb, condensing lens, colored filter, and a blower cooling unit to assure constant normal temperature of the lamp house assembly.

Photography. The use of photography and photographs for studying surface detail and other phenomena will not be discussed in this book except for the panoramic camera.[6] The permanent record provided by a photograph is often useful for future comparison and reference, as in the study of corrosion. For further details of these techniques the reader is referred to such books as given in references 7 to 9. However, it should be mentioned that color photography, high-speed photography, and closed-circuit television will find more and more use in the non-destructive testing field, especially in helping to determine and study the causes of material or component failures in service.

A schematic diagram of the panoramic camera is shown in Fig. 2-12. This camera was developed for inspecting reactor fuel element sub-assemblies. These subassemblies consist of a number of parallel spaced plates welded together to form a set of long thin channels, accessible only from one end. In this camera, light from the ribbon-filament lamp is focused on slit 1, which then acts as the source of illumination for the rest of the system. The illuminating-system lens focuses an image of slit 1 on the wall of the channel in the specimen at point *A*, the light having been reflected from the beam splitter and the mirror supported within the channel. This illuminates an area of the wall. The light, reflected from and modified by the surface, is again reflected from the

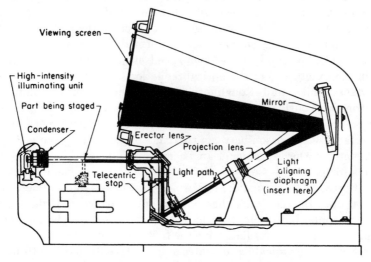

FIG. 2-11. Optical comparator. (*Courtesy Jones and Lamson Machine Company.*)

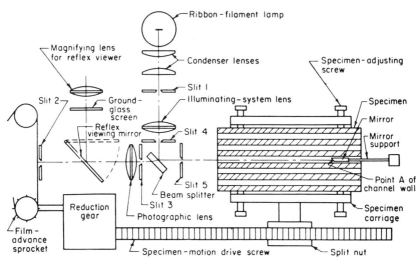

Fig. 2-12. Schematic diagram of panoramic camera. (*Courtesy American Society for Testing Materials.*)

mirror, passes through the beam splitter and the photographic lens which focuses an image of the illuminated channel wall on the photographic film. Slit 2 is adjusted to exclude all but the image-forming rays reflected by the mirror. If the specimen and film are kept stationary, a narrow strip of the wall is photographed. In order to obtain an image of the entire channel wall, the specimen and film are moved in synchronism while the mirror and other camera parts remain stationary. Under these conditions the image point corresponding to a given point on the object will be fixed relative to the film, and the exposure time will be the time required for the image point to cross the opening of slit 2. Thus an image of the channel wall is formed as the specimen moves. The ratio of film speed to specimen speed must be maintained constant and in the exact ratio of the optical reduction.

Surface Finish and Roughness. Measurements of the flatness or smoothness of surfaces are conveniently done using the principles of the interference of visible light. Using the theory of interference between flat surfaces, comparison between two gage blocks or between a standard block and another specimen can be made. The working standard of flatness for practical use is the optical flat. Flats are made of ordinary glass, Pyrex, or fused quartz, in various thicknesses and diameters, and may be optically flat on either one or both surfaces. Ordinary glass flats are somewhat more transparent, but Pyrex has better thermal stability. Fused quartz flats have the best thermal stability and wearing properties. The optical flat deviates only by 0.00001 to 0.000001 in. from absolute flatness across the entire width of the flat.

The action of the optical flat in producing an interference pattern or "fringes" is shown in Fig. 2-13. The flat is tipped at an angle to the work surface so that a narrow "air wedge" is formed. Light from a monochromatic light source is directed from the top. That portion of light which is reflected from the top surface of the flat, or is transmitted or absorbed at the work surface, does not affect the interference pattern and consequently is not shown. A part of the light will be reflected from the "glass air" boundary. Another part of the light will be transmitted across this boundary, across the air wedge, and be reflected from the work surface. These two reflected rays will ordinarily be separated slightly, but will both be received simultaneously by the eye. In Fig. 2-13, the rays are shown crossing the air wedge at a point where the wedge is exactly one-half wavelength thick. When this condition exists, the two reflected rays are said to be 180° out of phase and will cancel, marking the center of a dark band or "fringe." This condition for interference will exist wherever the thickness of the air wedge is an integral number of half wavelengths in thickness. Accordingly, a series of dark fringes will appear across the field of view. The condition of reflection as regards phase is different for the two rays shown in Fig. 2-13. A useful rule to remember is that, when the light is reflected without a change of medium, there is no change of phase. In the above figure, the ray reflected in the glass suffers no phase change. The ray reflected at the air–work surface boundary is changed in phase by 180°. This phase concept is consistent with the existence of a dark fringe at all points where the flat and the work surface are in contact. Figure 2-14 shows the interference pattern for various conditions using a rectangular specimen.

Surface finish is often of importance because of its influence on metal fatigue and friction. In many cases, the performance and service life of a product depend largely on the roughness or "smoothness" of bearing races, O-ring grooves, gear teeth, or other surfaces. When the surfaces are too rough, performance suffers. Sometimes a surface that is too smooth performs as poorly as one that is too rough. Thus, by specifying and producing the proper surface roughness, better products are obtained. A rough estimate of surface smoothness can be made by rubbing the

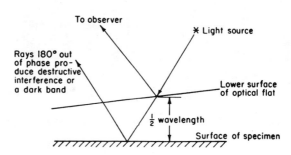

FIG. 2-13. Optical interference.

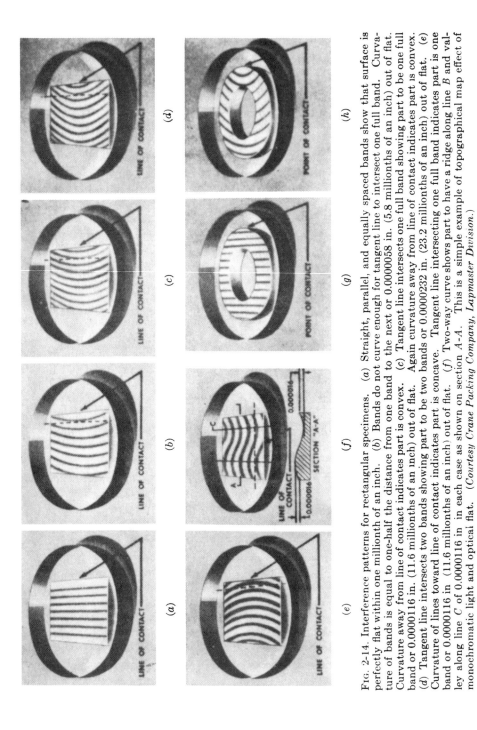

Fig. 2-14. Interference patterns for rectangular specimens. (a) Straight, parallel, and equally spaced bands show that surface is perfectly flat within one millionth of an inch. (b) Bands do not curve enough for tangent line to intersect one full band. Curvature of bands is equal to one-half the distance from one band to the next or 0.0000058 in. (5.8 millionths of an inch) out of flat. Curvature away from line of contact indicates part is convex. (c) Tangent line intersects one full band showing part to be one full band or 0.0000116 in. (11.6 millionths of an inch) out of flat. Again curvature away from line of contact indicates part is convex. (d) Tangent line intersects two bands showing part to be two bands or 0.0000232 in. (23.2 millionths of an inch) out of flat. (e) Curvature of lines toward line of contact indicates part is concave. Tangent line intersecting one full band indicates part is one band or 0.0000116 in (11.6 millionths of an inch) out of flat. (f) Two-way curve shows part to have a ridge along line B and valley along line C of 0.0000116 in in each case as shown on section A-A. This is a simple example of topographical map effect of monochromatic light and optical flat. (Courtesy Crane Packing Company, Lapmaster Division.)

30

surface lightly with the fingernail. Mechanical methods of measuring surface finish are refined methods of feeling the surface.

The methods of designating surface roughness originate through the use of an instrument that will either produce a magnified graph of the roughness profile or operate electrically or mechanically to provide a number in some way descriptive of the dimensions of the irregularities. Microscopic inspection of the surface generally gives only the plan view of the specimen and consequently does not yield a measurement. The Zeiss interference microscope[10] is an exception, however, and is useful for detailed study of certain types of smooth surfaces having roughness irregularities between about 1 and 20 μin. in total height.

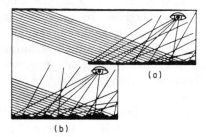

With any surface, a human observer sees only those portions of the surface which reflect light into the eye. Thus, the appearance of any machined, ground, or finished surface depends entirely on the way in which light is reflected from the surface irregularities. This, in turn, depends on the shape of the irregularities or, more specifically, on the number and spacing of those facets which are at the proper angle to reflect light into the eye. With the two surfaces shown in Fig. 2-15, the contours are such that the number and spacing of light rays reflected into the eye are the same for both surfaces. Thus, as far as the eye can tell, these surfaces are identical in roughness, whereas one surface b is actually several times as rough as the other surface a. In the same way, the shape of the specimen often affects the apparent roughness contour as surface b above; yet it reflects light into the eye from fewer points, and thus appears to be considerably smoother.

In addition, when the angle of lighting is changed, some parts of

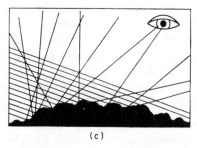

FIG. 2-15. Reflections from rough surfaces. With machines, ground and finished surfaces, only a relatively few facets reflect light into the eye at any one time. Those facets are the only portions of the surface that are seen; so the appearance of a surface depends on the number and spacing of those facets.

With the two magnified surfaces shown here, the number and spacing of such facets are the same; so these surfaces look alike, though they differ greatly in roughness. Lines indicate light rays.

The curved surface c is of substantially the same roughness as the flat surface b above; but because of its curvature, it reflects fewer light rays into the eye. Thus surface c looks much smoother than surface b. (*Courtesy Micrometrical Manufacturing Company.*)

irregularities that were invisible will be at the proper angle to reflect light into the eye, and will thus be seen; and some parts that were visible will no longer reflect light into the eye, and will disappear from view. This is the reasoning behind the common procedure of cocking a surface at different angles "to get a better light." In short, with any given angle of lighting, the eye sees only a small per cent of the roughness irregularities—and that is much too sparse and spotty a selection from which to judge the character of the surface.

It is common practice for shopmen to back their visual judgment of surface roughness by running a fingernail over the surface and comparing the "feel." In addition, "human" inspection methods do not indicate either the range of roughness along the surface or point-to-point variations in roughness, which are important to production control. With many common surfaces, such as narrow shoulders, small holes and slots, recesses, deep holes, and gear teeth, the shape or dimensions of the workpiece prevent using either the eye or the fingernail to get a roughness rating. In order to reach the bottom of most roughness irregularities, it has been found that a tracer point of not more than 0.0005-in. tip radius is required. It is readily demonstrated that with a duller point accurate roughness ratings cannot be obtained. Consequently, the procedure of comparing surfaces by the fingernail-feel technique is very crude and inaccurate.

The American Standards Association has designated that stylus-type instruments used to determine a surface roughness shall meet the following specifications.

A cone shaped stylus with a tangential spherical tip shall be considered standard unless otherwise specified by the design engineer. For some special applications, the chisel point may be desirable. A 500 microinch spherical tip radius is considered standard. Other radii can be used only in unusual circumstances, where the 500 microinch radius may not provide the information desired. When other radii are used they shall be chosen from the series 100, 500, 5,000, 50,000, etc., microinches. The tip radius for an instrument in satisfactory condition shall be within 30 per cent or 100 microinches of nominal value, whichever is greater. The nominal included angle of the conical stylus shall be 90°. The stylus support shall be such that under normal operating conditions no lateral deflections sufficient to cause error in the roughness measurement will occur. For the standard tip radius of 500 microinches the stylus force is not to exceed $2\frac{1}{2}$ grams at any point within the displacement range of the stylus. For special tip radii the maximum force shall not exceed that given by the following relation:

$$\text{Maximum force (grams)} = 0.00001 - \text{tip radius microinches} \qquad (2\text{-}4)$$

The minimum stylus force shall be sufficient to maintain contact with the surface under conditions of maximum irregularity amplitude, maximum tracing speed, and minimum wavelength for which the instrument is designed.

A number of instruments for measuring surface roughness have been devised which employ optical, acoustical, mechanical, and pneumatic principles. However, the large majority of instruments in use for measuring surface roughness depend on electrical amplification of the motion of a stylus perpendicular to the surface over which the stylus is traversed. The American Standards Association[11] has prepared a standard which deals with the geometric irregularities of surface of solid materials, physical specimens for gaging roughness, and the characteristics of instrumentation for measuring roughness.

For specifying surface finishes, the terms arithmetical average and root mean square average are used. The arithmetical average is defined as the mean line or center line about which roughness is measured in a line parallel to the general direction of the profile throughout the roughness width cutoff length such that the sum of the areas contained between it and those parts of the profile which lie on either side of it are equal. The arithmetical average deviation from the mean line is

$$Y = \frac{1}{L} \int_{x=0}^{x=L} |y|\, dx \qquad (2\text{-}5)$$

where Y is the arithmetical average deviation from the mean line, y is the ordinate of the curve of the profile, $|Y|$ refers to the absolute value of Y, and L is the length over which the average is taken. The root mean square average (rms) is defined by the following expression:

$$Y = \left[\frac{1}{L} \int_{x=0}^{x=L} y^2\, dx \right]^{\frac{1}{2}} \qquad (2\text{-}6)$$

where the symbols have the same meaning as above except that Y is now the rms deviation from the average line.

An approximation of the average roughness may be obtained by adding the y increments shown in Fig. 2-16 without regard to sign and dividing the sum by the number of increments taken:

$$\text{Arithmetical average } Y = \frac{y_a + y_b + y_c + \cdots y_n}{n} \qquad (2\text{-}7)$$

$$\text{rms average} = \left(\frac{y_1^2 + y_2^2 + \cdots + y_n^2}{n} \right)^{\frac{1}{2}} \qquad (2\text{-}8)$$

Roughness measuring instruments calibrated for rms average will be found to read 11 per cent higher on a given surface than those calibrated for arithmetic average. Instruments originally calibrated to read rms average can, in many cases, be adjusted to read arithmetic average. This 11 per cent difference is in such a direction that continued use of drawings which originally specified rms measurements

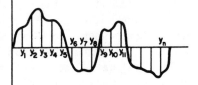

FIG. 2-16. Surface profile divided into increments.

and the inspection of surfaces to those values with instruments calibrated to arithmetical average, as called for in the standard, would theoretically allow the acceptance of surfaces with up to 11 per cent greater roughness than originally intended. This difference, however, is of lesser magnitude than the variations from point to point on a machined surface and is considerably less than the variation expected from one piece to another. For these reasons and because the absolute limit of roughness for satisfactory functioning is usually even more indefinite, many manufacturers consider it practical to adopt arithmetical average ratings without changing the roughness values given on part drawings.

Two commercially available instruments for measuring surface roughness are the Brush Surface Analyzer and the Profilometer. The Brush Surface Analyzer is a roughness measuring device which will produce both a profile record and a meter reading in microinches rms. The tracing device is motor driven to produce a length of trace of $\frac{1}{16}$ in. for one complete back and forth oscillation requiring 10 sec. The tracer point (stylus) is a diamond with a tip radius of 0.0005 in. This radius is suitable for all meter readings but may result in some rounding of the peaks and sharpening of the valleys of the profile records. The recording will be faithful on irregularities up to about 0.004 in. roughness width, whereas meter readings will include the contributions of irregularities up to about 0.012 in. roughness width. The Brush Surface Analyzer is provided with electrical means for calibration, and a specimen is provided for checking the condition of the tracer point. Measurements can be taken inside $\frac{1}{4}$-in.-diameter holes and on other surfaces large enough to accommodate the tracer point and skid and still allow the $\frac{1}{16}$-in. stroke.

The Profilometer is a mechanical electronic instrument built for shop use. Its principal working parts consist of a mechanical tracer and an amplimeter. The Profilometer is an electronic averaging type of instrument giving a meter reading of average deviation from the mean surface in terms of microinches rms. Several types of tracers are available to suit the size, shape, and placement of the surface to be measured. Basically, the instrument consists of a tracer, which is moved either by hand or mechanically over the surface being measured, and an electronic amplifier and amplitude measuring unit, connected to the tracer by a cable. The instrument is provided with a microinch meter that shows the average height of the roughness irregularities as the tracer moves along the specimen. Readings are either arithmetical or rms average, as selected by a switch on the panel.

This instrument utilizes a differential transformer. As the tracer moves over the specimen, the point moves up and down to follow the contour of the surface irregularities. These vertical movements of the tracer point are transmitted to a coil which moves in the field of a per-

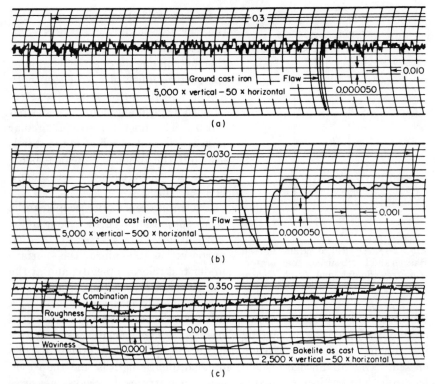

FIG. 2-17. Typical surface records. (a) This Proficorder chart shows the total profile on a ground surface (at left) and end milled surface. (b) This chart shows the same paths of trace but with waviness eliminated so as to show only the roughness irregularities. (c) These two charts again show the same paths and trace but with waviness eliminated to show only the waviness contour. (*Courtesy Micrometrical Manufacturing Company.*)

manent magnet. A small fluctuating voltage is produced which is proportional to the height of the surface irregularities. This voltage is supplied to an electronic amplifier and integrated so as to actuate the direct-reading microinch meter. The greatest roughness width measured depends on the speed of the trace and with manual tracing will exceed 0.033 in. The mechanical tracer supplied by the manufacturer has a tracing speed of 0.3 in./sec which will satisfactorily measure roughness having widths of 0.020 in. As the tracer moves over the specimen, the instrument automatically computes the mean line, shown in Fig. 2-17, for the roughness profile of the surface and then computes the average height of the irregularities. This information is presented on a meter which shows the average height values in microinches. This measuring, computing, and averaging process is continuous, so the instrument shows the variations in average roughness height that occur from place to place in the specimen.

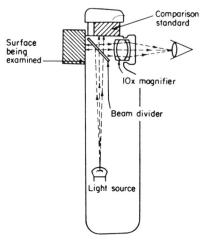

Surface being examined —

Comparison standard

10x magnifier

Beam divider

Light source

FIG. 2-18. Optical diagram of the Bausch and Lomb surface comparator. (*Courtesy Bausch and Lomb Optical Company.*)

Bausch and Lomb has developed the portable surface roughness comparator shown in Fig. 2-18. With this instrument the operator compares the surface finish of a specimen with a commercial roughness standard strip or with a specimen having the desired finish. The operator sees a circular image, one half of which is the standard and the other the surface being checked. Both surfaces are magnified ten times.

The interference microscope can be used to measure surface roughness. For surface testing it serves a dual purpose: it magnifies the surface under test microscopically, and at the same time it reveals the surface structure in readily measurable form by a "contour line" representation. With this microscope, roughness depths of about 80 to 1.2μ in. can be measured. The principle used in this instrument is the same as that used in the optical flats previously discussed. One advantage of using monochromatic light is that the wavelength is not subject to change, and the instrument always measures correctly and does not need calibration. In monochromatic light, interference bands are sometimes hard to trace when the deflection of the bands is greater than the band width; when white light is used, differently colored bands result which clearly show the course of deeper scratches. Measurement of the surface contour, however, is always made using the monochromatic light. Such an interferometer has been combined with an electronic counter, which gives a direct reading of the light fringes. The counter reading can be converted into dimensions by multiplication of a constant factor. Such an instrument permits measurements to 0.000001 in.

Tolansky[12] discusses how precision optical interference methods can be used for the study of the topography of reasonably smooth surfaces including metals, crystals, and plastics, as well as the properties of thin films and certain optical properties of metals. These methods are capable of high precision, and measurements on properties of molecular dimensions can be carried out.

Light Sensitive Devices. Photocells, phototubes, and other light sensitive devices can sometimes be used to advantage for replacing the human eye or to monitor processes where light energy can be used or is available. In the chapter on penetrant inspection such a device is mentioned for scanning specimens for defects indicated by penetrant inspection. Figure 2-19 shows the construction of a phototube and a typical circuit in which such a tube can be used. Figure 2-20a shows the output of such a tube as the light intensity is varied and Fig. 2-20b the output as a function of wavelength of the incident radiation. It should be pointed out that such detectors can be made to operate on infrared radiation as well as visible radiation.

Pinholes in a metal strip can be detected by passing the strip between a light source and a phototube. When a pinhole passes through the scanner, light shines through the hole into the detection chamber, causing the phototubes to produce a signal which when amplified can be used to operate a meter, marker, or rejection mechanism. J. R. Burns[13] used this technique to study the porosity of magnesium castings. Strips of the porous castings were passed slowly in front of a photo-

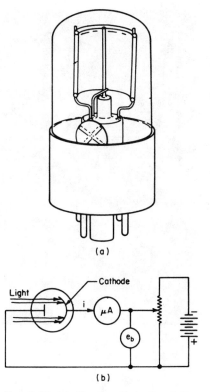

(a)

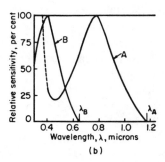

(b)

Fig. 2-19. (a) Construction of a phototube; (b) test circuit for phototube. (*Courtesy of Radio Corporation of America.*)

tube having an aperture of ⅛ by ⅝ in. connected to a vacuum tube amplifier and milliammeter.

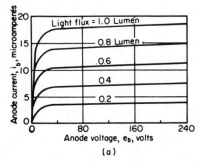

FIG. 2-20. Phototube properties. (a) Anode characteristics for a high-vacuum phototube; (b) spectral sensitivity characteristics. (*From Ralph Benedict, "Introduction to Industrial Electronics," pp. 129 and 131, copyright 1951 by Prentice-Hall, Englewood Cliffs, New Jersey. Reproduced by permission of the publisher.*)

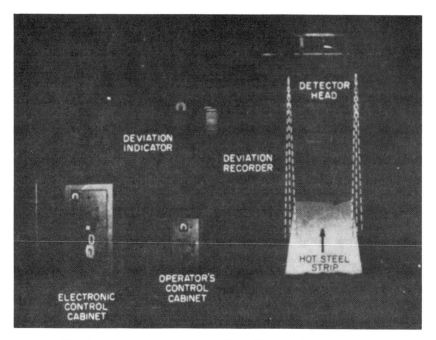

Fig. 2-21. Width gage for hot strip steel mills.   (*Courtesy General Electric Company.*)

An optical means has been used in a noncontacting type of gage to measure the width of hot metal strip.[14] The gage to be described measures, indicates, and can provide a record of strip width. The accuracy of the gage is better than $\frac{1}{8}$ in. for strip ranging in width from 10 to 96 in. The width indication is independent of reasonable lateral and vertical motion of the strip as it bounces rapidly along a rolling mill table. The indication is also independent of strip temperature for a 300°F change within the range of 1500 to 2100°F. Since this gage observes the edges of the strip by optical means, it is possible to locate the detector head unit 15 ft above the hot strip, as shown in Fig. 2-21. At this distance, the unit is largely unaffected by extreme ambient conditions such as temperature, moisture, fumes, and dirt of the mill near the strip itself.

The light radiated from the hot strip edges is used to obtain signals for measuring width. The gage operates by scanning an optical image of each edge of the strip to be measured as reproduced in the scanning units of the detector head. A block diagram of the gage is shown in Fig. 2-22. The position of these two scanning units is adjusted by a motor-driven screw in order to place the two units directly above the nominal position of the edges of the strip. The lens at the bottom of each scanning unit focuses the image of the edge of the hot strip onto a disk. Behind the scanning disk is a phototube, which gets its illumination by the radiation from the hot strip. The optical image for each edge of steel is converted into an electrical signal by the phototube. The rotating slotted disk provides a means for repeatedly scanning across the image of the edge of the strip at right angles to the direction of strip travel. Each unit scans approximately 10 in., nominally 5 in. off the edge of the strip and 5 in. on the strip. The scanning field is thus wide enough to allow for a certain amount of sidewise motion of the strip as well as for normal changes in width. The scanning action causes each phototube to generate a rectangular wave shape of voltage versus time in which the percentage pulse width is directly proportional to the position of the edge. The two sets of amplified signals, one from each edge of the strip, are transmitted to the amplifiers in the electronic control cabinet. The amplified signal gives the width of the strip as seen by the phototubes. This width must be added to a value determined by the position of the phototubes to give the total width of the strip.

Bujes[15] has used visible light in conjunction with a cadmium selenide crystal detector to measure the thickness of the inhibitor on rocket pro-

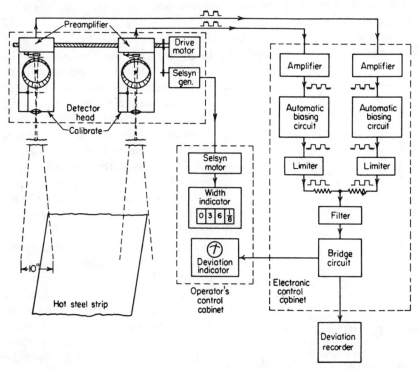

Fig. 2-22. Block diagram of width gage shown in Fig. 2-21. (*Courtesy General Electric Company.*)

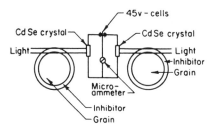

FIG. 2-23. Inhibitor thickness gage. (*Official United States Navy photograph.*)

pellant grains. Figure 2-23 shows the arrangement for this measurement. A collimated light beam falls on the inhibited grain, and part of it is absorbed by the inhibitor and grain. The transmitted beam is detected by the cadmium selenide crystal. The cadmium selenide crystal is a semiconductor, and its operation depends on electronic conduction in a solid. The resistance of the crystal is a function of the intensity and the wavelength of the radiation falling upon it. Thus, the crystal has the highest resistance when no radiation falls upon it and diminishing resistance as the radiation increases. Figure 2-24 shows the response curves for a cadmium sulfide and a cadmium selenide detector. It was found by experiment that by using a cadmium selenide crystal operating at 90 volts the variation in the inhibitor thickness could be detected without amplification of the current, using a microammeter. With inhibitor thicknesses of 0.058, 0.048, and 0.034 in., the respective microammeter readings were 18, 6, and 0.5 μamp.

Figure 2-23 shows the paradoxical fact that the intensity of the light transmitted through the inhibitor increases with the thickness of the inhibitor. Since the primary beam is traversing not only the inhibitor but also the grains, more grains will be in the path of the light beam. Because the density of the grain is about 1.6, as compared with about 1.3 for the inhibitor, under identical geometrical conditions, the grain will absorb more light than the inhibitor. It might be of interest to explain the superior effectiveness of the cadmium selenide crystal versus a photocell for this particular application. The light sensitive area of the photocell is, in general, many times larger than that of a crystal (3 by 1 mm), and assuming equal specific sensitivity for both detectors, a beam of 3 by 1 mm will affect the crystal 100 times more than the photocell. The very steep slope of

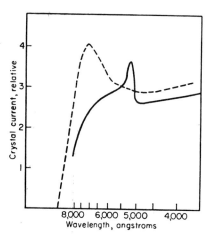

FIG. 2-24. Spectro response on cadmium sulfide crystal (solid curve) and cadmium selenide crystals (dashed curved) to visible position on the spectrum runs from about 3,800 to 7,600 A. (*Courtesy General Electric Company, X-Ray Division.*)

the spectral curve in the infrared and red regions, as shown in Fig. 2-24, makes the cadmium selenide crystal very sensitive to filament lights glowing at low temperature; so a small bulb operating at low voltage can be used.

Tolansky, Stephens, and Heavens[16] discuss some optical methods for nondestructive testing. These techniques include two-beam interferometry, multiple-beam interferometry, the light profile microscope, optical shadow casting, total reflection, optical rotation of plane polarized electromagnetic radiation, optical reflectivity, and phase contrast microscopy. These techniques can be used to study surface roughness, properties of thin films, surface inhomogeneity and mechanical strain, surface flaws, and shape and distribution of etch and corrosion pits; to examine the flow of metal surrounding hardness indents, depth of hardness indentations, topography of grain boundaries in alloys, directional hardness; and to study electrodeposits and the rate of material removal by chemical and discharge etching. They suggest the use not only of visible light but also infrared and ultraviolet light and microwaves for special applications. For example, the use of infrared radiation in studying surfaces would tend to eliminate the effect of the smaller surface blemishes.

GENERAL REFERENCES

Jenkins, Francis A., and Harvey E. White: "Fundamentals of Optics," 3d ed., McGraw-Hill Book Company, Inc., New York, 1957.
Jones, R. Clark: Performance of Detectors for Visible and Infrared Radiation, in "Advances in Electronics," vol. 5, pp. 1–96, Academic Press, Inc., New York, 1953.
Monk, George S.: "Light Principles and Experiments," McGraw-Hill Book Company, Inc., New York, 1937.
Monk, George S., and W. H. McCorkle (eds.): "Optical Instrumentation," McGraw-Hill Book Company, Inc., New York, 1954.
Robertson, John Kellock: "Introduction to Physical Optics," D. Van Nostrand Company, Inc., Princeton, N.J., 1941.

SPECIFIC REFERENCES

1. Peterson, F. A., R. A. Jones, and A. W. Allen: A New Method for Studying Fractures of Porcelain Enameled Specimens, *J. Am. Ceram. Soc.*, vol. 31, pp. 186–193, July, 1948.
2. A New Method for Surface Flaw Detection, *Ind. & Welding*, vol. 25, pp. 79–80, April, 1952.
3. Hunter, M. S., and Wm. G. Fricke, Jr.: Metallographic Aspects of Fatigue Behavior in Aluminum, *Proc. ASTM*, vol. 54, 1954.
4. Craig, W. J.: An Electron Microscope Study of the Development of Fatigue Failures, *Proc. ASTM*, vol. 52, pp. 877–889, 1952.
5. Monk, George S.: "Light Principles and Experiments," pp. 186–188, McGraw-Hill Book Company, Inc., New York, 1937.
6. Cocks, G. G., and C. M. Swartz: A Panoramic Camera for Inspecting Wall of Deep Narrow Slots, *ASTM, Spec. Tech. Pub.* 223, 1958.

7. James, Thomas Howard: "Fundamentals of Photographic Theory," John Wiley & Sons, Inc., New York, 1948.

8. Mees, C. E. K.: "Theory of the Photographic Process," The Macmillan Company, New York, 1948.

9. Doucher, Paul E.: "Fundamentals of Photography," D. Van Nostrand Company, Inc., Princeton, N.J., 1947.

10. Kehl, George L.: Interference and Profile Microscopy, to be published by American Society for Metals, Cleveland, Ohio.

11. American Standards Association, Surface Roughness, Waviness, and Lay, *ASA, Bull.* 46.1, 1955.

12. Tolansky, S.: "Multiple-Beam Interferometry of Surface Films," Oxford University Press, London, 1948.

13. Burn, Jay R.: Susceptibility of Four Magnesium Coating Alloys to Microporosity and Its Effect on the Mechanical Properties, *AIME Inst. Metals Div. Tech. Pub.* 1955, 1946.

14. Sampson, E. S.: Width Gage for Hot Strip Rolling Mills, *Natl. Electronics Conf.,* vol. 5, p. 341, 1949.

15. Bujes, J. I.: Testing of Ordnance Products by Electromagnetic Radiation, *J. Soc. Nondestructive Testing,* vol. 14, no. 5, pp. 28–31, September–October, 1956.

16. The Institute of Physics: Physics of Nondestructive Testing, *Brit. J. Appl. Physics, Suppl.* 6, 1957.

CHAPTER 3

PRESSURE AND LEAK TESTING

In pressure and leak testing, defects are revealed by the flow of gas or liquid into or through the defects. The simplest and most commonly used pressure test is a hydrostatic test. In this test the pressure within the test object (hollow) is made greater than the external pressure. A simple example is the technique used at service stations to find leaks in an inner tube. In this test the tube is filled with gas at a pressure greater than the surrounding air, and leaks are located by immersion in water with bubble formation. Water, oil, or air or other gases can be used for building up the pressure. The expansive force of compressed air or other gases is relatively great. Since there is always a possibility of a failure under test, the use of air or other gas for testing should be discouraged except under extraordinary circumstances.

For some hydrostatic testing, clean water or water containing dye is used. The temperature of the water should be no lower than that of the surrounding atmosphere; otherwise sweating will result which makes examination difficult and may obscure defects. The hydrostatic pressure should be applied gradually. Test pressures to be used are often stipulated in codes or specifications. Leaks can be detected by water or gas seepage. The presence of leaks can also be revealed by changes in pressure of the liquid or gas being used for applying the internal pressure. Bubble formation when the test object is covered with a soap solution or immersed in a liquid such as water or kerosene can also be used for locating defects. The bubble test is a rather insensitive test compared with the helium leak test to be described later in this chapter.

Hydrostatic Testing. Hydrostatic testing is often required of welded pressure vessels, piping, or valve and fitting sections that can be sealed at their open ends. Such containers are subjected to water pressure of about 1½ to 2 times the working pressure. Back seepage usually is indicated on a limewash coating applied over the surface within 24 hr after the hydrostatic test. Although sometimes near-penetrating defects may enlarge sufficiently to seep water, only the larger defects are generally revealed by this method, such as center-line cracks in welds and pinholes. Fine check cracks may not be indicated. Rules for hydrostatic testing of Unfired Pressure Vessels are given in the ASME Boiler and Pressure Vessel Code[1,*] and those for Pressure Piping are given in the American Standard Code for Pressure Piping.[2] Additional test pressures for piping and tubing are given in the applicable ASTM specifications.

Air pressure testing is primarily used on welded vessels. Air pressure is introduced into the vessel, and leaks are observed by means of soap solutions applied to the outside of the weld or by immersing small vessels under water. Chemical indicator testing is based upon detecting seepage of gases introduced inside the vessel by means of sensitive solutions or gases. Among various methods used, the ammonia phenolphthalein method probably is the best known. The technique consists of cleaning the weld joint surface and applying the white phenolphthalein indicator. An air-ammonia mixture is then introduced into the dry vessel at approximately ½ psig ammonia and 2 psig air pressure. Leakage of the gas through the weld causes the indicator to turn pink. The indicator usually is made of 5 parts 2 per cent phenolphthalein in alcohol, 2 parts distilled water, 10 parts glycerin, and sufficient titanium oxide powder to thicken the solution to thin paint consistency. Another material occasionally used is ammonia–sulfur dioxide.

Bubble Test. The immersion of containers in water and pressurizing with air is used as a crude test in some cases to check for leak tightness; however the sensitivity can be made very high if very high pressures are used. Biram[3] made an investigation of the sensitivity of the bubble test. His work showed that with the right liquid the bubble method can be made extremely sensitive. For comparison purposes, Table 3-1 lists

TABLE 3-1. TYPICAL SENSITIVITIES FOR LEAK FINDING METHODS

Method	Sensitivity, $l\mu sec$
Rate of rise of pressure	Unlimited
NH₃ and indicator	10^{-2}
Freon technique	3×10^{-3}
Freon technique with d-c amplification	10^{-5}
Mass spectrometer	10^{-10}
H₂ Pirani gage	10^{-4}
Ion gage, and variations, e.g., H₂-palladium system	$10^{-3}–10^{-6}$
Air–soap bubbles	10^{-2} is given
H₂-ether or H₂-alcohol	5×10^{-4}, 1 atm gage

the various standard methods employed. Their limiting sensitivity is expressed in the usual vacuum technology unit, the micron-liter per second ($l\mu sec$) which is approximately equivalent to a leak rate of 1 cm³/sec at atmospheric pressure. It will be seen that the limit for the bubble method with the best combination of liquid and gas is 5×10^{-4} to

* Superscript numbers indicate Specific References listed at the end of the chapter.

10^{-3} lμsec for low gage pressures. This is sufficiently sensitive for most practical purposes, including continuously pumped vacuum systems.

One main reason for the insensitivity of testing in water is that comparatively large bubbles are formed. Such bubbles take so long to appear that they can easily be missed. In liquids of low surface tension the bubbles formed are seven or eight times smaller in diameter. Hence they are emitted several hundred times more frequently; they move more slowly and appear as a vertical stream which is clearly visible, under the proper viewing conditions. Also at a given pressure the limiting size of hole which can just emit bubbles, and hence be detected, is smaller for the liquids of lower surface tension. The vacuum leak rate for the smallest detectable hole is found to depend upon the cube of the surface tension of the test fluid.

The best combination was found to be hydrogen and ether, the ether having a low surface tension and the hydrogen a fast flow through small leaks. Both materials are hazardous, and alternatives can be used with some loss of sensitivity, e.g., methanol and helium, which is as good as hydrogen for all but the smallest leaks. Before using such materials the inspector should become familiar with the hazards involved and the necessary safety precautions. With a further loss of sensitivity, compressed air can be substituted, provided that it is obtained from a cylinder. Compressed air lines are not recommended as dirt and oil can block small leaks temporarily. Soap solution, although it has a low surface tension, is also not to be recommended, as it has a similar effect. A pressure of 15 to 20 psi (gage) is convenient and was used in most of the tests, but lower pressures will yield valuable information. For pressure up to about 50 psi suitable plastic or rubber tubing can be used to connect a small gas cylinder, with reducing head, to the specimen. For higher pressures, armored cable should be used and suitable valves and clips.

Pressure should be applied to the test unit before immersion, so that fluid will not enter small leaks. Greater pressure is needed to detect a leak once it has been filled with fluid. The unit may then be immersed in alcohol, acetone, or ether and a careful inspection made for bubbles. If the leak is small, the bubble may be difficult to see until the eye is adapted. A reading glass having a magnification of two or three times at 4 to 5 in. from the specimen will be found of great assistance. There is a minimum size of bubble, so the glass does not introduce a new strain by revealing smaller bubbles. Good lighting is, however, essential, and a dark background may be helpful. A small stream of bubbles may be more easily detectable from above than from the side. If large vessels have to be tested, immersion may be impossible, but channels can often be built around suspected areas to contain alcohol.

The bubble test has been used by Moore[4] in his investigation of surface cracks. The specimen in which a crack is suspected is sprayed with gasoline. The gasoline seeps into any small crack and, as repeated stress is applied, a bubbling mixture of gasoline and air is forced out of the crack. The line of bubbles formed locates the crack. This method is most suitable where the repeated stress can be applied at a rate of at least 10 to 20 cps. Another version of this technique is described by Kiner[5] who suggests pouring a small amount of soap solution on the suspected area of the fatigue specimen, with the test either running or stopped. When the test is continued, any crack present will be indicated by a line of soap bubbles. It is reported that minute cracks can be detected by this technique.

The advantages of bubble testing are that it is cheap, can be done by inexperienced personnel, is rapid, gives accurate location of leak, and the whole specimen is inspected simultaneously. This technique cannot locate very small leaks. In some cases leaks have been known to pass gas in one direction only, and if this is inward the bubble technique will not locate them. The ideal liquid for bubble testing would have a low surface tension and a low viscosity. The bubble size depends on the viscosity of the liquid, pressure, diameter of leak, and degree of vibration.

U.S. Patent 2,751,780[6] describes an apparatus for testing a sheathed object for suspected leaks through the sheath. The specimen to be tested is coated with a liquid having a low viscosity and low surface tension, such as a soap solution or kerosene. The liquid used must wet the sheath and must not boil at reduced pressure. The specimen is then placed in a transparent container and the container evacuated. In case of leakage through the sheath, bubbles will be formed through the liquid because of the lower external pressure surrounding the sheath.

McGonnagle, McLain, and Wood[7] describe a technique, called the vacuum bubble test, for finding small leaks in the fusion welds in the end closures of aluminum-silicon alloy bonded fuel elements that contain an air space between the uranium and the end closure. The welded end of the specimen is inserted part way into a vacuum chamber and sealed with a collarlike gasket. While the fuel element is standing in a horizontal position, a vacuum of about 1 in. of mercury is drawn in the chamber and then the apparatus is rotated 90° to pour a special oil from a reservoir over the weld. As the gas in the air space seeps out through any leaks in the weld, it forms a thin stream of bubbles in the oil covering the weld. This test was found to be most sensitive for the detection of weld leaks when the weld is free from dirt and oxides. Better results were obtained when the chamber was evacuated before the oil was poured over the weld. The oil found most suitable had, at room temperature, a specific gravity of 0.787, a dynamic viscosity of 0.0086 poise, and a surface tension of 25.8 dynes/cm^2.

The Boeing Aircraft Company uses a binaural stethoscope for detection of pressurization leaks in high-altitude bomber components. The testing is carried out in a pressurized area because all leaks must be pinpointed at their source. With an outward pressure differential, several leaks at points on the low-pressure side might originate from a single leak on the high-pressure side.

Pressure and leak tests can be made more sensitive by adding a fluorescent material such as fluorescene to the water and illuminating the surface of the specimen with black light. Dye and fluorescent penetrants can be used to detect through leaks in tubing and similar type vessels. This method of nondestructive testing will be discussed in Chap. 4. Gases such as Freon, ammonia, and helium can be used to make pressure and leak tests more sensitive. Radioactive liquids and gases can be used to improve sensitivity greatly.

Techniques Using Radioactive Material. The Reed-Curtis Nuclear Industries has developed a leak testing technique for inspection of small hermetically sealed parts, called the Radiflo.[8] In the Radiflo technique the specimen to be tested is placed in a large tank, and the tank is then sealed. A radioactive gas is applied to the interior of the tank at approximately 1 atm of differential pressure with respect to the internal pressure of the specimen. The specimens are left in the gas for a sufficient period of time so that in any specimen which has a through hole a sufficient amount of gas will accumulate. After removal from the tank, the specimens are cleaned to remove any radioactive contamination. Each specimen is then placed in front of a radiation detector. The counting rate determined by the radiation detection is directly proportional to the amount of radioactive gas in the part. The radioactive gas is mixed with any desirable gas such as air or nitrogen. The sensitivity of the test can be varied by adjusting the gas pressure, the dilution factor, and the time in the pressure tank. Using this technique it is reported that it is possible to detect, in small hermetically sealed parts, a leak of 1 standard cubic centimeter in 500 years. In theory, this limit can be extended by factors of 10 or more. Bell Laboratories[9] reports the use of a similar technique to test repeater amplifiers for leak tightness.

Putman and Jefferson[10] discuss the use of radioactive sodium for measuring flow and leakage in hydraulic systems. In one technique a length of water main is filled uniformly with a radioactive solution. All outlets from the main are closed, and the pressure is maintained for about half an hour to allow some of the radioactive solution to pass through the leaks to the surrounding soil. The radioactive solution is pumped out, and a check is made of the activity in the earth around the main, particularly at pipe junctions. Leak rates of more than 100 ml/hr are found routinely, and leaks as small as 2 ml/hr have been positively identified.

To localize small leaks in long mains another technique is often used. Both the inlet and outlet valves are closed, and pressure is maintained by supplying water at an opening in the middle of the suspected section. When pressure equilibrium is established, a small volume of radioactive material is injected into the water supply. The radioactive material flows mainly toward the greatest leak, and counters placed on each side of the supply point indicate the direction of the leak. Long lengths of main in which high leak rates occur can be tested by measuring the velocity of water in the main when all the outlets except the leaks are closed. In this technique a small volume of radioactive material is injected into the closed main. Radioactive detectors are placed at convenient points along the main, and by timing the passage of the radioactive material between successive points, the velocity of the water is found. When a point of leakage is passed, an abrupt change in velocity occurs, and by comparing the apparent velocity in the section containing the leak with the velocities in the previous and succeeding sections, the position of the leak can be found.

For detecting leaks in very long mains or oil pipelines, still another technique is used. A quantity of radioactive solution is injected into the pipeline from the input end. After sufficient time has elapsed for the injected solution to have traveled approximately $1\frac{1}{2}$ miles, a battery operated radiation detector and recorder are put into the line. This unit is carried along by the flowing liquid. Positions along the line where the radioactive solution has leaked out are located by the radiation detector. Small gamma-ray sources are placed at convenient intervals against the outside of the pipeline as distance markers. The detection unit is removed at the other end of the line, and from the record the leakage points can be located.

Division of any liquid stream from one channel to another can be readily detected using radioactive material. Under the proper conditions such leaks can be measured quantitatively. Leakage between cross streams in a heat exchanger, Fig. 3-1, offers an excellent example of this technique.[11] The radioactive material is injected in a short surge into the heating stream inlet. Radiation detectors are attached to the exit pipes of both the heating medium and the process streams. Indication of radiation by the detector on the process stream indicates a leak. Leak size is measured by the number of counts on this detector compared with that on the other detector. Leaks of a few tenths of 1 per cent are detectable.

Verkamp and Williams[12] have used radioisotopes for testing leak tightness of nuclear plants. Their technique is to fill the system with a radioisotope solution having an initial concentration of about 0.01 microcure per milliliter of

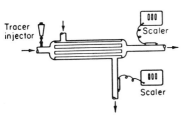

FIG. 3-1. Leak testing with tracers. (*Courtesy Nucleonics.*)

water. All welds, valve bonnets, and vent plugs are wrapped with absorbent tape. The pressure on the system is raised to 1.5 times the design pressure and held for 6 to 8 hr. Then the system is cycled several times, the system is depressurized, and the tapes are removed and counted. From the counting rates obtained, small leaks can be located. In using radioactive material, care must be taken to prevent contamination of the test specimen and/or excessive exposure to personnel. Precautions to be observed in handling radioactive materials will be discussed in Chap. 7. The U.S. Atomic Energy Commission and the National Bureau of Standards have issued information on the handling of radioactive materials.

Halogen Leak Test. The commercial halide torch used for leak testing consists of a tank for gas and a brass plate. The gas is burned and heats the brass plate. In the presence of halogen gas, the flame color changes owing to the formation of copper halide. The flame is also used to inspirate gas through the probe, which is a length of laboratory hose.

The General Electric Company[13] has developed a supersensitive leak detector shown in Fig. 3-2, which is sensitive to and recommended for use only with most gases that contain chlorine, fluorine, bromine, or

Fig 3-2. Halide detector   (*Courtesy General Electric Company.*)

iodine. This includes Freon gas,* Genitron gas,† and the halogen gases. The Freon family of gases and the Genitron family of gases include sulfur hexafluoride, trichloroethylene, and carbon tetrachloride. The halogen gases are chlorine, fluorine, bromine, and iodine. This instrument can detect a leak so small that in a year only 0.01 oz of Freon will pass through the opening. This corresponds to a halogen gas concentration of 1 part per million. The basic principle used in this detector is the fact that positive ion emission from a heated platinum surface is increased in the presence of halogen gases and most of their compounds. A suction fan is used to draw a continuous sample of gas over the heated halogen sensitive element (temperature approximately 1470°F). One precaution which must be observed in using this instrument is to keep the area in which the testing is being done free of vapors from halogen containing compounds; otherwise the presence of leaks may not be detectable because of background contamination. Cigarette smoke will often affect such a device. This leak detector is not a quantitative device and can give only an approximate indication of the size of the leak.

Helium Leak Detector. Helium gas used in conjunction with a commercially available mass spectrometer helium leak detector gives a very sensitive leak test. The commercial helium leak detector can detect the presence of less than 1 part of helium in 10 million parts of air. This corresponds to helium leakage rates of less than 1 cm³ at standard temperature and pressure (STP) per year. The helium leak detector is a portable mass spectrometer especially designed to be highly sensitive to helium gas. A mass spectrometer is an instrument for separating or sorting atoms of different mass. Gas molecules entering the mass spectrometer are bombarded by electrons emitted by a heated filament. This ion beam produced by the electron bombardment is accelerated in the form of a narrow beam by means of an electric field. The ions then pass between the pole pieces of a permanent magnet. The magnetic field deflects the ions in circular paths. The radius of curvature of the path depends upon the mass of the ion. The radius of the path R taken by a particular ion is given by the following expression:

$$R = KV \frac{M}{H} \qquad (3\text{-}1)$$

where K is a constant, V the velocity of the ion, M the mass of the ion, and H the magnetic field intensity. Ions having equal mass will all emerge from the magnetic field at a certain position. A helium leak detector is adjusted so that only helium ions are collected. The flow

---

* Freon is a registered trade mark of the Kinetic Chemicals Divisions of E. I. du Pont de Nemours & Co.

† Genitron is a registered trade mark of the General Chemical Division of the Allied Chemical and Dye Corporation.

of helium ions to the collector constitutes a minute electrical current which can be detected, amplified, and used to activate an electrical meter to control the pitch of an audio signal generator. Figure 3-3 shows a schematic diagram of a helium leak detector and the associated vacuum system. Helium is usually used for leak detection because it is an inert gas and does not react with other gases and materials in the system. Helium is not present in any significant quantity in the atmosphere, thus causing little interference in sensitive leak detection work. Helium, having a light mass, passes through small leaks more readily than heavier gases.

Four different techniques using a helium leak detector will be described: the probe technique, the sniffer technique, the accumulation technique, and the pressurization technique. In all these techniques it is necessary

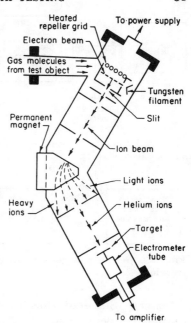

FIG. 3-3. Schematic of helium leak detector. (*Courtesy Veeco Vacuum Corporation.*)

to have clean test specimens since dirt, moisture, scale, and oil may easily seal comparatively large leaks. There is sufficient moisture in the human breath to plug a small leak. The first technique to be discussed, the probe technique, is illustrated in Fig. 3-4. The test specimen is continuously evacuated by the auxiliary pump or pumps. A continuous sample of gas is passed into the leak detector through the throttle valve. The throttle valve is used to control the amount of gas being sampled, in order not to exceed the operating pressure of the leak detector, which is recommended by the instrument manufacturer. High operating pressures can be tolerated but will result in some loss of sensitivity. To detect leaks, a fine jet of helium, such as that obtained from a hypodermic needle, is passed over the exterior surface of the specimen. Helium gas will be drawn into any opening through the walls of the specimen and register on the leak detector as a visible or audible indica-

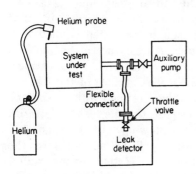

FIG. 3-4. Helium leak testing, probe technique.

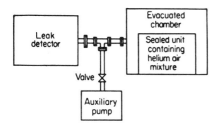

FIG. 3-5. Helium leak testing, envelope vacuum technique.

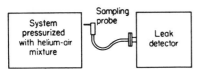

FIG. 3-6. Helium leak testing, sniffer technique

tion. By using a small stream of helium, it is possible to locate precisely the position of an opening or openings in the specimen and get a quantiative measure of its size. Normal air contains an insignificant amount of helium and will not cause any difficulties. However, if the air surrounding the specimen contains a large amount of helium, the presence or position of a leak may not be distinguished because it is masked by the high background indication. The size of leaks found using this technique can be determined by using a calibrated leak. Calibrated leaks can be obtained from commercial suppliers of helium leak detectors. Standard leaks are customarily calibrated in standard cubic centimeters per second. If a pressure differential of 15 psi is applied across a 10^{-6} leak, it would take 6 months to fill a 1 in.^{3} volume with helium. A 10^{-7}, 10^{-8}, or 10^{-9} leak under the same pressure differential would require 5, 50, and 500 years, respectively, to fill the same volume with helium. The pressure in the specimen and the probing speed affect the sensitivity of the test. For example, increased pressure in the test specimen increases the sensitivity of the test.

Sometimes it is desired to determine only the presence of leaks or the total magnitude of all the leaks. In such a case the specimen can be surrounded with a helium atmosphere by putting it into some sort of gastight chamber, such as shown in Fig. 3-5. A plastic bag often makes a satisfactory chamber.

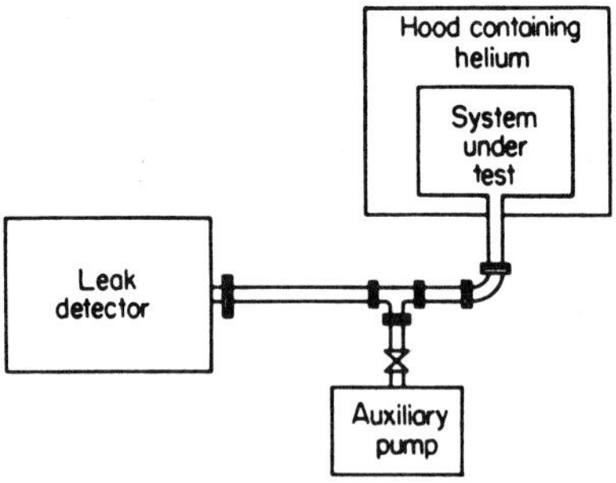

The third technique is shown schematically in Fig. 3-6. The specimen to be tested is filled with helium or a mixture of helium and air to a pressure greater than atmospheric. The surface of the test object is then scanned with a "sniffer" connected to the leak detector. Helium flowing out through any openings will be sucked into the leak detector system by the sniffer. A variation of this technique is to fill the test specimen with helium

FIG. 3-7. Helium leak testing, envelope pressure technique.

at any pressure. The specimen is then placed in a chamber connected to the leak detector, as shown in Fig. 3-7. The latter chamber is partially evacuated. Helium will flow through any leaks into the evacuated chamber and then to the leak detector. This latter variation has the advantages that it gives the over-all leak rate of the specimen. In using the sniffer, the presence of an excessive amount of helium in the atmosphere surrounding the specimen may mask indications and position of leaks. In the accumulation technique a static supply of helium is contained in the specimen. The specimen is placed in a glass chamber and the chamber evacuated. The helium leakage from the specimen is allowed to accumulate for a given length of time. Then a gas sample from the accumulation chamber is analyzed for helium. This technique increases the sensitivity of the test.

If the specimen to be tested is to be used at elevated temperatures, the leak testing should be carried out at the operating temperature. Testing at elevated temperatures can be carried out with slight modifications of the techniques previously described.

The fourth technique, the pressurization technique, can be used to test specimens in which there is no way to attach the leak detector or a source of helium gas. This technique has found application in the nuclear reactor field for detecting minute holes, cracks, and fissures in the cladding and end weld closures. In applying this technique, the specimen is first placed in a helium pressurizing vessel and exposed to a helium atmosphere, Fig. 3-8, detail *A*. The pressure and the pressurization are not

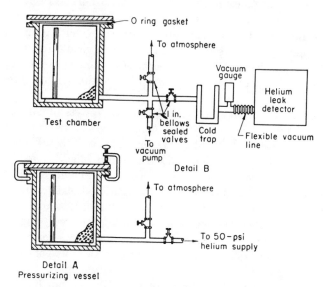

FIG. 3-8. Helium leak testing, pressurization technique. (*Courtesy Argonne National Laboratory.*)

critical. The specimen is removed from the pressurizing vessel and transferred to a second chamber which is connected to a vacuum pump and helium leak detector, Fig. 3-8, detail B. The latter chamber is first evacuated to a pressure of about 150 microns, then the throttle valve on the leak detector is opened. The relative quantity of helium released is read on the leak indicating meter. Typical results for a group of 128 welded fuel elements are given in Table 3-2. For comparison, a clean

TABLE 3-2. TYPICAL HELIUM LEAK TEST RESULTS

Leak detector reading (arbitrary units)	Number of samples in range
0.0–0.5	48
0.6–1.0	5
1.1–1.5	18
1.6–2.0	19
2.1–3.0	30
3.1–4.0	3
4.1–5.0	3
12	1
14	1

aluminum rod gave a relative response of 0.10. The two specimens which gave excessive response were destructively tested and found to have defective welds. It has been found possible to detect the presence of one which has an excessive response in a small group with normal response. Dirt, moisture, oil, and scale on the specimen cause a large response, but specimens that have had an acid dip are sufficiently clean for testing. A large open cavity (such as a toolmark which is visible to the eye) does not produce a noticeable response because helium is not retained long enough after the specimen is removed from the helium pressurizing vessel.

The relative sensitivity of the helium leak test is illustrated by the following. Pappin[14] of the Knolls Atomic Power Laboratory reports that a 30-ft section of 2-in. pipe was first tested with a halogen sniffer and soap bubble test. A helium leak test found 28 additional leaks. In another system, reported to be leak-tight after being checked with a halogen sniffer and soap bubble test, 68 additional leaks were found.

Thermal Leak Detector. Minter,[15] at the Naval Research Laboratory, has developed a thermal conductivity leak detector. Four tungsten filaments mounted in a metal block are connected in the form of a Wheatstone bridge. The bridge circuit is connected to a source of d-c potential. The output of the bridge is connected to a sensitive microammeter or to a null point inductance with a d-c amplifier. The author has found that leaks detected by the mass spectrometer are detected by this instrument.

Sonic Leak Detection. In this technique the component or system is pressurized to its operating pressure or evacuated. The escaping gas or the flow of gas into the system produces detectable sound. The background noise affects the size of the leak that can be found. The sensitivity of the technique can be increased by use of electronic devices but the practical limit of sensitivity is a leak rate of 1 cc/sec.

SPECIFIC REFERENCES

1. ASME Boiler and Pressure Vessel Code, section VIII, Unfired Pressure Vessels, American Society of Mechanical Engineers, New York, 1956.
2. American Standard Code for Pressure Piping, *ASA, Bull.* 31.1, 1951.
3. Biram, J. G. S.: Bubble Leak Detection, *Atomic Energy Research Establ. (G. Brit.)*, March, 1957.
4. Moore, H. F., and F. C. Howard: A Metallographic Study of the Path of Fatigue Failure in Copper, *Univ. Ill. Eng. Expt. Sta. Bull.* 176, Mar. 15, 1927.
5. Kiner, G. B.: Early Detection of Fatigue Cracks, *Metal Prog.*, vol. 45, no. 1, p. 89, January, 1944.
6. Plott, Robert F.: Leakage Testing Apparatus, U.S. Patent 2,751,780, June 26, 1956.
7. McGonnagle, W. J., Stuart McLain, and E. C. Wood: Applications of Nondestructive Testing to Fuel Elements for Nuclear Reactors, *J. Soc. Nondestructive Testing*, vol. 15, no. 2, pp. 86–90, March–April, 1957.
8. Reed, Clifton W.: Atomic Energy in Quality Control of Hermetically Sealed Parts, *J. Soc. Nondestructive Testing*, vol. 15, no. 5, pp. 266–268, September–October, 1957.
9. Bell Labs Test Leaks with Cs^{134}, *Nucleonics*, vol. 14, no. 7, p. 82, July, 1956.
10. Putman, J. L., and S. Jefferson: Application of Radioisotopes to Leakage and Hydraulic Problems, *Proc. Intern. Conf. Peaceful Uses Atomic Energy*, Geneva, 1955.
11. Leak-Testing, *Nucleonics*, vol. 13, no. 4, p. 21, April, 1955.
12. Verkamp, J. P., and S. L. Williams: Testing Nuclear Plant Leak Tightness, *Nucleonics*, vol. 14, no. 6, pp. 54–57, June, 1956.
13. General Electric Company: Leak Detector Type H-1, *Bull.* GEC-233E, 1954.
14. Pappin, W. H.: Helium Leak Detection Techniques, *ASTM, Spec. Tech. Pub.* 223, 1958.
15. Minter, Clarke C.: Thermal Conductivity Leak Detector, *Rev. Sci. Instr.*, vol. 29, no. 9, pp. 793–794, September, 1958.

CHAPTER 4

LIQUID PENETRANT INSPECTION

Penetrant inspection is a nondestructive testing method that can be used for the detection of surface discontinuities or flaws which extend to the surface of the test specimen. The use of penetrants may be considered as an extension of visual inspection. Very few discontinuities are revealed by penetrants that cannot be found by a well-trained visual inspector. Penetrants, however, delineate a discontinuity to a much greater extent, making the inspection much less dependable on the human element. This makes the method more adaptable to production testing by increasing the general reliability and speed of inspection. Penetrant techniques are applicable to all metals as well as to glazed ceramics, plastics, and other nonporous materials. A special penetrant is used for locating flaws on a porous surface such as unfired ceramic materials. The penetrant method is applicable to either magnetic or nonmagnetic materials where magnetic particle inspection cannot be used. Liquid penetrant inspection has the advantage of being a quick, easily applied, relatively inexpensive and reliable test. The primary limitation and disadvantage of penetrant inspection is that only surface defects or defects which open to the surface are revealed. All defects found by use of penetrants give only an approximate indication of the depth and size of the flaw. After considerable experience, some inspectors can make fairly reliable estimates of the depth and size of defects.

Penetrant techniques can be used to locate grinding cracks, welding cracks, casting cracks, fatigue cracks, shrinkage, blowholes, seams, laps, pores, cold shuts, porosity, lack of bonding, pinholes in welds, through cracks, forging laps and bursts, gouges, tool marks, and die marks, provided that the defects are open to the surface. The defects must behave as capillaries to draw in the penetrant and retain it after the excess material has been removed. Shallow or wide open flaws are difficult to detect because the penetrant is easily removed. Fortunately, such open flaws are usually located during a visual examination. Surfaces must be free of any material which will block the openings. In the case of small tubing, the use of penetrants is limited to the outer surface of the tubing because of the inaccessibility of the interior.

Oil and Whiting Technique. The simplest and oldest penetrant test is the oil and whiting test. A penetrating oil such as kerosene is applied to the surface to be inspected. After allowing the oil sufficient time to penetrate any surface defects, the excess oil is completely removed and the surface thoroughly dried. A thin coating of whiting (calcium carbonate) either as a dry powder or mixed with alcohol is applied to the surface. When dry, the whiting has nearly the same refractive index as the oil. After a period of time the oil seeps out of the defects into the whiting, causing an appreciable reduction in whiteness. The oil and whiting technique was widely used to inspect railroad axles in railroad shops. After the coat of whiting was dry, the axle was rotated and struck with a hammer to help force the oil out of the minute cracks. Hot oil is sometimes used because of the lowered surface tension and viscosity of the oil. In addition, the heat causes the crack to expand slightly. Ouwerkerk and Binkharst[1,*] indicate from their studies that this technique is limited to cracks in excess of 0.02 mm in width.

A variation of the oil and whiting test is the oil vapor blast test. The surface is first thoroughly cleaned and then soaked in kerosene for a few minutes. After soaking, the surface is vapor blasted with clean grit (approximately 100 mesh), washed in water to remove any extra grit, and allowed to air dry. Indications will appear on the vapor blasted surface because of seepage of the kerosene. However, these indications are not too permanent and may disappear in a relatively short period because of vaporization of the kerosene. In another technique of this general type the sample is first etched with dilute hydrochloric or sulfuric acid to enlarge any cracks which may be present. The sample is covered with a 3 per cent solution of tannic or gallic acid which is allowed to dry. After the sample is polished, cracks will appear black on a bright background. Flaws produced during brazing and grinding of carbide tips can be found by blasting the tip with fine abrasive in a fluid medium. The abrasive and hydraulic actions combine to force fine sintered material out of invisible cracks to become visible defects.

A satisfactory penetrant can be made from a mixture of Du Pont bleeding red dye, oleic acid, and fuel oil (diesel). A developer which can be used with this penetrant is desiccated magnesium dissolved in alcohol or talc in alcohol.

The New York Naval Shipyard compound the following materials to make a penetrant for their use: diesel fuel oil, 870 cm^3; cutting oil, 80 cm^3; oleic acid, 50 cm^3. One ounce of red alizarin dye in powdered form is added to the 1,000-cm^3 mixture and stirred until the dye has gone completely into solution. Commercially procurable talcum powder dissolved in alcohol is used as the developer.

The oil and whiting test and the oil vapor blast test are less sensitive than the penetrants which are now available commercially. Those now in use include Zyglo, Zyglo-Pentrex, and Spotcheck (sold by Magna-

* Superscript numbers indicate Specific References listed at the end of the chapter.

flux Corporation), Met-L-Chek and Flaw Findr (sold by Met-L-Chek), Dy-Chek and Chek-Spek (sold by Turco Products), and Dyeline (sold by Zaco Laboratories). Table 4-1 gives a summary of the commercially available penetrants and some of their properties. Glo-Crack (fluorescent) and Hyglo are used in Great Britain and Europe.

TABLE 4-1. COMMERCIALLY AVAILABLE LIQUID PENETRANTS

Trade name	Dy-Chek	Met-L-Chek	Spotcheck	Zyglo
Manufacturer	Turco Products, Inc.	The Met-L-Chek Co.	Magnaflux Corp.	Magnaflux Corp.
Penetrant type	Liquid red dye	Liquid red dye	Liquid red dye	Fluorescent liquid
Applied by	Brush, spray, or dip	Brush, spray, or dip	Spray can	Dip, brush, or spray
Temperature range, °F.	70–150	70–130	Above 60	Above 60
Removed by	Solvent	Water	No. 1 cleaner	Water
Developer	White powder suspension	White powder suspension	Volatile liquid suspension of white powder	Powder or liquid
Indication visible by	White light	White light	White light	Near ultraviolet light
Auxiliary apparatus	None	Source of water	None	Sources of water and electric power

Commercial Penetrants. Penetrants can be divided into two kinds, dye penetrants and fluorescent penetrants. In the case of dye penetrants, a dye is dissolved in the liquid penetrant. The color of the dye is selected to give a high color contrast between the developer and penetrant. In the fluorescent penetrants a fluorescent material is dissolved in the penetrant. All petroleum products fluoresce to a certain degree, so that fluorescent penetrants have an oil base. A fluorescent additive is added to increase fluorescence. Fluorescent penetrants must be viewed under near ultraviolet or black light of approximately 3,650 A. A minimum of 90 to 100 ft-c should be used. One of the most recent developments in the field of penetrants is the introduction by the Magnaflux Corporation of Zyglo-Pentrex. It differs from Zyglo in that the penetrant has been separated from the emulsifier. The penetrant is thus designed for pure penetrating ability, and penetration is more rapid.

Technique. The technique used in penetrant inspection is essentially the same for all penetrants. The basic steps in their application are:

1. Cleaning the surface of the specimen
2. Applying the penetrant

3. Removing the excess penetrant
4. Applying the developer
5. Inspection and interpretation

With the postemulsification type of penetrant an additional step is required in the procedure. To make the penetrant water washable, an emulsifier must be added when it is time to remove the excess penetrant. The postemulsification type of penetrant will locate smaller defects than other types of penetrants because the addition of the emulsifier to the liquid before applying to the test surface increases its surface tension and viscosity.

Cleaning the Surface. It is absolutely essential that the surface of the specimen to be tested be free of dirt, lint, wax, paint, grease, scale, or any other material which will fill or clog the surface openings. Penetrant held by the dirt or other material on the surface can also give false indications. Liquid solvents, vapor blasting, vapor degreasing, and acid etching can sometimes be used to prepare the surface properly. Suppliers of commercial penetrants often provide a suitable solvent. Sandblasting is not recommended for cleaning the surface, since it has the tendency to close up small surface openings. Scale is best removed by vapor blasting, which will not close up or cover existing defects. Buffing and certain other surface-finishing operations tend to bridge over the flaws. Such operations should not be performed before the metal is inspected. Surface treatment involving use of strong acid or caustic solution tends to reduce the fluorescing characteristics of fluorescent penetrants. Such treatment should be postponed until after the inspection has been made if a fluorescent penetrant is to be used.

Applying the Penetrant. The penetrant is applied to the surface of the test object by immersion, brushing, or spraying. The latter method undoubtedly gives the more uniform coat of material. Because of the low surface tension, penetrant is readily drawn into small surface openings by capillary action. The action of the penetrant is shown in Fig. 4-1. If the surface of the specimen is warm when the penetrant is applied, better results will be obtained because the surface openings will be slightly expanded and therefore more easily penetrated. The rate of penetration may also be increased by raising the temperature of the penetrant. However, some suppliers do not recommend heating their particular penetrant because of the relatively low flash point of the penetrant. Each supplier of commercial penetrants recommends the temperature best suited for his penetrant, see Table 4-1. Specimens to be tested are sometimes heated and then

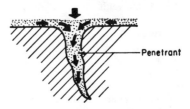

Fig. 4-1. Action of penetrant. The penetrant can seep only into flaws that reach the surface.

immersed in the cooler penetrant.   The drop in temperature causes a low-pressure area in the defect.   This pressure differential may aid in drawing penetrant into the defect.   Striking or vibrating the test object helps to open up the defect for penetration.

The penetrant is allowed to remain on the surface for a period of time which depends on the material, type and size of defect, and penetrant used.   The penetrant must be allowed to remain on the surface long enough so that a sufficient quantity enters the defect to obtain subsequent development of flaw indications.   To detect very fine flaws, the penetrant may be applied more than once to the same surface.   The exact penetration time can be determined only by experimentation on the particular specimen which is to be inspected.   By varying penetration time, the optimum time for a particular size and kind of defect can be determined.   Table 4-2 shows typical penetration times for Dy-Chek and Spotcheck as recommended by their manufacturers.   Table 4-3 shows typical penetration times for Zyglo-Pentrex.   To find the proper penetration time, a typical specimen should be carried through the entire procedure, allowing a fixed time in the penetrant.   If indications did not develop, the procedure should be repeated, varying the penetration time, until indications of the defect is obtained.   The proper penetration time may be excessive for a particular part, in which case a compromise must be made.

TABLE 4-2. TYPICAL PENETRATION TIMES FOR TWO DYE PENETRANTS

Material	Type of defect	Penetration time, min	
		Dy-Chek (60–90°F)	Zyglo
All..............................	Heat-treatment cracks	3–5	2
	Grinding	7–10	10
	Fatigue	7–10	10
Plastics.........................	Cracks	3–5	1–5
Ceramics.......................	Cracks	3–5	1–5
	Porosity	3–5	1–5
Aluminum welds.................	Cracks and pores	3–5	10–20
Steel welds.....................	Cracks and pores	7–10	10–20
Forgings.......	Cracks	7–10	20
	Laps	7–10	20
Metal rollings...................	Seams	7–10	10–20
Die castings....................	Surface porosity	3–5	3–10
	Cold shuts	3–5	10–20
Metal-permanent mold casting.....	Shrinkage porosity	3–5	3–10
Carbide tipped cutting tools........	Poor braze	3–5	1–10
Cutting tools....................	Cracks in steel	3–5	1–10
	Cracks in tip	3–5	1–10

TABLE 4-3. TYPICAL PENETRATION TIMES FOR ZYGLO-PENTREX

Material	Penetration time, min
Aluminum:	
Castings	5–15
Forgings	30
Welds	30
Fatigue	30
Magnesium:	
Castings	15
Forgings	30
Welds	30
Fatigue	30
Stainless steels:	
Castings	30
Forgings	60
Welds	60
Fatigue	30
Brass and bronze:	
Castings	10
Forgings	30
Brass	15
Fatigue	30
Plastics	5–30
Glass	5–30
Carbide tipped tools:	
Poor braze	5
Cracks	30
Reground	10

Removing the Excess Penetrant. All traces of penetrant must be removed from the surface of the specimen, using tap water or a solvent recommended or supplied by the manufacturer. Removing the excess penetrant is a critical operation. Too much washing may remove the penetrant from the flaws, and insufficient washing will leave penetrant on the surface. Penetrant remaining on the surface will give false indications or mask true indications. The specimen is then dried by a hot-air dryer, by a blast of clear dry air, or by standing in air.

In the postemulsification technique the emulsifier is applied over the penetrant. The emulsifying agent diffuses into the penetrant film, rendering it water soluble. The time interval allowed for this diffusion process controls the ability of the penetrant to reveal broad shallow defects. If the diffusion time is short, these defects are more likely to be revealed since with longer times the emulsifier would diffuse into the part of the penetrant film. The emulsified penetrant and the excess emulsifying agent are removed by a water wash.

Applying the Developer. Developer is applied to the dried surface. The developer usually consists of finely divided powder such as talc. The purpose of the developer is to draw the penetrant from the defect

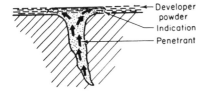

FIG. 4-2. Action of developer. When the developer dries, it draws the penetrant out of the crack onto the surface on each side of it, increasing the size of the indication.

by a blotting action and spread it on the surface of the test object for a small area around the defect. The action of the developer is shown in Fig. 4-2. This magnifies the indication and makes it possible to locate small defects. The rate of seepage out of the defect may be improved by heating the part. Striking or vibrating the specimen aids in forcing the penetrant out of the defects.

Two kinds of developers are used, a dry developer and a wet developer. Both types have about equal sensitivity. In using either type of developer, it is desirable to apply it in as light a coat as possible. Dry developer may be applied with a spray bulb or powder gun or by dipping small parts in the powder. Dry powder has a tendency to build up in any complex contours of the specimen.

Wet developer consists of a powder suspended in a volatile or quick-drying liquid. Wet developer may be applied by dipping or spraying. Care must be used in applying wet developer. If the developer is applied too sparingly, it will not cover all surfaces and will not adequately perform its intended function. If too liberally applied, it will wash out the penetrant. Hot air is recommended for drying the wet developer. The drying serves the twofold purpose of evaporating the developer solvent and heating the specimen which helps bring the penetrant to the surface.

Inspection and Interpretation. The fifth step in the procedure, inspection and interpretation, is the most important operation in the entire procedure. The time between developing and inspection may vary from a few minutes to several hours. The type and thickness of the developer coating determine the minimum defect size that is detectable. If a thick coating is used, small defects will not be revealed, since there is not sufficient penetrant to diffuse completely through the coating. If a dry developer is used, even the finest and smallest discontinuity will be revealed. The type of indication depends on the type of penetrant used. If a fluorescent penetrant is used, defects will show up as glowing yellow-green dots or lines against a dark background. In dye penetrants, defects are indicated as red dots or lines against a white background. Figures 4-3 to 4-5 show typical defect indications. In the oil and whiting method the penetrant will darken the whiting. In the kerosene sandblast method the seepage of kerosene will darken the surface.

The interpretation of the characteristic patterns indicating the types of flaws is of extreme importance. A crack or cold snut will be indicated by a line of penetrant. Dots of penetrant indicate pits or porosity. A series of dots indicates a tight crack, cold shut, or partially welded lap.

(c)   (d)

FIG. 4-3. Defect indications. (a) Zyglo detects and marks defects in aluminum copilot balance control. (b) Forging crack in helicopter landing-gear cap detected with Pentrex after anodizing. (c) Brazing and grinding cracks in carbide tools shown with Zyglo. (d) Welding porosity and cracks in railroad diesel valves indicated by Zyglo and Pentrex. (*Courtesy Magnaflux Corporation.*)

A rough estimate of the opening may be estimated by the width of the indication or spreading of the penetrant on the developer. Proper interpretation can only be learned by experience and consequently should only be done by experienced inspectors. The manufacturers' instructions should be followed carefully.

Test Blocks. Test blocks for gaining experience with penetrants can be made. A flat piece of oil-quenched tool steel is heated to approximately 1500°F and then quenched in water. A large number of cracks will be produced. The Magnaflux Corporation suggests the following test block. The test block is made of 2024 aluminum, 2 by 3 by ⅛ in. The blocks are heated to 950°F and then quenched in cold water. This

FIG. 4-4. Defect indications. (*Courtesy Turco Products, Inc.*)

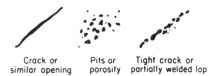

Crack or similar opening   Pits or porosity   Tight crack or partially welded lap

Large crack or opening	Crack or cold shut	Fatigue crack· or lap	Porosity or pits

Fig. 4-5. Defect indica-
tions. (*Courtesy The Met-
L-Chek Company.*)

produces a heat-treating crack pattern that runs both around the hot spot and lengthwise on the test block.

A publication from The Central National Council for Applied Scientific Research in the Netherlands (T.N.O.) describes a test specimen in which cracks ranging from very small to fairly large can be produced. A tapered steel rod was made, and 10 transverse holes were drilled along the length of the rod. The rod was loaded on a torsion bench and a torque applied to produce fatigue. A diagram of the rod and the cracks produced is shown in Fig. 4-6. Because of the taper of the rod, the shearing stress increased along the length of the rod. Thus, when a load was applied, an increase in stress occurred along the edges of the holes. This increase in stress is dependent on the ratio of the hole diameter to the rod diameter. Because of the differing stresses, cracks of different sizes are produced. Table 4-4 shows the sizes of cracks found. The holes are numbered 1 to 10 in the order of decreasing stress. The edges of the holes are marked A on one side and B on the other wide of the rod.

Equipment. Completely mechanized automatic or semiautomatic equipment for penetrant inspection is sometimes used in industry. Figure 4-7 shows such an automatic conveyorized Zyglo processing unit for inspection of castings. Figure 4-8 shows an installation for the

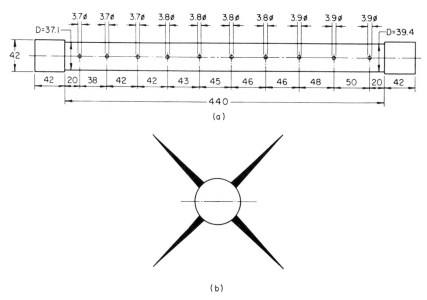

(a)

(b)

Fig. 4-6. Rods with cracks.

TABLE 4-4. CRACK LENGTHS BY MET-L-CHEK

Hole	Crack size, mm				Hole	Crack size, mm			
	I	II	III	IV		I	II	III	IV
A 1	17.1	5.8	10.9	4.6	B 1	4.0	5.3	4.2	5.0
A 2	2.7	5.7	2.5	2.8	B 2	3.6	3.8	3.9	3.4
A 3	1.9	1.7	1.9	1.8	B 3	2.9	2.8	2.8	2.5
A 4	2.0	2.0	2.5	1.0	B 4	1.0	2.6	2.5	2.5
A 5	1.3	1.5	1.3	1.4	B 5	0.5	2.0	1.0	1.9
A 6	0.5	1.4	0.8	1.5	B 6	...	0.5	2.0	1.6
A 7	1.0	0.5	0.5	0.5	B 7	2.0	1.5	1.9	1.5
A 8	1.0	0.6	1.2	1.4	B 8	2.0	1.3	1.5	1.9
A 9	1.3	0.7	0.4	1.5	B 9	0.5	1.0	1.3	1.2
A10	1.0	...	0.2	1.0	B10	0.5	1.5	1.0	1.0

penetrant inspection of aircraft components. The Battelle Memorial Institute[2] has developed a mechanical device for scanning specimens for fluorescent penetrant indications.

Precautions. The manufacturers' precautions concerning the safe handling of the particular penetrant should be explicitly followed. Some of the penetrants are volatile, and others have a low flash point. The vapors arising from the penetrant may be toxic; so adequate ventilation for the inspection personnel should be provided at all times.

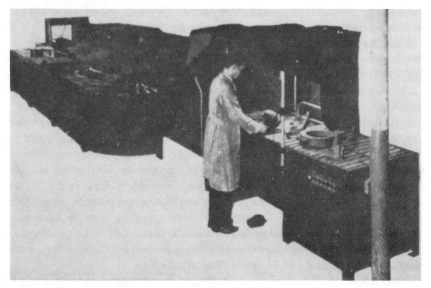

FIG. 4-7. Automatic Zyglo unit for castings. (*Courtesy Magnaflux Corporation.*)

Fig. 4-8. Penetrant installation for aircraft components. (*Courtesy Magnaflux Corporation.*)

The oil base of the penetrants may cause irritations of the skin. The skin can be protected by wearing a Neoprene glove. Washing with soap and water will remove penetrant from the skin or clothing. Dry developing powders are usually not toxic. Special precautions should be taken to prevent breathing of the powder by the operator.

Black lights are noninjurious to the eyes or skin, but such sources should not be used directly. The liquid in the eyeball glows in black light but this is not harmful. The fluid produced by the liquid in the eye gives the viewer the sensation that a cloud or mist is formed over the eye.

Applications. Dolan[3] has used penetrant techniques to detect fine fatigue cracks in boiler plate specimens. He used a moist coating method in which the coating is made with ordinary white lead thinned with a very low-viscosity mineral oil to retard tendencies to drying on the surface. The coating is usually brushed on with a fine camel's hair brush, with the final brush marks in a direction perpendicular to the axis of the expected cracks. A somewhat better coating can be obtained by spraying. With this solution, Dolan reports that cracks approximately $\frac{1}{32}$ in. long can be detected by the darkening of the coating as the solution seeps in and out of the surface cracks.

Application of this method to rotating beam fatigue specimens has been made by workers at the University of Minnesota. With a coating of the proper consistency of white lead and mineral oil, cracks of the order of 0.005 in. were observed, with the coating brushed on reversed bending specimens. No indication is found in cracked specimens unless alternating strain is applied for several cycles. A major advantage of the technique is that the coating can be applied at the start of a fatigue test and then visually examined for crack indications at regular intervals throughout the test. Only one application is necessary. The coating on the specimen must be protected from dirt and lint during the test for clear evidence of cracking. Comparative tests indicate that continuous application of penetrant facilitates detection of fatigue cracks at an early stage in their development during the course of a fatigue test. A practical use in crack growth studies is to locate the points of cracking and then make plastic replicas of the surface for examination at high magnification. Crack growth may be followed by making subsequent replicas. Proper precautions should be taken when employing lead-containing coatings.

Radioactive Penetrant. A method of surface crack detection utilizing radioactivity is described by Mehl.[4] The specimen is immersed in a radioactive liquid and subjected to a pressure sufficient to cause the fluid to penetrate into the cracks. After cleaning the surface of the specimen, the amount of radioactive fluid retained by the open defects may be determined either by a photographic method or by a suitable radiation detector. Kaiser[5] and Kaindl and Mathiaschitz[6] also describe a technique using radioactive material.

Makin[7] has used oil containing radioactive Pd^{109} for determining porosity in magnesium alloys. The specimens are covered with hot oil, and pores which lead to the surface of the specimen are filled with the radioactive oil. Autoradiographs are taken of the surface of the specimen. The density distribution on the photographic film shows the extent of the pores in the surface of the specimen. Pores extending from a surface area 1.5 by 10^{-5} cm^2 and within 0.4 mm of the surface have been detected.

Filtered Particle. Filtered particle[8–10] inspection can be used to locate defects in porous surfaces such as unfired dried clay, certain fired ceramics, concrete, some powdered metals, carbon, and partially sintered tungsten and titanium carbide. In this method a fluid containing suspended particles is applied to the porous surface. The liquid is absorbed in the specimen. Because of the extra absorbent area provided by a defect, the fluid is absorbed to a greater extent at the defect areas. As the fluid is absorbed by the porous surface, the suspended particles being larger than the defect opening are filtered out and left behind on the surface to form an indication. The suspended particles can give a visible indication, or particles coated with a fluorescent pigment can be used to give higher contrast.

A variation of this basic method is to incorporate a dye in the solution. Thus when a defect is located it is possible to trace the extent of the flaw by the staining of the side walls of the crack. This method requires that the surface be porous. However, even on very porous material the differential absorption effect is not pronounced enough to detect flaws. For example, Staats[10] points out that the type of porosity found in a 60-mesh grinding wheel does not exhibit sufficient differential absorption. The method just begins to operate on the type of porosity found in 100-mesh grinding wheels. The commercially available materials consist of a light petroleum distillate carrying in suspension a wide range of particle sizes in the micron range. A filtered particle method is sold under the trade name Partek by the Magnaflux Corporation.

In the chapter on leak testing, it was mentioned that penetrant inspection could be used for detection of through leaks. The procedure is to apply the penetrant to one side of the specimen. The other side is examined for visible indications of the dye or for fluorescent material under black light. The use of a developer can be used to intensify the indications. The penetration time will depend on the type and thickness of the object. Coarse porosity in the leakage path slows penetration because the capillary action is reduced. This method is particularly effective on thin-walled vessels, but can be used for walls up to about $\frac{1}{4}$ in. thick. It will find defects in thin-walled vessels that are not found by the usual air test with pressures of 5 to 20 psi. Trapped moisture or dirt in the leakage path will prevent penetrations.

SPECIFIC REFERENCES

1. Ouwerkerk, L. V., and D. J. Binkharst: Verglerch der Kalmilchprufung mit dem Magnetpulverfahren zum Nachweiss von Hoarrissen, *Wärme*, vol. 64, pp. 357–359, 1941.
2. Wenk, S. A., K. D. Cooley, and R. M. Kimmel: Photoelectric Scanning of Fluorescent Indications, *J. Soc. Nondestructive Testing*, vol. 11, no. 1, pp. 28–31, Summer, 1952.
3. Dolan, T. J.: Basic Concepts of Fatigue Damage in Metals, Fatigue, a series of lectures presented to the American Society for Metals during 35th National Metal Congress, Cleveland, Ohio, 1953.
4. Mehl, R. F.: Nondestructive Testing with Gamma Rays, *Iron Age*, vol. 127, p. 1651, 1931.
5. Kaiser, H. F.: Possible Use of Radioactive Substances in the Testings of Metals, *Trans. ASM*, vol. 27, pp. 403–427, 1939.
6. Kaindl, K., and A. Mathiaschitz: Nondestructive Crack Detection Method Using Radioactive Indicators, *Werkstoffe u. Korrosion*, vol. 2, pp. 368–369, October, 1951.
7. Makin, S. M.: Porosity Detection in Magnesium Alloys Using Paraffin Containing Pd[109], *Intern. J. Appl. Radiation and Isotopes*, vol. 4, pp. 253–255, 1959.
8. Betz, C. E.: Two New Testing Methods for Ceramic Products, *J. Soc. Nondestructive Testing*, vol. 7, no. 2, pp. 22–26, Fall, 1948.
9. Staats, Henry N.: The Testing of Ceramics, *J. Soc. Nondestructive Testing*, vol. 10, no. 3, pp. 23–26, Winter, 1951–1952.

10. Staats, Henry N.: Filtered Particle Inspection of High Tension Insulators, *J. Soc. Nondestructive Testing*, vol. 11, no. 3, pp. 21–24, January, 1953.

ADDITIONAL REFERENCES

Carson, H. L., W. G. Cook, and R. Schnurmann: Symposium on Principles of Penetrant Methods—Physics of Nondestructive Testing, *Brit. J. Appl. Phys. Suppl.* 6, pp. 518–523, 1957.

Met-L-Chek, Visible Penetrant Dye Method Finds Defects in Jet Parts Faster at Less Cost, *Western Metals*, December, 1954.

Migel, Hamilton: Magnetic Particle, Penetrant and Related Inspection Methods as Production Tools for Process Control, presented at SAE Summer Meeting, June, 1954.

Migel, Hamilton: Activities of Subcommittee III on Magnetic Particle and Penetrant Testing, *ASTM Spec. Tech. Pub.* 213, p. 11, 1957.

Schnurmann, R.: Penetrant Methods of Inspection, *Research Applied in Industry (London)*, pp. 254–257, July, 1958.

CHAPTER 5

THERMAL METHODS

The basic principle utilized in thermal tests is to apply heat to the test specimen and measure or observe the resulting temperature distribution. Flaws alter the temperature distribution on or in the specimen. The heat may be applied by direct thermal contact with a heat source, by electrical current heating, by induction heating, or by infrared heat sources. The resulting temperature distribution may be detected by use of temperature indicating substances. For example, wax, stearin, frost, Tempilstik, Tempilaq, Thermocolor, Thermochron, formation of characteristic oxides, temperature sensitive phosphors, infrared film, thermocouples, bolometers, resistance thermometers, photoconductive materials, or the color of thin oil films may be used.

The term "temperature" is often used loosely and occasionally interchangeably with the word "heat." Before discussing nondestructive thermal testing, it is perhaps advisable to define the term temperature. The dictionary defines it as the "degree of coldness or hotness measured on a definite scale." Several different scales of temperature measurement are now used. However, only three of these will be mentioned here. The oldest temperature scale was that devised by Fahrenheit in 1714. It used as its zero point the temperature of a mixture of snow, salt, and sal ammoniac. The high end of the scale was determined by the temperature of a normal human body. Almost universally accepted now for scientific use is the centigrade temperature scale formulated in 1742 by the Swedish astronomer Celsius. In this scale, the freezing and boiling points of pure water at normal atmospheric pressure have been assigned the temperatures of 0 and 100°C, respectively. Lord Kelvin formulated an absolute temperature scale using the previously defined centigrade scale. On the absolute or Kelvin scale, the freezing point of water is 273.2°K, and the boiling point is 373.2°K. Other important fixed points on the international temperature scale are shown in Fig. 5-1. Since all radiation phenomena are a function of the absolute temperature, the absolute or Kelvin scale is the one of importance in radiation measurements.

A change in temperature of a body is associated with the transfer of heat energy to or from the body. This energy transfer is commonly accompanied by other observable effects. These may include a change in dimension, electrical conductivity, electromotive force, pressure in a closed system, intensity of emitted radiation, and peak wavelength of emitted radiation. Others include change in physical state or in color due to chemical reaction or molecular rearrangement and changes in physical properties such as hardness, elasticity, and crystal structure. Ways have been devised for measuring the magnitude of these effects, and some of these have been used for nondestructive testing.

Frost Test. One thermal test that has been widely used in testing nuclear reactor fuel is the so-called "frost test." A reactor fuel element consists of a core of fissionable material such as uranium entirely encased or clad by a protective material. The uranium is bonded to the clad by diffusion or by the introduction of a third material. Areas of the clad which show relatively lower heat transfer rates because of lack of bonding or flaws may be detected by the frost test.[1,*] Diphenyl and acenaphthene (melting point, 200°F) are suitable frost materials.

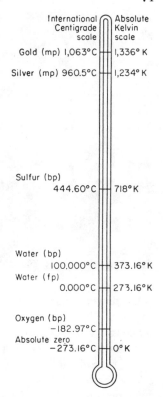

Fig. 5-1. International temperature scale.

In the application of this test the clad fuel element is sprayed with a 14 per cent solution of acenaphthene or a 40 per cent solution of diphenyl in CCl_4. This leaves a rough or "frosted" surface. The specimens are then moved at a uniform velocity through an induction heating coil by a conveyor system. Eddy currents are induced in the outer portions of the element, particularly in the cladding. The major portion of the heat flows inward through the clad and bond into the uranium core which serves as the thermal sink. An unbonded area or a flaw in the bond produces a thermal barrier. The cladding over this area is heated above the melting point of the frost. The frost melts to form a smooth surface outlining the defective area. Further, the organic material remains in place. Visual inspection may be carried out at a convenient time even up to several days. Figure 5-2 shows typical frost test results. For cylindrical fuel elements a single turn inductor can be used. For flat fuel elements a rectangular shaped inductor can be used. It was necessary carefully to adjust the power input to the specimen. If too

* Superscript numbers indicate Specific References listed at the end of the chapter.

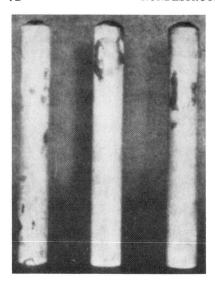

Fig. 5-2. Frost test results. (*Courtesy Argonne National Laboratory.*)

much power is used, complete melting of the frost occurs. If the power used is too low, the frost will not melt evenly over the defective areas. It is necessary to position the specimen properly in the induction coil. It was also found necessary accurately to control the initial temperature to attain reproducible results.

The frost test has limitations in its ability to find defects. Defects must present a thermal resistance sufficiently great to make the surface temperature over the defect rise above the melting point of the acenaphthene. A defect 0.1 to 0.2 in. in cross section may offer only a small thermal resistance though it may be an inch long. Likewise, large non-bonded areas in which there are small well-bonded areas may present little in the way of thermal resistance. This test can locate defects of approximately 0.1 in.² area which have a minimum width or length of 0.2 in. Cases have been found in which even a large defect did not present a sufficient thermal resistance. Unfortunately this test has not been found entirely dependable. Smaller thermal resistance can be detected by increasing the applied power; however, above a critical power level, false indications may occur because of the fuel element being off center as it passes through the inductor or the uranium being eccentrically positioned in the clad.

Thermography. A more sensitive thermal test is the thermographic method in which a temperature sensitive phosphor is used. The phosphor is so temperature sensitive close to a transition temperature that localized small increases in temperature reveal themselves as relatively dark areas. Dr. F. Urbach of the Eastman Kodak[2] Company has published the characteristics of a number of such thermal sensitive phosphors. A zinc–cadmium sulfide phosphor, when irradiated with ultraviolet light, will change its emissivity by 20 per cent when its temperature is changed by 2°F. This particular phosphor has its maximum sensitivity over the range of 40 to 130°F. These phosphors are also sensitive to the intensity and wavelength of the ultraviolet radiation impinging on its surface; the optimum wavelength for this phosphor is 3,650 A. There is also an optimum intensity of the ultra-violet radiation.

The phosphor is applied as a paint on the test specimen. The paint vehicle must be a material that will not fluoresce when subjected to ultra-violet light, because such fluorescence might obscure the heat sensitive properties of the phosphor pigment. Beetleware 227, a liquid plastic, is a vehicle that meets this requirement. To make a suitable paint vehicle, the plastic is thinned in these proportions: Beetleware 227, 680 g; butanol, 75 cm³; and xylene, 45 cm³. The pigment and vehicle are mixed in the proportions of 100 g phosphor to 100 g vehicle. This mixture can be sprayed or brushed on the test specimen.

The specimens are coated with the phosphor to an approximate maximum thickness of 0.001 in. If heavier coatings are employed, the sensitivity of the test is diminished because the paint is a heat insulator. The test specimen must be alternately heated and cooled by subjecting it to a flow of hot water at 130°F and cold water at approximately 40°F, passing it through an induction heater, or heating by infrared radiation and cooling by a blast of air. The disadvantage of this technique of testing is that the color change is reversible; so no permanent record is obtained as in the frost test. The fact that the phosphor coated surface must be illuminated with ultraviolet light is also a disadvantage. If a permanent record is desired, a movie camera or some type of photocells must be used.

This technique was used as a test to differentiate between areas of intimate metal-to-metal bond and a fusion bond.[3] The experimental arrangement for this test will be briefly described. A 250-watt Switzer ultraviolet lamp filtered to 3,650 A was set up 24 in. from the phosphor coated sample at an angle of approximately 30°. A 16-mm motion picture camera was set up directly in front of the sample at a distance of 30 in. XX Super-panchromatic film was used. Exposures were made at 8, 16, and 32 frames per second. The following technique was used for this particular problem. A stream of cold water (32°F) was sprayed over the back of the sample for several seconds, establishing an approximate temperature equilibrium. At this temperature the color of the phosphor surface was bright and even. The cold water flow was then turned off, and hot water (130°F) was suddenly applied. Within a few seconds the sample was in equilibrium with the hot water. During the short transient period, the bright fluorescence was more persistant over spots of poor thermal conductivity. The procedure was then reversed, the hot water being replaced suddenly by the cold water. This time the regions of poor conductivity became visible momentarily as dark streaks or spots on the more rapidly brightening background. The color changes of the phosphor were observed visually and recorded with motion pictures.

In order to evaluate the usefulness and sensitivity of this technique it was necessary to have samples containing known defects. Known defect blocks were made in an endeavor to get various degrees of metal-to-metal contact. These blocks were made from zirconium. A defect block was made as follows: holes were drilled into zirconium plates perpendicular to the edge of the plate, the holes were filled with zirconium wire of approximately the same diameter as the hole, and the plates were reduced in thickness by rolling. This type of synthetic defect was believed to approach closely the desired condition of metal-to-metal bonding without fusion bonding. One such defect block was made from a piece of zirconium 0.250 in. thick. A hole 0.113 in. in diameter was drilled in the block perpendicular to the edge, a zirconium wire 0.110 in. in diameter placed in the hole, and the block warm rolled to a thickness of 0.112 in. Examination of the edge revealed that, in rolling, the wire flattened out into an elongated ellipse (major axis, 0.0225 in., minor axis, 0.035 in.), giving metal-to-metal contact between the zirconium wire and the block. In addition, the hole collapsed at one end of the ellipse, giving an additional metal-to-metal contact along a line 0.007 in. long. Figure 5-3 shows typical results using the thermographic technique.

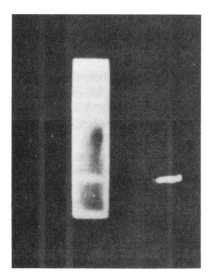

Two defect blocks similar to the type just described were made especially for comparison purposes. Defect block A was 0.500 in. thick before rolling and had two holes drilled into it perpendicular to the edge. One hole was 0.020 in. in diameter and the other 0.040 in. in diameter. Zirconium wire slightly smaller than 0.020 in. in diameter was placed in each hole and the block warm rolled to 0.250 in. thickness. Figures 5-4a and b show how the wall of the hole completely collapsed about the wires after rolling, giving metal-to-metal contact of approximately 0.095 and 0.045 in., respectively, for the large and small defects. This known defect block, after testing at a thickness of 0.250 in., was subsequently

FIG. 5-3. Thermographic results. (*Courtesy Argonne National Laboratory.*)

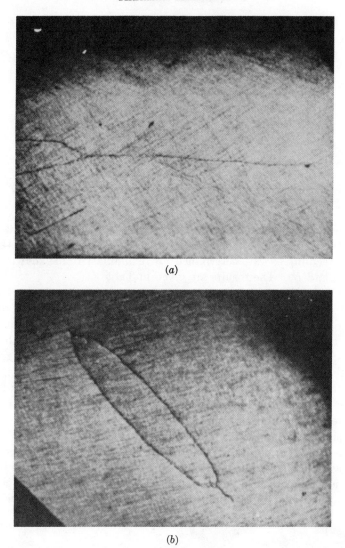

(a)

(b)

Fig. 5-4. Micrographs of test specimen. (a) Section cut through large hole filled with zirconium wire; (b) section cut through small hole filled with zirconium wire. (*Courtesy Argonne National Laboratory.*)

machined in steps to thicknesses of 0.160, 0.120, 0.090, 0.060, and 0.030 in. and tested after each machining operation.

Defect block B was similar to defect block A. A hole 0.013 in. in diameter was drilled perpendicular to the edge and filled with zirconium wire 0.010 in. in diameter. The block was warm rolled to a thickness of 0.250 in. Examination of the edge showed the wall of the hole com-

TABLE 5-1. THERMOGRAPHIC METHOD RESULTS USING DEFECT BLOCKS *A* AND *B*

Thickness of block, in.	Block *A*		Block *B*
	Large defect	Small defect	
0.250	NR*	NR	NR
0.160	NR	NR	NR
0.120	NR	NR	NR
0.090	PR†	NR	NR
0.060	R‡	NR	NR
0.030	R	NR	NR

* NR = not revealed.   † PR = partially revealed.   ‡ R = revealed.

pletely collapsed about the wire after rolling. This block was machined and tested at the same thickness as *A*. The sensitivity of the thermographic technique was determined by examination of known defect blocks *A* and *B*. The results are given in Table 5-1.

The United States Radium Corporation has developed new thermal phosphors of high sensitivity. Figure 5-5 shows discontinuities in the bonding of stainless steel sheet to brass. The Corporation reports that in the temperature range of 68°F zinc cadmium sulfide with silver activator and a trace of nickel as "poison" is most useful. Figure 5-6 shows the effect of temperature on the brightness at four levels of ultraviolet illumination. This corporation has developed a lacquer for use with the phosphor which has a minimum fluorescence. A spray gun is recommended for applying the material.

Tempilstik. Tempilstiks are temperature sensitive crayons having calibrated melting points. There are 60 different Tempilstiks commercially available, covering the range from 113 to 2000°F, each indicating a specified temperature, within a tolerance of plus or minus 1 per cent of its rating. For most applications the specimen is marked with the

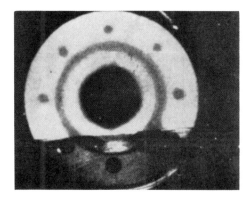

FIG. 5-5. Contact thermography showing discontinuities in bonding of stainless steel to brass. (*Courtesy United States Radium Corporation.*)

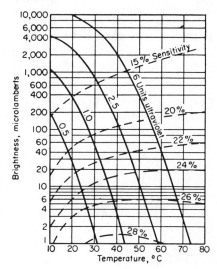

FIG. 5-6. Performance of radelin thermographic phosphor No. 1807. Solid curves show brightness against temperature at four levels of ultraviolet excitation. Dashed curves show sensitivity against brightness and temperature. (*Courtesy United States Radium Corporation.*)

appropriate Tempilstik before heat is applied. On subsequent heating the crayonlike mark on the Tempilstik melts as soon as its temperature rating is reached. The mark made by a Tempilstik gradually changes in color as heating progresses. The change in color, or in color intensity, is never to be interpreted as a temperature indication. The Tempilstik mark melts when its temperature rating is reached, and only this change from the dry to the liquid condition is the significant temperature signal. In some applications this simple technique may not suffice because of evaporation of the Tempilstik mark on prolonged heating, or its absorption by the surface at high temperatures may leave too little residual substance for an unambiguous observation. In such cases stroking or touching the specimen with the Tempilstik at regular intervals during the heating process is recommended. Obviously, the Tempilstik will leave a dry mark at temperatures below its rating and a liquid streak when its temperature rating has been reached or exceeded.

On rapidly moving objects, such as magnesium or aluminum sheet during spinning operations, it is impossible to see whether the applied Tempilstik leaves a dry or melted mark. However, with a little experience operators quickly learn to sense the smooth gliding of a Tempilstik over a surface which is hot enough to melt it on contact, as contrasted with the frictional drag on a cooler surface. If the top of the Tempilstik is brought in contact with the moving metal surface for a brief moment, and with light pressure only, the heat caused by friction is trivial enough to be ignored. An alternative procedure consists of lightly sandpapering the tip of the Tempilstik, touching the spinning object, and then examining the tip to see whether it has become fused by encounter with a sufficiently hot surface.

The above-described technique may also be employed on hot radiating surfaces where the bright background makes it difficult to distinguish whether or not the Tempilstik mark has melted. However, a radiating surface will appear relatively dark under intense illumination from a.. external source, and this fact can be utilized to improve recognition of the temperature signal against a red-hot background. The use of Tempil Pellets or Tempilaq instead of Tempilstiks will further improve visibility under these conditions. An appropriate series of Tempilstik marks, applied to the specimen or surface under investigation before heating begins, will provide a record of the maximum temperature attained during a process or operation. Subsequent examination will show that all Tempilstik marks up to a certain temperature rating had melted, while those of a higher rating had not. The maximum attained temperature must consequently lie somewhere in the interval between the highest melted and the lowest unmelted Tempilstik rating.

The temperature distribution and isothermal boundaries of surfaces such as furnace walls, melting ladles, cement kilns, etc., can be established by using an appropriately chosen series of Tempilstiks to draw a pattern of lines (parallel, concentric, radial, etc.) on the area under investigation. The heat conduction along a surface can be effectively studied by the progressive melting of suitably chosen Tempilstiks. Caution must be exercised not to permit the mark made with one Tempilstik rating to cross or overlap the mark of a dissimilar Tempilstik, as such a mixing of different Tempilstik marks would destroy their accuracy. Tempilstiks in the range above 600°F will not function properly in strongly reducing atmospheres, such as pure hydrogen, cracked ammonia, or water gas, although under milder reducing conditions many of the Tempilstik ratings will function satisfactorily. Tempil products will give good results in induction heating and in ionized air, as well as in the presence of static electricity about electrical equipment, where electrical means of measuring temperatures often function erratically.

Tempilaq is a temperature-indicating lacquer, which consists of materials very similar in nature to the Tempilstik, suspended in an inert volatile nonflammable liquid. The lacquer is applied to the specimen by dipping, spraying, or brushing. When dry the lacquer gives a dull opaque coating. On heating the dried lacquer melts as soon as its temperature rating is reached. On cooling the melted lacquer solidifies with a characteristic glossy crystalline or transparent appearance distinctly different from the original coating. The lacquer must be applied before heating begins. It is useful for application to smooth surfaces such as glass, plastics, or polished metal. Temperature-indicating pellets are available in the temperature range from 113 to 2500°F and are intended for use for testing specimens where observations must be made at some distance as, for example, heat-treating operations. The pellets are placed on the specimen before heating begins, and they melt at the rated temperature. The liquid residue will remain for a long period of time.

Temperature Sensitive Pigments. Temperature-indicating pigments covering the temperature range from 104 to 2912°F and known by the trade name Thermocolor have been developed by Badischeanilin-Osoda-Fabrink in Germany and are sold in the United States by the Curtiss-Wright Corporation. These pigments have the characteristic of changing their original color when certain temperatures are reached. These color changes are not merely into shades of the original color but are definite and therefore distinctly visible. Some of the Thermocolors undergo one color change whereas others may undergo two, three, or four changes in color. For example, one of the Thermocolors is pink at room temperature, becomes light blue at 149°F, yellow at 293°F, black at 347°F, and olive-green at 644°F. Most of the color changes are permanent; however, atmospheric moisture may cause certain of the pigments when cooled to return in a period of several hours to their original color.

The temperatures given for the various pigments by the manufacturer are accurate for heating periods of approximately 30 min, timed from the beginning of the warm-up to the color change. If the change in color occurs in less than 30 min, a somewhat higher temperature is indicated; if the change takes place after a longer period, this indicates a slightly lower temperature. For special cases involving sudden heating (heating periods of 5 to 30 sec), information on the color-change temperature is available. When testing temperatures using the 30-min heating period, the accuracy is plus or minus 5°C (plus or minus 9°F), so that the measurements are precise enough for all practical operating purposes. Most Thermocolor pigments do not recede upon cooling; others recede so slowly that, even after cooling, conclusions can be reached as to the temperatures attained. However, because of the equalization of the temperature gradient by conduction, the isothermal patterns may continue to change after the heating has stopped. Therefore when carrying out certain investigations, it may be necessary to freeze the colors by rapid cooling.

Thermocolor paints may be applied with a brush or by spraying. The specimen to be painted must be clean; traces of oil are best removed with an appropriate solvent, followed by alcohol. Thermocolor brands work best on uncoated surfaces. Thermocolors may be used not only on metals, but also on ceramic materials, bricks, insulating stone, porcelain, plastics, wood, and glass. The coats dry within a few minutes. It is advisable to allow approximately one-half hour for drying to eliminate blistering and peeling. The coat may be removed by washing with alcohol, brushing with hot water, scraping, or sand blasting. If especially easy removal of the Thermocolor is desired, the surface should be well rubbed with a slightly oily rag before painting; in this case a particularly thin coat is applied. Certain Thermocolor pigments can be used for testing in steam, carbon dioxide, sulfur dioxide, ammonia, and hydrogen sulfide atmospheres. Figure 5-7 shows the application of Thermocolor. In using Thermocolor for nondestructive testing, the rate of change of color may be useful. Figure 5-8 shows the reaction time of one Thermocolor in relation to temperature and time.

On specimens of moderate thickness, for instance, sheet metals, measuring the heat conductivity vertical to the surface of the specimen can give useful information. The heat conductivity of air is 100 times less than that of steel. Therefore, an extended fold of 0.001 in. in a steel sheet acts like an additional thickness of 0.1 in. Thermocolor can be used to locate this type of flaw. The specimen is coated with Thermocolor on one side and is uniformly heated from the other side. On the painted side, above a sound cross section, the change in color caused by rise in temperature appears sooner than above a faulty one, for example, a cross section with a fold. By observation of the time intervals in which the color changes appear and their uniformity, conclusions as to the presence of faults can be drawn. This technique has been used to determine if the bonding between steel cylinders and the bronze rings drawn around them is good. For this test, the sleeve was painted on the outside with the proper material. After drying, the specimen was filled with hot oil. After about 30 sec, the color change appeared on the outside. If the change was uniform, the metal bond was judged as good. If the irregularities in the time of the change of the color were more or less temporary, the points of delayed change were marked as probable faults.

The spread and degree of heat in the metal adjacent to the weld can be watched by means of Thermocolors. The smoothness of the isotherms obtained by applying Thermocolor along the length of the weld gives a

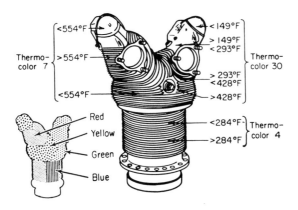

FIG. 5-7. Application of Thermocolor. (*Courtesy Curtiss-Wright Corporation.*)

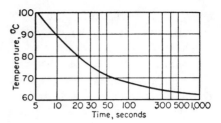

Fig. 5-8. Reaction curve of a color change. (*Courtesy Curtiss-Wright Corporation.*)

record of the rate with which the weld was run. When the isotherms are irregular or bulge on the side of the weld, it has been proved that oxidation or other undesirable conditions exist in the weld. When preheating is required for welding or any other process, Thermocolors serve to control and record the uniformity and degree of heating. Thermocolor has also been used to inspect castings.

In this application the casting is coated on one side with Thermocolor, and an even heat is applied to the opposite side. When the surface reaches the given temperatures, it will change color. There will be no immediate color change wherever there is porosity in the casting. This technique will show not only the defects but also the location and approximate size of the defects.

Infrared. In addition to the light which we can see, there also exists invisible radiation at both ends of the visible spectrum. Beyond the violet in the spectrum is radiation which is called the "ultraviolet," which is of short wavelength. It is invisible but has strong action on photographic materials, making it easy to detect by photographic means. At the other end of the spectrum, at wavelengths longer than the red, there exists radiation called the "infrared," meaning "below the red." The infrared region extends out indefinitely from the end of the visible region. As the wavelength increases, the radiation merges into heat waves and finally into the radio waves. Even though the infrared extends far out, it is the region only quite near the visible red which is of interest photographically. It can be recorded on plates and films which have been specially prepared. The longest wavelength of radiation photographically recorded is about $1,350$ μ (micron), but in general infrared work, the regions between 700 and 860μ are used.

Infrared photography[4] extends the vision of the camera beyond the limits of the human eye, affording a new photographic and pictorial dimension, as well as a medium for making record and technical photographs not otherwise possible. Infrared photography penetrates haze and renders the sky, green foliage, fabrics, etc., in a manner entirely different from ordinary photography. Infrared film and plates are as simple and dependable in practice as the more usual photographic materials. The usual photographic plates and films, even panchromatic ones, are not sensitive to infrared. By treatment of the emulsion with special dyes, sensitivity to infrared can be achieved. The infrared sensitive materials currently available may be used in almost exactly the same manner as panchromatic material. Although the speed of these films and plates permits their exposure in hand-held cameras, best definition is obtained by the use of a tripod and the smallest lens opening feasible. Best quality in infrared pictures results from critically correct exposure; the latitude of these materials is less than that of the usual panchromatic films. Photographs can be made with infrared-sensitive film without a filter in an emergency, but the resulting photograph will be similar to that of a blue-sensitive film. The quality will be less satisfactory than that produced by either orthochromatic or panchromatic film. Infrared materials can be handled in any darkroom if all white light is excluded.

There is no fundamental difference between the practice of infrared photography and that in which visible light is used. Any photographer, equipped for work with panchromatic films and plates, can make infrared photographs without additional equipment other than a suitable filter for use on the camera lens. There are, however, a few precautions which should be observed. Infrared rays, because of their longer wavelengths, in the case of many lenses do not focus in the same plane as visible light rays. It is therefore necessary to make an increase in the lens-to-film distance to correct for the focusing difference between infrared and visible light rays.

The value of photography by infrared lies in the fact that infrared radiation and visible light often are reflected and transmitted quite differently by common objects. For example, chlorophyll in live green foliage absorbs a large percentage of the visible light which falls upon it but does not absorb the invisible infrared radiation. This is reflected almost entirely by the leaf structure and, therefore, is recorded by means of the infrared sensitive material. Foliage thus appears white in an infrared photograph. Painted materials which match chlorophyll in color but which do not reflect strongly in the infrared will appear dark in an infrared photograph. Most photographic materials render blue sky relatively light, but, since little infrared radiation is present in the blue sky, the infrared materials render it dark. Infrared radiation is freely transmitted through atmospheric haze; so distance scenes can be recorded with greater clarity than they can be seen with the eye. Many dyes which appear bright colored to the eye do not absorb infrared and therefore record as white. The skin is somewhat transparent to infrared radiation, and therefore an infrared photograph is sometimes of value in medicine in diagnosing abnormal conditions immediately beneath the surface of the skin. Infrared photography has been used successfully in criminology, photomicrography, botany, and other scientific fields.

Infrared-sensitive materials are valuable for studying the temperature distribution of hot bodies which are just below red heat, such as cooling ingots and castings, stoves, engine parts, high-pressure boilers, flatirons, and other subjects similar in temperature. A hot electric flatiron can be photographed in about 1 to 5 min at $f/4.5$. At temperatures below 750°F, the exposures are very long, about 6 hr at $f/5.6$ being required at 625°F. If the plates are hypersensitized, the exposure is reduced to one-half or one-third. Photography of hot objects must be done in a completely darkened room; otherwise a photograph is obtained by reflected, not emitted, infrared. Proper precautions should be taken to prevent the surroundings from burning if the object is to be heated for a long exposure.

Ektron Detectors. The Eastman Kodak Company has developed sensitive infrared detectors which are sold under the trade name of Kodak Ektron detector. These detectors consist of a thin layer of lead sulfide sprayed on glass. The manufacturers of these particular detectors claim an increase in sensitivity of 10,000 times compared to the other detectors using the same material. The response of the Ektron detectors extends from 0.25 (1 μ = 10^{-3} mm = 10^{+4} A) to 3.5 μ, with maximum sensitivity at 2.2 μ as shown in Fig. 5-9. The detector is more sensitive and has a greater output at lower temperatures.

Lead sulfide is a photosensitive semiconductor. A semiconductor is a material whose conductivity is much poorer than that of most metals but much better than that of insulators. Such materials include selenium, germanium, copper oxide, zinc oxide, cadmium sulfide, and lead sulfide. The conductivities of these semiconductors increase with rising temperature primarily because of the presence of impurities. The conductivity value is profoundly affected by the degree of purity and the manner of preparation of the material. Photosensitive semiconductors or photoconductors have a different electrical conductivity in darkness and in light. In darkness they are virtually insulators, but in the presence of light they become moderate conductors. Semiconduc-

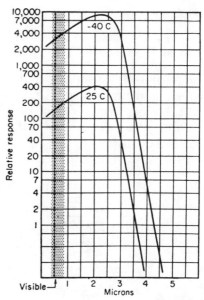

FIG. 5-9. Spectral response of Eastman Ektron detector. (*Reproduced with permission from the copyrighted Kodak booklet "Kodak Ektron Detector."*)

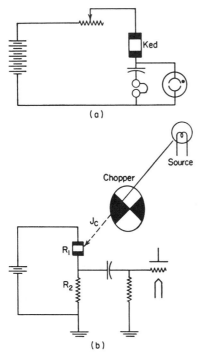

(a)

Chopper

Source

(b)

FIG. 5-10. Ektron detector as a photo-cell. (*Reproduced with permission from the copyrighted Kodak booklet "Kodak Ektron Detector."*)

tors exhibit varying sensitivity to light of different wavelengths. Lead sulfide has a maximum sensitivity at about 2 μ in the infrared region, but has useful response in the visible and ultraviolet region of the electromagnetic spectrum.

In addition to their exceptional spectral response, Kodak Ektron detectors are characterized by high signal-to-noise ratio, time constants in the range of 400 to 1,000 μsec, and dark resistance of 0.2 to 0.8 megohm, and a negative coefficient of resistance, that is, a decrease in resistance with an increase in temperature which is in the order of 2 per cent per degree Fahrenheit. An increase in signal may be obtained by employing an optical system to collect the radiation and focus it on the detector. However, care must be used to select materials that will transmit the desired radiation; for example, ordinary glass is useful only to 2.7 μ whereas quartz can be used over the entire range of usefulness of lead sulfide. Ektron detectors can be used to detect radiation, as shown in Fig. 5-10. The Ektron detectors may be made part of a bridge circuit, with one receiving the radiation. As the charging resistor is a relaxation oscillator, Fig. 5-11, the unit can be used to detect hot spots.

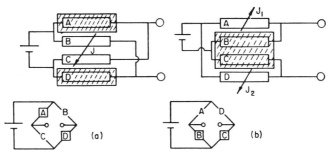

FIG. 5-11. Circuits for Ektron detectors. (*Reproduced with permission from the copyrighted Kodak booklet "Kodak Ektron Detector."*)

Remote temperature measurement by means of infrared radiometry is possible because the radiation emitted by a body is a well-known function of its temperature. A radiometer is an optical instrument which collects radiation from a narrow field of view and converts the received energy into an electrical signal. The amplitude of the electrical signal can be measured with great accuracy. Since the characteristics of radiation are well established by basic laws and relationships, it is possible to calibrate the electrical output in terms of temperature for a given source. Thus, a practical means has

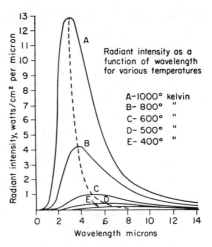

FIG. 5-12. Black body radiation curves.

been established for measuring temperature without physical contact. In the following section a review of the basic characteristics of radiation will be made. Any object with a temperature above that of absolute zero emits energy over a wide band of frequencies. Most of this radiation is emitted at or near a specific wavelength which is dependent upon the temperature of the object, Fig. 5-12. The wavelength at which maximum radiation occurs becomes shorter and shorter as the temperature increases. Black body radiation is defined as radiation emitted by an object which absorbs all incident radiation without reflection. Such radiation is greater in total energy per unit area than that emitted by any other type of object at the same temperature.

Calculations to determine the various characteristics of infrared radiation are based upon three laws, each thoroughly discussed in the literature. These three fundamental relationships are the Stefan-Boltzmann law, Wien's displacement law, and Planck's empirical equation. According to the Stefan-Boltzmann law the total radiation from an object is

$$W = \epsilon \sigma A T^4 \qquad (5\text{-}1)$$

where W = total radiation, watts

σ = 5.672 $\times$ 10^{-12} watt/(cm^2)($^\circ$K^4)

A = area of source or aperture, cm^2

T = temperature, $^\circ$K

Since physical objects do not meet the requirements of an ideal black body, the emissivity factor ϵ is necessary to correct for the surface characteristics of the object. The ϵ factor varies with the material and equals 0.02 for a silver mirror surface and 0.95 for lampblack. Wien's displacement law gives the wavelength λ at which maximum radiation is emitted for any specified temperature and is generally expressed as follows:

$$\lambda = \frac{K}{T} \tag{5-2}$$

The surface constant K has the value of 2960 for a black body. If T is in degrees Kelvin, then the wavelength of maximum radiation will be expressed in microns.

The radiation emitted by a black body as a function of wavelength may be computed from Planck's equation. For this purpose the expression may be written as follows:

$$J_{\lambda T} = C_1 \lambda^{-5} (e^{c_2/\lambda t} - 1)^{-1} \tag{5-3}$$

where $J_\lambda t$ = radiation power emitted as a function of wavelength, watts/(m²)(μ)

$C_1 = 3.74 \times 10^8$

$c_2 = 1.44 \times 10^4$

λ = wavelength, μ

T = temperature, °K

e = base of Napierian logarithms (2.718)

The actual radiation distribution for gray bodies ranging from $-150°$F to 1830°F is shown in Fig. 5-13. This depicts graphically the results of the fundamental radiation laws. The bar charts clearly show that 80 per cent of the radiant energy emitted by objects within this temperature range lies between 1.8 and 40 μ in wavelength. This is the range of wavelengths over which it is desirable to measure the total infrared radiation in order to determine the temperature of objects below 1830°F.

From the basic radiation laws it can be seen that the radiation emitted by objects is uniquely related to their temperature and that almost all the energy is in the infrared portion of the spectrum. Most common objects have nearly constant emissivity in the infrared and hence have radiation distribution similar to black bodies of the same temperature. The temperature of objects can hence be determined through measurement of the emitted infrared radiation. All radiation measurements

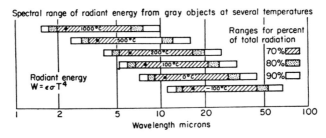

FIG. 5-13. Spectral range of thermal energy at various temperatures.

involve comparison with a primary or secondary reference standard. The reference level of radiation is often contained within the radiometer.

The radiometer focuses the infrared radiation from the heated specimen onto an infrared radiation detector. The radiometer optical system must be equally efficient at all infrared wavelengths, which generally requires the use of reflecting optics. The detector must have an electrical output proportional to the magnitude of the incident radiation for all wavelengths of infrared radiation. It must hence be a "black" or "thermal" type infrared detector.

These instruments are often classified as d-c infrared radiometers. This terminology is applicable since the instrument measures a change in d-c electrical properties of a thermoelectric or bolometric infrared sensing element. The radiation difference measured by this type of instrument is the difference between the source energy focused by the radiometer optics on the infrared detector and the radiant energy level of the detector itself. The emissivity and temperature of the infrared detector determine the reference radiation level of this instrument and are of great importance.

Recent advances in high-speed infrared detectors have led to the development of new types of radiometric instruments. These instruments are called a-c infrared radiometers, since they employ a-c chopping techniques in which the radiation is periodically interrupted. In this system the infrared detector alternately sees the image of the source and the blackened chopper. The radiation signal measured is the difference between the source and chopper radiation. The reference radiation level in this instrument is determined by the emissivity and temperature of the blackened chopper. An a-c electrical signal is developed by such system detectors.

Thermistors. Thermistors (thermally sensitive resistors) are semiconductors composed of oxides of manganese, nickel, and cobalt. When suitably processed, mixtures of these oxides form a stable structure which has a large negative temperature coefficient of resistance. Thermistors are unique for their speed of response, ruggedness, flexibility, and ease of electronic amplification. Equipped with special infrared transmitting windows they are uniformly sensitive to 40 μ. The thermistor infrared detectors are used in the infrared d-c and a-c infrared radiometers previously described. These materials have a high resistivity so that they can be connected directly into the grid circuit of a vacuum tube. Figure 5-14 shows a typical thermistor bolometer assembly. In a thermistor bolometer a very thin sheet in the order of 10 μ is mounted on a heat sink. The assembly consists of an active element and a matching compensating element which is shielded from the incoming radiation. The shielded matched compensating element, mounted in the housing, is shown in Fig. 5-15. The detector is operated in a bridge circuit.

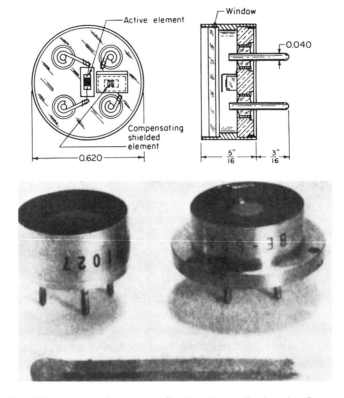

FIG. 5-14. Thermistor bolometer. *(Courtesy Barnes Engineering Company.)*

Blackening of the semiconductor increases the absorption of the incident energy. The compensating element is used to minimize changes in the characteristics of the detector assembly due to variations in ambient temperature.

Evaporograph. The Evaporograph provides a two-dimensional heat picture from infrared radiation emitted by a specimen. This instrument will show if there is a temperature difference in the field of view. No

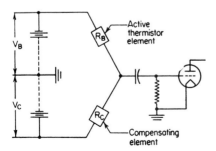

FIG. 5-15. Thermistor bolometer circuit. (*Courtesy Barnes Engineering Company.*)

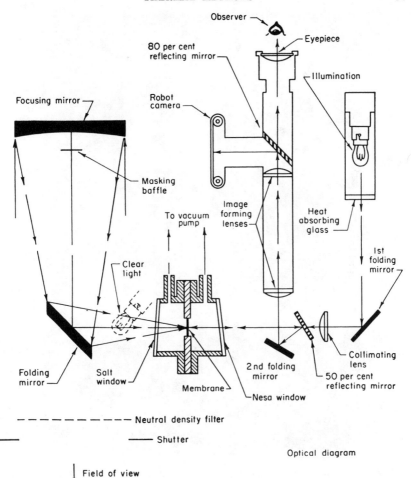

FIG. 5-16. Diagram of Evaporograph. (*Courtesy Baird-Atomic, Inc., Cambridge, Mass.*)

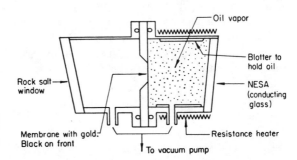

FIG. 5-17. Evaporograph cell. (*Courtesy Baird-Atomic, Inc., Cambridge, Mass.*)

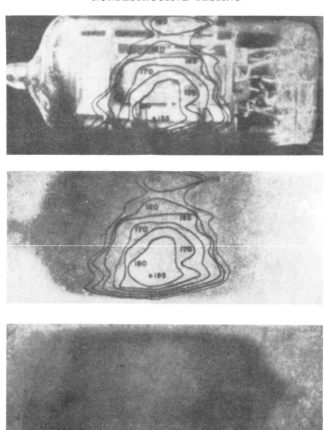

Fig. 5-18. Temperature contours in electronic tube. (*Courtesy Baird-Atomic, Inc.,* *Cambridge, Mass.*)

physical contact with the specimen or auxiliary illumination is required. Either visual or photographic pictures of temperature variations can be made in daylight or in total darkness. The Evaporograph uses a nitro-cellulose membrane so thin that it reflects white light as a color, like the familiar oil film on water. This membrane, about 4 μin. thick, is located in an evacuated two-compartment cell between an optical system and a viewing system.

Figures 5-16 and 5-17 show the construction of the instrument and details of the Evaporograph cell. The operating principle of the Evaporograph is briefly described as follows: The optical system gathers infrared radiation from the field of view and forms an image of the field on the membrane. The coating on the front surface of the membrane absorbs this radiation. The membrane temperature varies from point to point, depending on the amount of radiation received. The membrane temperature alters the oil film thickness on the rear side of the membrane, depending on the local temperature. The various oil film thicknesses over the membrane cause white light to be reflected as different colors, thus giving a colored isothermal image of the field of view.

The radiation from the field of view can be measured in two ways, visually and photographically. The results can be converted to temperatures by use of radiation-temperature tables based on the Stephan-Boltzmann law. Figure 5-18 shows the various temperature contours found in an electronic tube. Temperature differences as small as a few tenths of a degree Fahrenheit can be measured with this equipment. A high degree of accuracy for absolute temperature measurements as well as good reproducibility of results is claimed for this equipment.

SPECIFIC REFERENCES

1. Brown, George H., Cyril N. Hoyler, and Rudolph A. Bierwirth: "Theory and Application of Radio-Frequency Heating," D. Van Nostrand, Company, Inc., Princeton, N.J., 1947.
2. Urbach, F.: Thermography, *Phot. J.*, vol. 90B, pp. 109–114, 1950.
3. McGonnagle, W. J., J. H. Monaweck, and W. G. Marburger· Methods of Bond Testing, *J. Soc. Nondestructive Testing*, vol. 13, no. 2, pp. 17–22, March–April, 1955.
4. Clark, Walter: "Photography by Infrared," John Wiley & Sons, Inc., New York, 1946.

CHAPTER 6

X-RAY RADIOGRAPHY

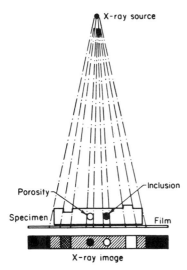

FIG. 6-1. Differential absorption.

Radiography is a method of non-destructive testing which uses X-ray or gamma radiation. In order to distinguish this type of radiation from other types of radiation such as visible light and sound, X rays and gamma rays will be referred to in this book as "penetrating radiation." Radiography is one of the oldest nondestructive tests, having been used since the early 1920s. Today it is one of the most widely used of all nondestructive tests. The intensity of the penetrating radiation is modified by passage through material and by defects in the material. The phenomenon of differential absorption illustrated in Fig. 6-1 serves as the basis for the use of radiography in nondestructive testing. The contrast (difference in density) on the developed film between the image of an area containing a defect and the image of a defect-free area of the specimen permits the observer to distinguish the flaw.

X rays and gamma rays comprise part of the electromagnetic spectrum. This spectrum may be roughly divided, on the basis of wavelength, into the following categories:

TABLE 6-1. ELECTROMAGNETIC SPECTRUM

Gamma rays	0.1–0.005 A
X rays	5–0.0004 A
Ultraviolet	10–4,000 A
Visible	4,000–7,500 A
Infrared	7,500–4 $\times.10^6$ A
Radio waves	10^6–10^{13} A

The wavelengths are expressed in angstroms, an angstrom being equal to 10^{-8} cm. The divisions made between the various kinds of radiation

92

are somewhat arbitrary, and overlapping of the various regions will be evident as one reads further in the text.

X-rays and gamma rays have the following properties:

1. Are invisible electromagnetic radiation
2. Can penetrate matter
3. Are differentially absorbed
4. Travel in straight lines
5. Produce photochemical effects in photographic emulsions
6. Ionize gases through which they pass
7. Are not affected by electric or magnetic fields
8. Travel with a velocity of 3×10^{10} cm/sec or 186,000 miles/sec
9. Are capable of liberating photoelectrons
10. Cause some substances to fluoresce

The usual method of detecting X rays and gamma rays in the field of radiography is by means of a photographic emulsion. However, other means such as Geiger counters, semiconductors, photoconductors, and scintillation crystals are sometimes used.

X-ray Production. X rays are produced when high-speed electrons, called "cathode rays," strike a metal target. When the electrons strike the target, a transfer of energy occurs and X rays are produced. The basic requirements for producing a beam of X rays are (1) a source of electrons, (2) a means of accelerating the electrons to high velocities, and (3) a metal target. In X-ray tubes the electrons are accelerated by a difference of potential between the source of electrons, called the "cathode," and the target. This difference of potential is referred to as the "tube voltage." Since electrons in motion comprise an electric current, the number of moving electrons determines the magnitude of the electric current. The flow of electrons in an X-ray tube is referred to as the "tube current."

The energy of penetrating radiation is often expressed in terms of electron volts (ev), kiloelectron volts (kev), or million electron volts (Mev). The electron volt is the kinetic energy that an electron acquires when it is accelerated in an electric field produced by a difference of potential of one volt.

The velocity at which the electron strikes the target is determined by the tube voltage. It will be shown later that the energy of the X rays produced at the target is proportional to the square of the velocity of the electron, and the shorter the wavelength of the X rays, the greater their

TABLE 6-2. ENERGY RELATIONSHIPS

1 ev	1.6×10^{-12} ergs
1 kev	1,000 ev
1 Mev	10^6 ev

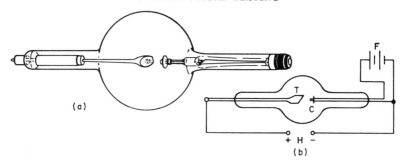

Fig. 6-2. Coolidge X-ray tube. *C*, cathode cup; *F*, filament battery; *H*, high-voltage source; *T*, target. (*Courtesy General Electric Company, X-Ray Division.*)

penetrating power. The velocity v acquired by an electron because of a difference of potential V can be calculated by setting the kinetic energy equal to the work done in accelerating the electron and is given by the following expression:

$$\tfrac{1}{2}Mv^2 = 1.6 \times 10^{-12}V \qquad (6\text{-}1)$$

where M = mass of the electron (9.1×10^{-28} g)

v = velocity, cm/sec

V = tube voltage, volts

There are two types of X-ray tubes which can be used for radiography, a gas tube and a Coolidge tube. The latter is used almost entirely today. A Coolidge tube is shown in Fig. 6-2. The tube is evacuated to the best possible vacuum, the metal parts in the tube being degassed by heating before the tube is sealed off. The cathode C consists of a spiral tungsten filament inside a small metal cup. A source of electrons is obtained by heating the tungsten filament to incandescence with an electric current. Emission of electrons from a heated filament is known as "thermionic emission." Experiment and theory both show that the electron current from a heated filament is a function of the temperature of the filament. The purpose of the cup is to shape the electromagnetic field and focus the electron beam. The electrons emitted by the heated filament strike only a small area on the target, called the "focal spot." The high voltage is applied between the target t and the cathode c.

The intensity of the X rays produced is directly proportional to the tube current and depends on the tube voltage raised to a power greater than 2.5. The efficiency of X-ray production is given by the following expression:

$$E = 1.4 \times 10^{-7}ZV \qquad (6\text{-}2)$$

where E = efficiency, per cent

Z = atomic number of target material

V = tube voltage

From this expression it can be seen that the efficiency of X-ray production is low at low voltages.   At 300 kv, only about 3 per cent of the energy of the electrons is converted to X rays.   The rest of the energy of the electrons at the anode appears in the form of heat.   The amount of heat generated at the target of a tube is proportional to the product of the tube voltage and tube current.   Consequently, it is necessary to cool the target.   This can be done in a number of ways.   Some tubes are made so that water or oil can be circulated in the interior of a hollow anode.   Other tubes are made with a long metal rod connected to the target which conducts the heat to cooling fins on the exterior of the tube. In still other cases the tube is so constructed that the anode can be made to rotate, and only a small area of the target is bombarded at any one instant.   Tungsten targets are sometimes embedded in a mass of copper, which will conduct the heat away faster.

In radiography it is desirable to have a source of X rays which approaches an ideal point source.   This can be achieved by proper design of the filament and focusing cup, but it does complicate the problem of heat removal.   In the usual case, a compromise has to be made on the size of the focal spot which can be used safely.   The finer the focal spot, the better will be the radiographic image.   Figure 6-3 illustrates how the effect of a small focal spot can be obtained and still have a larger area to dissipate the heat.   The target face is inclined at an angle to the cathode beam.   The projected focal spot at right angles to the beam has a size indicated by V, whereas the actual size of the focal spot is H.

As previously stated, the source of electrons for a Coolidge-type X-ray tube is a heated tungsten wire filament.   Most X-ray tube filaments operate in the range of 6 to 15 volts and use a current of 3 to 5 amp to heat the filament.   The necessary power is obtained by use of a step-down transformer.   The filament is usually not at ground potential. Consequently, a special step-down transformer in which the primary and secondary are isolated must be used.   The heating current is controlled by means of a rheostat in the primary circuit, as shown in Fig. 6-4. The high voltage for accelerating the electrons between filament and target of an X-ray tube is provided by a step-up transformer. In Fig. 6-4 an autotransformer is used to control the primary voltage of the high-voltage transformer.

X rays are produced only when the target is positive with respect to the filament.   Therefore, it is necessary to rectify a-c voltage from the secondary of the high-voltage transformer. The X-ray tube may be connected

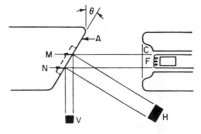

FIG. 6-3. Principle of line-focus X-ray tubes.

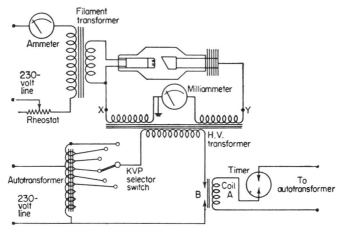

FIG. 6-4. Basic X-ray circuit.

directly to the secondary of the high-voltage transformer. In this case the target is positive during one half of each cycle of the alternating voltage and X rays will be produced only during this interval. During the other half of the cycle no X rays are produced. Thus the X-ray tube (being a form of diode) acts as its own rectifier. Care must be taken, however, to be sure that the target does not become hot enough to emit electrons, which would quickly destroy the filament by bombardment. Self-rectifying tubes are often used in portable X-ray units.

If the anode or target of a tube is at a certain potential when a current is flowing through the tube, the inverse voltage which occurs during the alternate half cycle will be considerably higher since no current is flowing and as a result there is no voltage drop in the transformer. Thus if the tube is to withstand the inverse voltage, the voltage which is used to accelerate electrons and produce X rays must be considerably reduced below the inverse voltage.

The rectifying action of an X-ray tube discussed above is utilized in special electronic tubes designed specifically for the purpose of rectifying high-voltage alternating current. This type of tube consists of a filament and plate or anode, as shown in Fig. 6-5. The filament is operated at about 12 amp and 12 volts. Electron flow between filament and anode takes place only during the half of

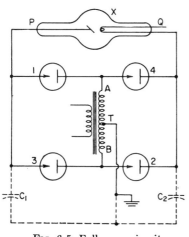

FIG. 6-5. Full-wave circuit.

the a-c cycle in which the anode is positive. Such tubes are so designed that when the filament is positive with respect to the anode no current flows even though the voltage may be as high as 200,000 volts. Using only one such rectifier tube, power

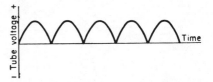

FIG. 6-6. Full-wave rectification.

is delivered to the X-ray tube only half the time; this is called half-wave rectification. Using the circuit shown in Fig. 6-5 it is possible to have power delivered to the X-ray tube at all times; this is called full-wave rectification. The operation of this circuit will be explained briefly. When point A on the high-voltage transformer is negative, point B is positive. Current will then flow in rectifier tubes 3 and 4. During the other half of the cycle, point A is positive and point B is negative. Current will then flow in rectifier tubes 1 and 2. During both halves of the cycle the target of the X ray is positive with respect to the filament. The waveform voltage appearing on the tube target can be seen in Fig. 6-6. It is possible by use of filter condensers to smooth out the voltage. Figure 6-7 shows how full-wave rectification can be obtained using only two rectifier tubes.

Instead of increasing the tube voltage by increasing the transformer voltage, a voltage doubling circuit is often used in industrial X-ray units. In voltage doubling circuits, the peak voltage applied to the X-ray tube is approximately twice the voltage supplied by the transformer, which is accomplished by charging condensers in parallel and discharging them in series. The electronic components required for such a voltage doubling circuit are readily available commercial items.

The Villard circuit shown in Fig. 6-8 is such a voltage doubling circuit. During the first half-cycle, connection A of the transformer secondary is negative with respect to the ground point while B is positive. C_1 is thus charged as shown by the current passing through K_1, and C_2 by the current through K_2. As the output voltage of the transformer falls

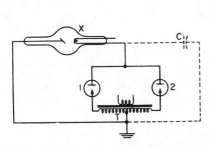

FIG. 6-7. Full-wave rectification circuit.

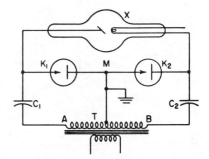

FIG. 6-8. Villard circuit.

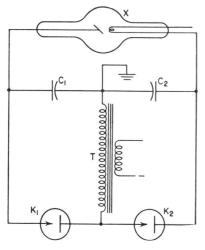

FIG. 6-9. Greinacher circuit.

to zero, so the voltage across the tube becomes equal to the peak voltage developed across the secondary. As connection A becomes positive and B negative, the voltage across the tube rises to twice the peak voltage of the transformer since the transformer secondary is effectively in series with C_1 and C_2 with the two rectifiers blocking. For satisfactory operation of this circuit the capacitance of C_1 and C_2 must be sufficiently high for their voltage to remain almost constant. Also the charging current through K_1 and K_2 must be considerably higher than the tube current. The resultant output of such a circuit will be pulsating and will have a frequency half that of the input voltage to the transformer. This makes the circuit unsuitable for use where very short exposure times are required since the output during the initial cycle will be distorted while the capacitors are being charged from zero. The circuit is, however, quite suitable for normal industrial work, where exposures are of a length which makes this irregularity negligible.

Another circuit of this voltage doubling type is the Greinacher circuit

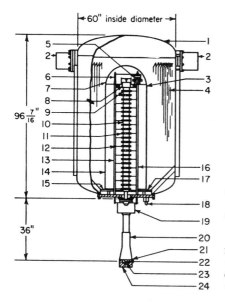

FIG. 6-10. Sectional drawing of 2-million-volt mobile X-ray unit. The numbers have the following significance: (1) steel tank; (2) cooler; (3) end turn filament coil; (4) laminated shield; (5) variable reactor; (6) spring for tie rod; (7) slotted brass shield; (8) cathode assembly; (9) first intermediate electrode; (10) glass envelope; (11) shields around X-ray tube; (12) tap lead; (13) glass tie rod; (14) secondary coils; (15) primary winding; (16) insulating filament control shaft; (17) laminated steel bottom; (18) filament control motor; (19) focusing coil; (20) lead shield; (21) water jacket; (22) extension chamber; (23) tungsten target; (24) lead diaphragm.

shown in Fig. 6-9. From the circuit diagram it can be seen that during the first half-cycle, if the free end of the winding becomes positive, C_1 will be charged through K_1 in the direction indicated, while K_2 blocks. During the next half-cycle, it becomes negative so that C_2 is charged as indicated, while K_1 blocks. The only discharge path for C_1 and C_2 is through the X-ray tube, and since they are effectively in series their potentials combine to give a tube voltage double that of the transformer output.

The resonant transformer, Figs. 6-10 and 6-11, has a low-voltage coil of rectangular wire and a high-voltage coil consisting of a number of thin flat sections spaced apart for cooling. The resonance principle of operation may be explained as follows: the high-voltage coil has a large inductance which is chosen to be in resonance with its terminal and distributed capacity at the operating frequency of 180 to 200 cps. To build the 30-kva resonant system up to the larger amplitude of oscillation of 1 million volts requires the relatively small input power of $3\frac{1}{2}$ kw, which equals the losses in the transformer plus the power for the X-ray tube and can therefore be supplied by the small primary winding. The inductance required for resonance can be obtained in a coil of this size without the use of an iron core, and the central space of the high-voltage coil is therefore available for the X-ray tube, thus facilitating the making

FIG. 6-11. 2-Mev resonance transformer removed from steel tank. (*Courtesy General Electric Company, X-Ray Division.*)

of connections to various tube electrodes and providing the benefit of electrostatic shielding by the transformer coil. The lower end of the high-voltage coil is grounded; the upper or high-voltage end is shielded by a rounded and radially slotted brass spinning. The 180-cycle power is derived from the 60-cycle line by means of a synchronous motor generator set which also renders operation independent of line-voltage fluctuations. To avoid the power loss which would ensue were the magnetic flux of the transformer to penetrate the surrounding steel tank, the latter is provided with an inner lining consisting of narrow overlapping silicon steel strips spot-welded to a thin shell and at the bottom with a ring of similar radial strips. The multisection high-voltage coil serves a quadruple purpose: to generate voltage, to grade the potential so as to prevent creepage, to shield the X-ray tube electrostatically from the grounded tank, and to support mechanically the rounded top shield.

Van de Graaff Generator. The Van de Graaff X-ray generator is shown in Fig. 6-12. This type of X-ray generator utilizes an electrostatic principle to build up the high voltage which is used to accelerate the electrons. In this generator an electric charge is sprayed on a rapidly moving insulated belt. The belt carries the charge to an insulated hemispherical high-voltage terminal which is insulated from the shell of the

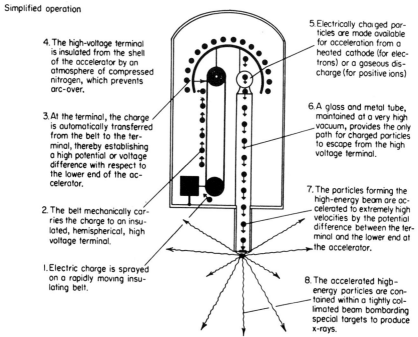

Simplified operation

4. The high-voltage terminal is insulated from the shell of the accelerator by an atmosphere of compressed nitrogen, which prevents arc-over.

3. At the terminal, the charge is automatically transferred from the belt to the terminal, thereby establishing a high potential or voltage difference with respect to the lower end of the accelerator.

2. The belt mechanically carries the charge to an insulated, hemispherical, high voltage terminal.

1. Electric charge is sprayed on a rapidly moving insulating belt.

5. Electrically charged particles are made available for acceleration from a heated cathode (for electrons) or a gaseous discharge (for positive ions)

6. A glass and metal tube, maintained at a very high vacuum, provides the only path for charged particles to escape from the high voltage terminal.

7. The particles forming the high-energy beam are accelerated to extremely high velocities by the potential difference between the terminal and the lower end of the accelerator.

8. The accelerated high-energy particles are contained within a tightly collimated beam bombarding special targets to produce x-rays.

Fɪɢ. 6-12. Van de Graaff X-ray generator. (*Courtesy High Voltage Engineering Corporation.*)

accelerator by an atmosphere of compressed nitrogen. The charge is transferred from the belt to the high-voltage terminal and builds up a voltage difference with respect to the lower end of the chamber. Electrons to be accelerated are produced by a heated filament. A glass and metal tube maintained at a high vacuum provides a path for the electrons. The electrons are accelerated by the difference of potential between the high-voltage terminal and the lower end of the chamber and produce X rays by bombarding a target. The Van de Graaff generator is a constant potential source. The constant potential is achieved by balancing the continuous charging current with the output current of the electron accelerating tube. By means of electrostatic

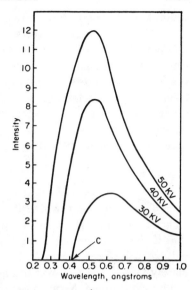

FIG. 6-13. Continuous X-ray spectrum from tungsten target.

beam focusing, the electron beam striking the X-ray target is sharply focused. Van de Graaff X-ray generators have focal spots less than 0.30 in. in diameter that produce a sharp image on the X-ray film.

Electromagnetic Radiation. Electromagnetic radiation may be thought of in either of two ways: as waves or as consisting of particles called "photons" or "quanta." Some phenomena are best explained by thinking of X rays or gamma rays as quanta, other phenomena by considering them as waves. In common with all wave motion, the relationship that exists between the frequency f, wavelength λ, and velocity V of electromagnetic radiation is given by the following expression (the velocity is 3×10^{10} cm/sec or 186,000 miles/sec):

$$V = f\lambda \qquad (6\text{-}3)$$

When the radiation from an X-ray tube is analyzed it is found to consist of two parts, the continuous X-ray spectra and the characteristic spectra. Figure 6-13 shows the intensity curves for the continuous X-ray spectrum for various tube voltages. The point of intersection of each intensity curve with the wavelength axis, i.e., point C for the 30-kv curve, is called the "short wavelength limit." The short wavelength limit is determined entirely by the tube voltage. X rays of the shortest wavelength are produced by the electrons having the highest velocity or most energy. The short wavelength limit can be calculated from the expression

$$\lambda = \frac{12,336}{V} \qquad (6\text{-}4)$$

where λ = wavelength, angstroms

V = tube voltage, volts

As the voltage across the tube is increased, Eq. (6-4) shows that the wavelength becomes shorter.   It will be shown later that this also means that the penetrating power or quality of the radiation increases with increased tube voltage.   Referring again to Fig. 6-13 it can be seen that the intensity of the radiation increases with increased voltage across the tube. It has been shown that intensity of the continuous spectrum increases with the square of the voltage:

$$I = KV^2 \qquad\qquad (6\text{-}5)$$

The K is a constant which is determined among other factors by the target material.   It should be pointed out in connection with Fig. 6-13. that if the tube current is increased the X-ray intensity for a given tube voltage will also increase.   The maximum intensity occurs at approximately two-thirds the maximum voltage.

The continuous X-ray spectrum and the short wavelength limit are independent of the target material.   There are at least three reasons for this continuous spectrum.   Some of the cathode rays penetrate into the target for a short distance before striking an atom, and as a result the X rays produced are filtered by the surface layers of the target. Variations in the voltage across the tube produce variations in X-ray energy.   X rays produced by a constant potential source have a higher percentage of short wavelength radiation than that produced by a self-rectified tube.   In the interaction between atoms of the target and the cathode rays, the amount of energy transferred varies.   The target material does affect the shape of the energy distribution curve.   The characteristic spectrum is determined by the target material.

Figure 6-14 shows the characteristic spectrum of tungsten.   The energy in the characteristic spectrum is small compared with that in the continuous spectrum.   The radiation which is characteristic of a par-

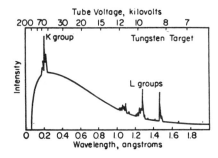

FIG. 6-14. Characteristic spectrum of tungsten.

TABLE 6-3. K SPECTRA

Element	Wavelength, A		
	α	β	γ
Fe	1.934	1.752	1.741
Co	1.787	1.617	1.605
Ni	1.656	1.497	1.485
Cu	1.539	1.389	1.378
Ag	0.560	0.496	0.496
W	0.211	0.184	0.179
Pb	0.168	0.146	0.141

ticular target material is not all of the same wavelength but is found to occur in groups of wavelengths or "lines" which are customarily designated by letters. The shorter wavelength group is designated as the K series, the next shortest L, M, etc. It is customary also to designate the wavelengths within a group according to their intensity, using the Greek letters alpha, beta, gamma, etc., as subscripts. The more intense wavelength is designated as K_α, the next K_β, etc. Table 6-3 shows the characteristic spectra of various materials. When the X-ray tube voltage is less than 70 kv for a tungsten target, the characteristic X rays are not present, as is shown in Fig. 6-14. This can be shown to be true by using Eq. (6-4). The characteristic spectra appear only when the tube voltage exceeds a certain minimum value, this minimum value being determined by the target material.

Absorption. The use of X rays in radiography depends upon differential absorption. Consequently, it is important to understand how X rays interact with matter. The first mechanism to be discussed will be absorption. How X rays are absorbed by material can be understood from the experiment shown in Fig. 6-15. It has been found from such an experiment that a homogeneous X-ray beam of intensity I passing through the thickness of material Δx undergoes a decrease in intensity ΔI. The decrease in intensity has been found to be proportional to the intensity of the incident beam and the thickness of the absorber, or in mathematical form

$$\Delta I = -\mu I \, \Delta x \qquad (6-6)$$

where μ is a constant of proportionality and the negative sign is used to indicate a decrease in intensity. This equation can be written in the differential form

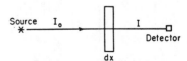

FIG. 6-15. Experimental setup for studying absorption.

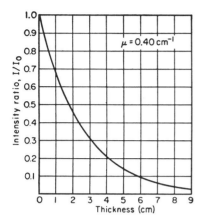

FIG. 6-16. Plot of absorption law, rectilinear coordinates.

and then integrated to give

$$dI = -I\mu\, dx \qquad (6\text{-}7)$$
$$I = I_0 \exp\,(-\mu x) \qquad (6\text{-}8)$$

where I_0 = intensity of incident radiation

I = intensity of transmitted radiation

x = thickness of specimen

μ = linear absorption coefficient for material

This expression is the basic law governing absorption of a homogeneous beam of X rays or gamma rays. Figure 6-16 shows a plot of this equation using rectangular coordinates, and Fig. 6-17 shows the straight line obtained when this equation is plotted on a logarithmic scale. The straight line has a slope equal to the linear absorption coefficient. The fact that the equation is exponential means that the intensity will be reduced by equal fractions for equal thicknesses of material. For example, if a given thickness of material reduces the intensity one-half, doubling the thickness will reduce the intensity one-fourth.

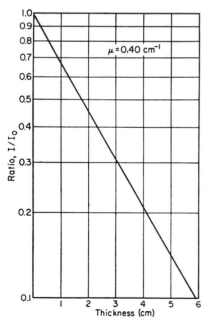

FIG. 6-17. Plot of absorption law, semilog coordinates.

The constant of proportionality μ is known as the linear absorption coefficient, and has the units per centimeter or per inch often written as cm^{-1} or $\mathrm{in.}^{-1}$ The linear absorption coefficient represents the fraction of energy absorbed per centimeter or per inch thickness of material transversed. The numerical value of μ depends upon the wavelength of the incident radiation and the material of which the absorber is made. Figure 6-18 shows how μ varies for steel as a function of energy. It is often more convenient to use the mass absorption coefficient μ_m which is related to the linear coefficient μ as follows:

$$\mu_m = \frac{\mu}{\rho} \qquad (6\text{-}9)$$

where ρ = density of material

Table 6-4 gives the linear and mass absorption coefficients for a number of different materials. The mass absorption coefficients as a function of energy are given for a number of materials by the National Bureau of Standards publication.[1,*] For values of mass absorption coefficients not found in Tables, Victoreen[2] gives an empirical method for calculating the necessary values. The mass absorption is independent of the physical or chemical state of the material. The intensity of an X-ray beam is not decreased as much in traversing 1 in. of steam as it is traversing 1 in. of water. However, when the density of the two materials is taken into consideration, the absorption per unit mass is the same. Thus the fraction of the X-ray beam absorbed by a given amount of water is the same whether the material is in the form of ice, water, or steam. Figure 6-19 shows how the mass absorption coefficient varies with wavelength.

The mass absorption coefficient of a chemical compound or mixture is an average of the mass absorption coefficients of the constituent elements, in proportion to the abundance of each element by weight. For example, the mass absorption coefficient for water (1 part H, 8 parts O) (H_2O) is given by

$$\mu_{H_2O} = \tfrac{1}{9}\mu_H + \tfrac{8}{9}\mu_0 \tag{6-10}$$

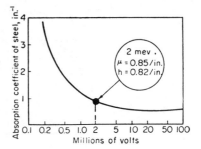

FIG. 6-18. Linear absorption coefficient of steel vs. energy.

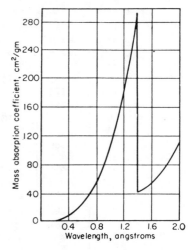

FIG. 6-19. Mass absorption coefficient vs. wavelength for copper.

where μ_H and μ_0 are the mass absorption coefficients for hydrogen and oxygen, respectively. Aluminum oxide (Al_2O_3) has an atomic weight of 102. The mass absorption coefficient of aluminum oxide for X rays having an energy of 0.5 Mev is calculated as follows:

$$\mu_{Al_2O_3} = 0.53\mu_{Al} + 0.47\mu_0$$
$$= 0.53(0.0840) + 0.47(0.0870)$$
$$= 0.085 \text{ cm}^2/\text{g}$$

* Superscript numbers indicate Specific References listed at the end of the chapter.

NONDESTRUCTIVE TESTING

TABLE 6-4. ABSORPTION COEFFICIENTS FOR X RAYS
($\lambda = 0.098$ A)

Element	At. No.	μ/ρ	μ	Element	At. No.	μ/ρ	μ
H	1	0.280		Ru	44	0.90	11.1
Li	3	0.125	0.067	Rh	45	0.95	11.8
Be	4	0.131	0.24	Pd	46	0.99	11.3
B	5	0.138	0.35	Ag	47	1.05	11.0
C	6	0.142	0.33	Cd	48	1.09	9.4
N	7	0.143		In	49	1.13	8.2
O	8	0.144		Sn	50	1.17	8.5
F	9	0.146		Sb	51	1.21	8.1
Ne	10	0.148		Te	52	1.25	7.8
Na	11	0.150	0.15	I	53	1.33	6.6
Mg	12	0.152	0.26	Xe	54	1.40	
Al	13	0.156	0.42	Cs	55	1.46	2.7
Si	14	0.159	0.37	Ba	56	1.52	5.7
P	15	0.162	0.32	La	57	1.60	9.8
S	16	0.166	0.33	Ce	58	1.68	11.6
Cl	17	0.176		Pr	59	1.75	11.4
A	18	0.184		Nd	60	1.81	12.6
K	19	0.191	0.16	Sm	62	1.95	15.2
Ca	20	0.200	0.31	Eu	63	2.02	10.5
Sc	21	0.208	0.52	Gd	64	2.08	12.3
Ti	22	0.217	0.98	Tb	65	2.13	17.7
V	23	0.227	1.4	Dy	66	2.23	19.2
Cr	24	0.238	1.7	Ho	67	2.33	21
Mn	25	0.250	1.8	Er	68	2.40	11.4
Fe	26	0.265	2.1	Tm	69	2.48	23
Co	27	0.287	2.5	Yb	70	2.55	14.0
Ni	28	0.310	2.8	Lu	71	2.63	26
Cu	29	0.325	3.0	Hf	72	2.72	31
Zn	30	0.350	2.5	Ta	73	2.80	47
Ga	31	0.380	2.3	W	74	2.88	56
Ge	32	0.41	2.2	Re	75	2.95	63
As	33	0.44	2.5	Os	76	3.02	68
Se	34	0.48	2.2	Ir	77	3.09	69
Br	35	0.52		Pt	78	3.15	68
Kr	36	0.56		Au	79	3.21	62
Rb	37	0.59	0.90	Hg	80	3.31	45
Sr	38	0.61	1.6	Tl	81	3.41	41
Y	39	0.66	2.5	Pb	82	3.50	40
Zr	40	0.71	4.6	Bi	83	3.57	35
Nb	41	0.75	6.4	Th	90	3.80	43
Mo	42	0.79	7.9	U	92	3.90	73

TABLE 6-5. K ABSORPTION LIMIT FOR SOME ELEMENTS

Element	Atomic No.	K abs limit
Cu	29	1.3774
Ag	47	0.4845
W	74	0.1782
Pb	82	0.1405

Figure 6-19 shows abrupt changes in the mass absorption coefficient as the wavelength changes. These discontinuities are known as characteristic absorption edges or critical absorption wavelengths. Comparing the values of critical absorption wavelengths given in Table 6-5 with the characteristic spectra given in Table 6-3, it can be seen that the

TABLE 6-6. DISTRIBUTION OF ELECTRONS IN THE ATOMS

X-ray notation	K	L		M			N				
Values of n, 1	1, 0	2, 0	2, 1	3, 0	3, 1	3, 2	4, 0	4, 1	4, 2	4, 3	
Spectral notation	1s	2s	2p	3s	3p	3d	4s	4p	4d	4f	
Element / Atomic number Z / First ionization potential, volts											Lowest spectral term
H 1 13.529	1	...	...	...	...	...	...	...	...	...	$^2S_{1/2}$
He 2 24.465	2	...	...	...	...	...	...	...	...	...	$^1S_0$
Li 3 5.37	2	1	...	...	...	...	...	...	...	...	$^2S_{1/2}$
Be 4 9.281	2	2	...	...	...	...	...	...	...	...	$^1S_0$
B 5 8.28	2	2	1	...	...	...	...	...	...	...	$^2P_{1/2}$
C 6 11.217	2	2	2	...	...	...	...	...	...	...	$^3P_0$
N 7 14.48	2	2	3	...	...	...	...	...	...	...	$^4S_{3/2}$
O 8 13.550	2	2	4	...	...	...	...	...	...	...	$^3P_2$
F 9 18.6	2	2	5	...	...	...	...	...	...	...	$^2P_{3/2}$
Ne 10 21.47	2	2	6	...	...	...	...	...	...	...	$^1S_0$
Na 11 5.12				1	...	...	...	...	...	...	$^2S_{1/2}$
Mg 12 7.61				2	...	...	...	...	...	...	$^1S_0$
Al 13 5.96	Neon			2	1	...	...	...	...	...	$^2P_{1/2}$
Si 14 8.12	configuration			2	2	...	...	...	...	...	$^3P_0$
P 15 10.9	or			2	3	...	...	...	...	...	$^4S_{3/2}$
S 16 10.3	10-electron			2	4	...	...	...	...	...	$^3P_2$
Cl 17 12.96	core			2	5	...	...	...	...	...	$^2P_{3/2}$
A 18 15.69				2	6	...	...	...	...	...	$^1S_0$

critical absorption wavelength is slightly shorter than the characteristic wavelength.

Atomic Structure. For a better understanding of absorption and characteristic spectra, a review of the structure of matter follows. Matter is composed of atoms, which consist of a central positively charged nucleus in which nearly all the mass of the atom is concentrated. Electrons revolve about the nucleus in orbits. In the normal atom, the number of electrons in the orbits is just equal to the net positive charge of the nucleus, leaving the atom uncharged. The nucleus is composed of protons and neutrons. The weight of the protons and neutrons in the nucleus is referred to as the "atomic weight." The positive charge on the nucleus or the number of orbital electrons is referred to as the "atomic number." Figure 6-20 shows the hydrogen, helium, tritium and deuterium atoms. All the elements can be placed in a series, in the order of increasing atomic number. As more and more orbital electrons are added to the atom, the development progresses in a manner determined by certain physical laws.[3] The orbit nearest the nucleus is referred to as the K orbit, the next nearest the L orbit, followed by the M and N orbits. Table 6-6 gives the distribution of electrons for

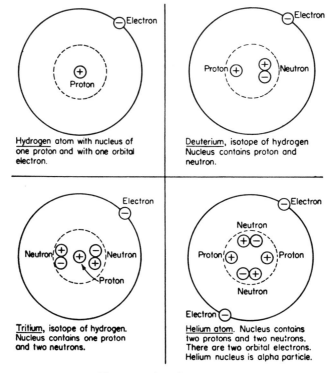

Hydrogen atom with nucleus of one proton and with one orbital electron.

Deuterium, isotope of hydrogen Nucleus contains proton and neutron.

Tritium, isotope of hydrogen. Nucleus contains one proton and two neutrons.

Helium atom. Nucleus contains two protons and two neutrons. There are two orbital electrons. Helium nucleus is alpha particle.

Fig. 6-20. Atomic structure.

a number of elements. The K electrons are more "tightly bound" to the nucleus than the L, M, or N electrons. If the voltage across an X-ray tube is sufficiently great, an electron will be removed from the K orbit. If the cathode ray does not have sufficient energy to remove a K electron, it may have sufficient energy to remove an L, M, or N electron. If a vacancy exists in the K orbit, an electron from one of the other orbits will fill the vacancy. The K series of spectral lines will be emitted when electrons go from the L, M, or N orbits directly to the K orbit. The L series of spectral lines will be emitted when electrons go from the M and N orbits to the L orbit. In the production of the K series of spectral lines a certain amount of energy must be supplied to the atom to remove the electron from the K orbit, after which the electron from an outer orbit going into the K orbit would emit a quantum of radiation. This latter quantum can be shown to always be less than the energy required to remove the electron initially.

Half-value Thickness. The half-value thickness t is the thickness of absorber required to reduce the intensity of a given monoenergetic radiation to one-half.

$$\frac{I}{I_0} = \tfrac{1}{2} = \exp\ (-\mu_m t \rho) \tag{6-11}$$

$$\log_e 2 = \mu_m t \rho \tag{6-12}$$

$$t = \frac{0.693}{\mu_m \rho} \tag{6-13}$$

In practical radiography neither Eq. (6-8) nor (6-13) is particularly useful since these equations are strictly true only for monochromatic X rays. However, by experimentally determining the half-value thickness, the corresponding value of the absorption coefficient in Eq. (6-8) can be determined. From a table giving absorption coefficient as a function of wavelength, the equivalent wavelength of the radiation can be determined. When the equivalent wavelength has been determined, the effective kilovoltage of the beam can be computed from the expression

$$\text{kv} = \frac{12.345}{\lambda\ (\text{equivalent})} \tag{6-14}$$

Scattering. In the experiment shown in Fig. 6-15, if the detector is moved either to the right or to the left, there will be an increase in reading when the absorber is placed in position. Some of the initial X rays emerge from the absorber traveling in directions different from that at which they entered. This radiation is called "scattered radiation," and when the test specimen has an appreciable thickness the radiation may be scattered more than once. The detector may receive scattered radiation as well as the primary radiation and cannot distinguish between the two. The measured absorption coefficient is composed of two parts, a true absorption and a scattering factor. The true absorption is characterized by the disappearance of an X-ray quanta, a transfer of energy from the X-ray quanta to the electrons of the material. The scattered radiation is characterized by a change of direction from that of the primary beam. The scattered radiation consists of modified and unmodified radiation. The modified radiation undergoes a change in energy, whereas the unmodified is not changed in energy.

There are three processes which take place as gamma rays pass through matter, namely, the photoelectric effect, pair production, and Compton scattering. These three processes are illustrated in Fig. 6-21. In the photoelectric effect, the photon is absorbed by an atom and one of the bound electrons in the atom is ejected. Electrons in the K and L shells account for most of the absorption by this process. The probability of this process occurring increases with increasing atomic number and decreases as the photon energy increases. In pair production a photon with an energy greater than 1.02 Mev is absorbed in the atoms and produces a pair of electrons. One electron is positively charged and the other negatively charged. The probability of this absorption process increases with the photon energy and with atomic number. In Compton scattering the photon is deflected by an atomic electron. Only part of the energy of the photon is lost, and the photon continues with lower energy in a direction different from its initial direction of travel. The atomic electron recoils out of the atom. The probability of this scattering process decreases with increasing photon energy and is proportional to the number of atomic electrons.

The total absorption coefficient is the sum of the photoelectric absorp-

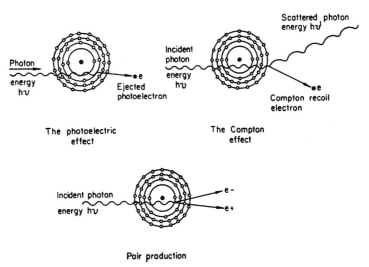

Fig. 6-21. Absorption processes.

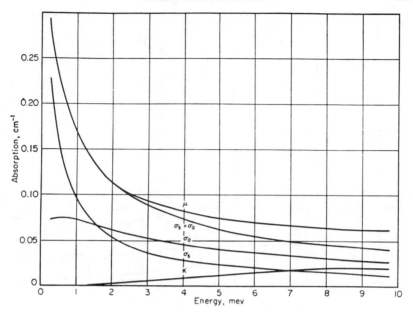

Fig. 6-22. Absorption coefficient of gamma rays in aluminum. μ, total absorption; σ_a, Klein-Nishina absorption; σ_s Klein-Nishina scattering; K, pair production.

tion coefficient, the Compton scattering coefficient, and the pair production coefficient. Figures 6-22 and 6-23 show the absorption of X rays as a function of energy. The individual absorption coefficients as well as the total absorption coefficient are shown. As can be seen from the figures, the Compton scattering makes a large contribution to the absorption in the 0.5- to 5-Mev energy range.

The change in wavelength of the modified radiation increases as the angle between the primary beam and scattered photon increases. The relationship is given by

$$\lambda' - \lambda = 0.02427(1 - \cos \theta) \tag{6-15}$$

where λ' = wavelength of modified radiation, A

λ = wavelength of primary radiation, A

θ = angle between incident radiation and scattered photon

The theory of Compton scattering may be visualized by means of Fig. 6-24 in which the photon is considered to be "bouncing off" the electron. The electron is displaced from its shell if the electron is free or loosely bound as in the case of light atoms. If the photon bounces off a firmly bound electron (such as an electron in the inner shell of a heavy atom), the electron is not knocked out of its shell, thus the unmodified scattering, in which the photon is deflected with no loss in energy. The probability for this process is large only for photons with low energy, that is,

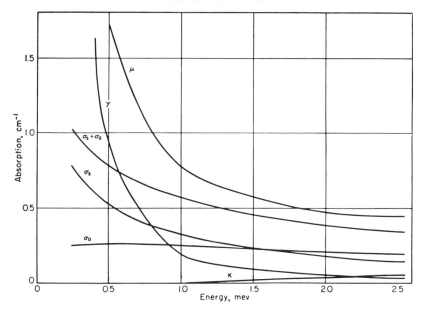

Fig. 6-23. Absorption coefficient of gamma rays in lead. μ, total absorption; σ_a, Klein-Nishina absorption; σ_s, Klein-Nishina scattering; τ, photoelectric absorption; K, pair production.

in the region where photoelectric absorption gives the main contribution to the total absorption coefficient.

In film radiography the scattered radiation should be reduced to a minimum since the scattered X rays do not contribute to the formation of the image on the film, but merely reduce the contrast on the film and tend to obliterate detail and reduce sensitivity, since the image forming radiation is that portion of the incident radiation traveling in the original direction after traversing the specimen. The X-ray scattering is roughly proportional to the object thickness and the density of the material. For example, with a steel specimen 5 in. in thickness the scattered X rays are five times that for a 1-in.-thick specimen. The scattering produced by a 5-in.-thick steel specimen is about the same as that produced by a 3-in.-thick lead specimen. By using a higher voltage, the scattering can be considerably reduced. Figure 6-25 shows the scattering factor per inch of steel as a function of tube voltage. The scattering factor is defined:

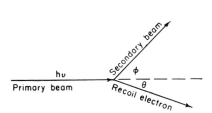

Fig. 6-24. Compton effect.

Scattering factor $=$

$$\frac{\text{scattered radiation}}{\text{direct radiation}} \quad (6\text{-}16)$$

With low-voltage X rays, as much as 80 per cent or more of the film darkening may be due to secondary radiation. As the X-ray voltage is increased, the secondary radiation decreases. For example, in radiographing a 4-in.-steel specimen with 20-Mev X rays, about 15 per cent of the film density is due to secondary radiation. With the same specimen using 10-Mev gamma rays, about 40 per cent of the film density is due to secondary radiation. From Fig. 6-25 it can be seen that the scattering factor tends to approach a constant value at about 3 or 4 Mev.

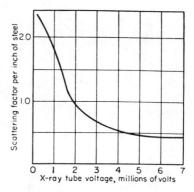

FIG. 6-25. Scattering factor as a function of energy.

X-ray Film Processing. X-ray film consists of a transparent cellulose acetate base material uniformly coated, usually on both sides, with a silver bromide gelatin emulsion. The silver bromide is in the form of tiny crystals and is distributed uniformly in the gelatin. The coatings are each approximately 0.001 in. thick. When X rays or gamma rays impinge on the emulsion, a photochemical change takes place in the silver bromide crystals. The amount of photochemical change which takes place is proportional to the intensity of the radiation and the duration of the exposure to the radiation. Before the photochemical changes which occur can be seen by the eye, the film must be treated with a chemical solution called the "developer."

The developer has a reducing action on the silver bromide and takes the bromide away from the exposed silver bromide crystal and deposits black metallic silver in the gelatin. The concentration of black metallic silver per unit area of emulsion depends on the exposure and determines the film density. It is also important that the developer does not affect crystals which have received no exposure. No chemical now known will leave an unexposed region indefinitely unchanged. The unexposed parts will begin to develop after a period of time, producing a condition known as "chemical fog." Figure 6-26 shows film defects. Developers generally contain metol and hydroquinone, which produce all the gradations of grays and the jet blacks. However, neither of the materials will develop when used alone. To produce any density on the film requires an alkaline solution; the alkali allows the developing agents to enter the pores of the emulsion. Sodium sulfite and potassium bromide are added to the solution to serve as a preservative and to suppress chemical fog. The rate of development is influenced by the temperature of the solution; a temperature of 68°F is usually recommended. Table 6-7 gives developing time at various temperatures.

FIG. 6-26. Film markings. (*Courtesy Eastman Kodak Company.*)

To stop development, the film is either placed in an acid short-stop bath or rinsed in water. The short-stop bath consists of a mild acetic acid solution to neutralize the alkali of the developer. The acid also protects the fixer solution which is slightly acid from the alkalies of the developer, thereby extending its useful life. If a stop bath is not used, films should be immersed in clear running water for at least 1 or 2 min. If the films are agitated, 20 sec is sufficient. After development, the emulsion contains all the unexposed and undeveloped grains of silver bromide. The undeveloped silver must be removed from the emulsion if the image is to be permanent. To accomplish this, a fixer is used. Sodium thiosulfate (hypo) or ammonium thiosulfate is used as a clearing agent to remove the undeveloped silver bromide in the film emulsion. Acetic and sulfuric acid in weak concentration are added to the fixer to neutralize the alkaline developer adhering to the film. A hardening agent such as potassium alum or aluminum chloride is added to the fixer

TABLE 6-7. DEVELOPING TIME AS A FUNCTION OF TEMPERATURE

Temperature, °F	Time, min
63	6
65	$5\frac{1}{2}$
68	5
70	$4\frac{1}{2}$
72	4
75	$3\frac{3}{4}$
80	$3\frac{1}{4}$

to harden the emulsion. When a film is removed from the developer, the undeveloped areas are swollen and yellow in appearance. After immersion in the fixer for a short period of time, this yellow becomes transparent. The time required for this change is known as the "clearing time." To fix a film adequately, it should be left in the fixer twice the clearing time.

Films are washed after fixing to remove any chemical present on the emulsion. Washing will never remove the last traces of fixer. The objective of washing is to remove enough of the fixer so that the radiograph may be kept without fading for a given period of time. The washing time should be at least 30 min. The final step in processing the film is drying. This may be done by hanging the films on racks and allowing air to circulate either by natural or forced convection. Hot-air dryers can be used to speed up the process. However, excessive heat should be avoided to prevent brittleness and buckling of the film.

Film Types. Various types of X-ray films are available which differ in speed, contrast, and grain size. Films having large grain size have higher speeds than those with a relatively small grain size. Likewise, high-contrast films are usually finer grained and slower than low-contrast films. A small-grained film is capable of resolving more detail than one having large grains. Two general types of X-ray films are used. One is intended for use with fluorescent intensifying screens, although it may be used without screens. The No Screen type is intended for use without any fluorescent intensifying screens, although lead-foil intensifying screens are often used. Table 6-8 lists some of these films and their characteristics as well as relative speed and contrast.

The completed radiograph consists of areas in which different amounts of metallic silver have been deposited. The density of the film is determined by the amount of white light transmitted through the film. The transmittance T is defined as

$$T = \frac{I_t}{I_0} \qquad (6\text{-}17)$$

where I_t = intensity of light transmitted through film
  I_0 = intensity of light incident on film
The film density D is defined as the logarithm of the reciprocal of the transmittance T of the film.

$$D = \log_{10} \frac{1}{T} = \log_{10} \frac{I_0}{I_t} \qquad (6\text{-}18)$$

A film which transmits half the incident light has a density of

$$\log_{10} 2 = 0.30$$

A film which transmits one-tenth of the incident light has a density of 1.0.

The density of the film depends on the intensity I of the radiation and the length of time t for which the exposure is made. The product of these two factors is known as the exposure:

$$E = kIt \qquad (6\text{-}19)$$

where E = exposure
   k = proportionality factor
   I = intensity of radiation
   t = exposure time

Radiographic exposure given by Eq. (6-19) can be written in terms of the tube current M as follows:

$$E = kIt = CMt \qquad (6\text{-}20)$$

where C = constant
   M = tube current, ma

TABLE 6-8. KODAK FILMS FOR INDUSTRIAL RADIOGRAPHY

Type of film	Relative speed and contrast				Development	
	Aluminum and other light alloys, no screens	Thin steel and brass, lead-foil screens	With million-volt X rays, lead-foil screens	With gamma rays, lead-foil screens	Normal development at 68°F (20°C), min	For max speed and contrast at 68°F (20°C), min
Type M:						
Relative speed........	30	40	30	30	5	8
Contrast.............	High	High	High	High		
Type A:						
Relative speed........	100	100	100	100	5	8
Contrast.............	High	High	High	High		
Type F:						
Relative speed........	200	200	200	360	5	8
Contrast.............	Medium	Medium	Medium	Medium		
Type K:						
Relative speed........	530	400	425	500	5	8
Contrast.............	Medium	Medium	Medium	Medium		
No Screen:						
Relative speed........	600	400	250	300	5	8
Contrast.............	High	High	High	High		
Blue Brand:						
Relative speed........	200	200	200	360	5	8
Contrast.............	Medium	Medium	Medium	Medium		
Type AA:						
Relative speed........	130	170	....	230		
Contrast.............	High	High	....	High		

The exposure at 10 ma for 2 min is four times the exposure at 5 ma for 1 min, all other factors being the same. If a number of strips of a particular film are given different exposures, each strip will have a different density. Figure 6-27 shows a graph of the film density as a function of the logarithm of the relative exposure. Such curves are called H & D or characteristic curves. The slope of the characteristic curve is changing along the entire length of the curve, the film contrast being greatest where the

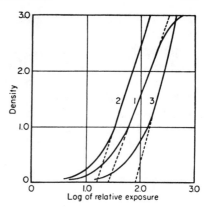

FIG. 6-27. Characteristic curves.

slope of the characteristic curve is greatest. The useful working range of the film is the straight-line portion of the curve, Fig. 6-27, and ordinarily lies in the density range between 0.7 and 2.0, with ordinary viewing facilities. A practical way to determine the required change in exposure for a given change in film density is illustrated by the following example. A radiograph is made on type F film with an exposure of 12 ma-min and has a density of 0.8. It is desired to increase the density to 2.0 for more contrast. The exposure required, using Fig. 6-27, curve 2, is

$$\text{For } D_1 = 0.8 \qquad \log E_1 = 1.30$$
$$\text{For } D_2 = 2.0 \qquad \log E_2 = 1.97$$
$$\log E_2 - \log E_1 = \log (It)_2 - \log (It)_1 \qquad (6\text{-}21)$$
$$0.67 = \log (It)_2 - \log (12)$$
$$(It)_2 = 56.12 \text{ ma-min}$$

or the exposure must be increased by 4.68 times.

Contrast may be defined as the difference in film density due to a change in thickness or density of the specimen. A defect is recognizable in a radiograph because of the contrast between the density of its image and the density of the image produced by the "good" material. The greater this difference, the easier it is to detect a flaw. From Fig. 6-27 it can be seen that the steeper the slope of the characteristic curve, the greater the contrast.

The total exposure of the radiograph comes from three sources: the primary radiation, the secondary or scattered radiation, and the "photographic fog." The latter may be inherent in the film or acquired during handling or processing. The secondary radiation is scattered in all directions. Consequently, unless some method is used to reduce the scattered radiation, it will cause a haziness or fog over the radiograph, reducing the contrast.

Geometrical Factors. Since a radiograph is a shadow picture, the image produced is affected by the relative positions of the specimen and film and by the size of the focal spot of the X-ray tube. The ideal condition would be a point source of radiation. Since the focal spot has definite size there will be a small amount of blurring or unsharpness around the edges of the specimen, as shown in Fig. 6-28. From the similar triangles in Fig. 6-28

$$\frac{FF'}{d} = \frac{U}{t} \tag{6-22}$$

$$U = \frac{(FF')t}{d} \tag{6-23}$$

where FF' = size of focal spot
 d = source-to-specimen distance
 U = unsharpness
 t = specimen-to-film distance

As can be seen from this equation, the smaller the focal spot, the less the unsharpness. Equation (6-23) also shows that the unsharpness is inversely proportional to the source-to-specimen distance. It is advisable to use the longest target-to-film distance possible consistent with a reasonable exposure time, the exposure time being directly proportional to the square of the distance. The unsharpness varies directly as the specimen-to-film distance. The film should be as close as possible to the specimen. The greater the d/t ratio, the better the definition on the radiograph. Definition or detail is the characteristic of a radiograph which describes the sharpness of the image. Deviation from the true shape of the specimen in the shadow picture is called "distortion."

Filters. A thin sheet of metal called a "filter" is sometimes placed between the source of radiation and the specimen to absorb the longer wavelengths. The shorter wavelengths do not produce so much scattering as the longer wavelengths. When radiographing specimens which vary in thickness, it is advisable to use a filter to reduce overexposure and scatter. The contrast is improved, which in turn permits a larger range of thickness to be radiographed on one film. Filters require an increase in exposure time to compensate for the addi-

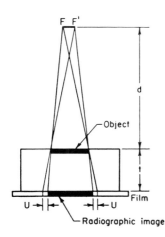

Fig. 6-28. Geometrical unsharpness.

tional material which must be penetrated by the X-ray beam. If the filter is placed between the specimen and the film, it becomes a screen.

Screens. Approximately 99 per cent of the X rays striking a photographic film do not produce a photochemical effect in the emulsion but pass entirely through the film. To decrease the exposure time and improve the quality of the radiograph, intensifying screens are often used. There are two kinds of intensifying screens, lead foil and calcium tungstate. Lead screens are made by mounting a thin lead foil on a cardboard base. The intensifying action of a lead screen is due to the emission of electrons from the surface of the lead when X rays impinge on the screen and by the action of the secondary X rays produced in the lead. The electrons in turn cause a photochemical reaction in the film. The electrons are more efficient in producing a photochemical change in the film than is the primary electromagnetic radiation. The intensifying screen should be placed in good contact with the film; otherwise the electrons emitted from the screen will cause fuzzy images. The lead screens also absorb the low-energy scattered radiation from the specimen being radiographed, giving a sharper radiograph.

Lead screen intensifying screens are often used in pairs, the film being placed between the screens. The screen placed closest to the X-ray source is commonly called the "front" screen, and the other is called the "back" screen. In industrial radiography the front screen is usually 0.005 in. thick. The rear screen is 0.005 to 0.015 in. thick. The front screen has a greater intensifying action since the electrons ejected from the lead tend to travel in the same direction as the primary radiation. Since the X rays must pass through the front screen before striking the film, the front screen acts as a filter and soft X rays are removed from the beam. At low kilovoltage the filtering action of the lead screen may be such as to increase the required exposure time. The rear screen tends to reduce the effect of backscatter on the film, since the screen tends to absorb the low-energy secondary radiation. The rear screen is made thicker than the front screen. The rear screen tends to reduce backscattered radiation from any object the radiation strikes after traversing the film, such as the film holder. The intensifying action increases as the energy of the photons increases. It requires 87.5 kv to eject a K electron from the lead atom. Consequently, the photon must have an energy in excess of 87.5 to produce photoelectrons to be of use in the photographic process. The absorption by the front screen reduces the photochemical effect of the X rays. The photoelectrons produced by the screen increase the photochemical effect. The intensification factor is defined as the factor by which the exposure is decreased by using a screen.

Pace[4] has made an investigation to determine quantitatively the intensifying effect of various thicknesses of lead screens. Figure 6-30

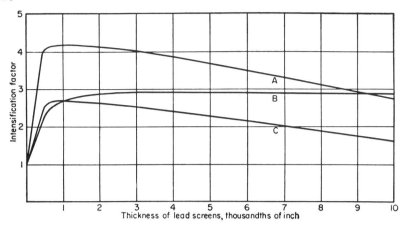

FIG. 6-29. Intensification vs. screen thickness for 1-in. steel, 250 kv. (*A*) double screens; (*B*) back screen; (*C*) front screen. (*Courtesy General Electric Company, X-Ray Division.*)

shows the intensification factor (double screen) for four different voltages. It can be seen that the maximum intensification for 175, 210, and 250 kv occurs at 0.001 in. or less of lead. The maximum intensification for 1,000 kv occurs at 0.003 in. of lead. This same investigation showed that the maximum intensification for 2 Mev occurred at 0.007 in. and for 10 Mev at 0.025 in. Figure 6-29 shows the intensification factor for various thicknesses of lead screens. The maximum intensification factor was obtained with 0.001-in.-thick front screen at voltages of 140 to 250 kw. From Fig. 6-29 the effect of the rear screen can be seen to reach a maximum at a thickness of 0.001 in. and remains the same up to 0.010 in. The intensification factor for lead screens (double) ranges from 2 at 140 kv to 4.7 at 1,000 kv. The back screen contributes the greater share of the radiographic density at the thickness of lead screen normally used. The photoelectrons from the back screen originate at the surface, and for thicknesses greater than about 0.001 in., the intensification factor is approximately the same. Lead screens are not very effective at voltages below approximately 140 kv; however, the back screen contributes appreciable density to the film even at 80 kv and should be used.

The intensification factor of lead screens is influenced by the thickness and kind of material being radiographed. For thin sections of steel, for example, the front and back screen intensification factor is greater than either the front or back screen alone. The intensifying effect of the lead screens reaches a maximum at about $\frac{1}{2}$ and $\frac{3}{4}$ in. of steel in the volt range of 140 to 250 kv.

Seeman[5] investigated other materials for intensifying screens and concluded that lead has the most desirable properties for such screens. The definition or graininess of the radiographic image is not affected by using lead screens. Since even thin layers of material absorb electrons readily, the screen must be in intimate contact with the film for maximum effect. The surface of the screen next to the film should also be kept free of foreign matter.

The other type of intensifying screen used is the calcium tungstate screen. The advantages of these screens are twofold: reduction in exposure time and the fact that radiography of thicker sections on X-ray machines of moderate power is facilitated. As in the case of lead screens, the film is often placed between a pair of such screens. When calcium tungstate is struck by X rays, it fluoresces with a bluish-white light. The intensity of the light emitted depends on the intensity of the incident radiation. The light emitted has a wavelength to which the X-ray film emulsion is sensitive. The screen is made by mixing finely divided power with a suitable binder, and a thin layer of the mixture is put on a cardboard backing. The intensification factor obtained when using calcium tungstate varies from 10 to 100. Figure 6-31 shows the effect of voltage on the intensification factor of calcium tungstate films. The coarser the grain, the more intense the fluorescence and the "faster" the film. The larger calcium tungstate crystals give the film a grainy appearance. The intensification factor of calcium tungstate screens is determined by the size of the crystals in the screen, the larger the crystal, the greater the amount of light emitted, and the greater the intensification. However, the size of the crystals affects the definition. A radiograph made using fluorescent screens has poorer definition than a radiograph made with or without lead screens. The poorer definition is due to the spreading of the visible light emitted by the screen. It is necessary to have good contact between the film and

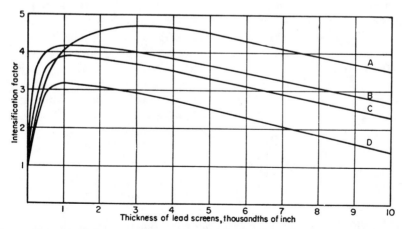

FIG. 6-30. Intensification vs. screen thickness for double screens. (A) 1,000 kv; (B) 250 kv; (C) 210 kv; (D) 175 kv. (*Courtesy General Electric Company, X-Ray Division.*)

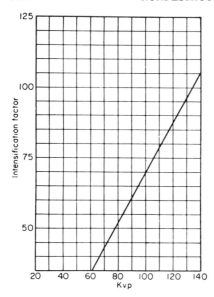

FIG. 6-31. Variation of intensification with voltage for calcium tungstate screen. (*Courtesy E. I. du Pont de Nemours & Company, Photo Products Department.*)

screen, because poor contact allows the fluorescent light to spread and produce a blurred image. For this reason calcium tungstate screens are seldom used except when a decrease in exposure time is required. Calcium tungstate intensifying screens are needed in the radiography of steel thicknesses greater than $1\frac{1}{2}$ in. at 200 kv peak and greater than 3 in. at 400 kv peak.

Some of the advantages of both lead foil and calcium tungstate screens can be obtained by using a lead screen on the side nearest the radiation source and the calcium tungstate screen on the back side. This technique reduces the exposure time without greatly affecting the definition. The latest development in intensifying screens is the use of lead barium sulfate (sold under the trade name of Hi-speed). Lead barium sulfate screens have a maximum response at 3,800 A (calcium tungstate response, 4,300 A) and consequently give a greater photographic effect. At 70 kv an exposure of 100 ma-sec with these screens can be made to give the same photographic effect as a 150-ma-sec exposure using calcium tungstate.

Inverse Square Law. Like other forms of electromagnetic radiation, X rays obey the inverse square law. How the intensity varies with distance can be seen from Fig. 6-32. Radiation spreads out from a point source S in a diverging beam. The cross-sectional area of the diverging beam depends upon the distance from

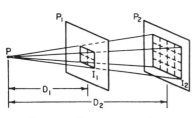

FIG. 6-32. Inverse square law.

the point source. At a point P_1, Fig. 6-32, the cross section of the beam is $ABCD$, and at P_2 the cross section is $A'B'C'D'$. The same amount of energy is present at the two positions (neglecting any absorption). The energy per unit area is called the "intensity." It can be shown by simple geometry that the area at point P_1 is proportional to D_1^2; the area at P_2 is proportional to D_2^2. If the intensity at P_1 is designated by I_1 and at P_2 by I_2, then

$$I_1 = \frac{KI_0}{D_1^2} \tag{6-24}$$

and
$$I_2 = \frac{KI_0}{D_2^2} \tag{6-25}$$

or
$$\frac{I_1}{I_2} = \frac{D_2^2}{D_1^2} \tag{6-26}$$

A useful expression for the radiographer is one that gives the relationship between the exposures and distances to produce the same film density.

$$\frac{E_1}{E_2} = \frac{D_1^2}{D_2^2} \tag{6-27}$$

This equation states that the exposure to produce a given film density is directly proportional to the square of the source-to-film distance. Suppose that the source-to-film distance for a 360-ma-sec exposure is 36 in. and that it is desired to make the film-to-source distance 48 in.; the exposure needed for equal blackening of the film can be calculated as follows:

$$\frac{360}{E_2} = \frac{(36)^2}{(48)^2}$$
$$E_2 = 640 \text{ ma-sec}$$

Detection of X Rays. When a charged atomic or nuclear particle passes through matter, it causes excitation and ionization of the molecules of the material. Charged particles produce ionization along their path by the direct action of their electric field on the orbital electrons of the atoms. X rays and gamma rays do not produce ionization directly. However, when a gamma ray or X ray is absorbed in matter, electrons are ejected from the atom. The electrons thus ejected have rather high energies and will produce ionization until their energy is expended.

When a charged particle passes through a gas contained in a chamber across which a small electric field is applied, the primary ions produced will move in opposite directions, each going to the electrode having the charge of opposite sign. A gas-filled chamber used for collecting the ions formed by incident ionizing radiation is called an "ionization chamber." Such a chamber consists of a cylindrical conducting shell with an insulated central electrode, Fig. 6-33. If an ionization chamber is

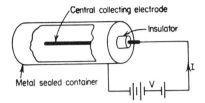

FIG. 6-33. Ionization chamber.

exposed to a constant source of ionizing radiation and the applied voltage
is increased, the current increases
until a certain value of voltage is
reached, after which there is no
further increase in current. This
current is known as the "saturation
current" and corresponds to the
condition where all the ions initially
produced travel to their respective electrodes with no recombination
of the ions taking place. However, if the applied field is increased
still more, a new phenomenon begins to take place. The primary
ions formed by the charged particles produce additional ions by
collision before collection at the electrodes. The size of the output
current pulse is now proportional to the energy of the atomic radiation.
Such a chamber is known as a "proportional counter." If the applied
field is increased further, a single ion pair in the chamber is sufficient to
cause a large ionization current pulse to pass. The magnitude of this
pulse is independent of the initial ionization. Such an instrument
is known as a Geiger-Müller counter. If the voltage on the tube is
increased further, the tube will break into a continuous discharge.
Figure 6-34 shows the size of the pulse (volts) produced as a function of
the voltage applied to the anode of the tube. The Geiger counter
exhibits a characteristic curve known as the "plateau curve." Figure
6-35 shows a typical plateau curve, which is made by plotting the counting rate as a function of the counter potential when a fixed source of
radiation is placed near the counter. Good Geiger counters commonly
have plateaus 200 to 400 volts in length.

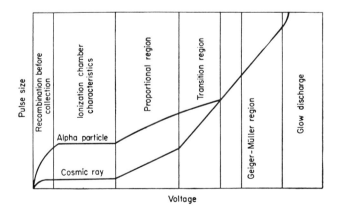

FIG. 6-34. Pulse size vs. voltage.

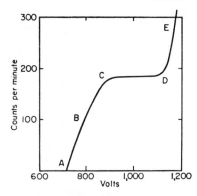

FIG. 6-35. Geiger plateau.

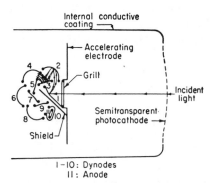

I–10: Dynodes
II: Anode

FIG. 6-36. Electron photomultiplier tube. (*Courtesy Radio Corporation of America.*)

Scintillation Counters. When X rays or gamma rays impinge on certain liquid or solid materials known as phosphors, visible or ultraviolet photons are emitted. The observation and measurement of the light "flashes" or scintillations produced by individual ionizing particles are the basis of scintillation counter instruments. The development of electron multiplier phototubes has made possible detection and counting of the scintillations by electronic means. The phototube converts the flashes into voltage pulses. Figure 6-36 shows the construction of a multiplier phototube. The basic processes involved in the detection and measurement of an ionizing radiation by a scintillation counter are shown diagrammatically in Fig. 6-37. The incident particle impinges on the phosphor, where it dissipates its energy in the ionization and excitation of the molecules. A fraction of this energy is converted into photons which are radiated in all directions. Some of these photons fall on the photosensitive cathode of the multiplier tube and eject a number of photoelectrons. These photoelectrons are accelerated by the potential applied between the cathode and the first electrode or dynode of the tube. On striking the first dynode, each photoelectron ejects several more electrons by secondary emission. This electron multiplication process is repeated at subsequent dynodes. Each dynode is at a higher potential than the preceding one. Finally after a multiplication of about 10^6 to 10^9, the electron

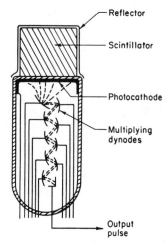

FIG. 6-37. Schematic diagram of a scintillation counter.

avalanche arrives at the collection plate. It produces a voltage pulse in the output condenser which is applied to an external pulse-recording circuit. Thus the initial energy of a single ionizing particle is transformed into a single voltage pulse. The whole system must be enclosed in a lighttight box, to eliminate effects other than those due to incident ionizing radiation.

When certain semiconductors such as cadmium sulfide and cadmium selenide are irradiated with X rays, their electrical resistance changes. There is a linear relationship between the photocurrent and the intensity of the X rays at relatively low levels of radiation, but at high levels saturation effects occur. The cadmium sulfide crystal itself amplifies the electron current by a factor of 10^4 to 10^6. One advantage of these cells is their high absorption compared to other detectors.

A semiconductor detector can be made to have an effective area of only a few square millimeters. Because of this small size it acts almost as a point detector. Frerichs and Jacobs[6] have used cadmium sulfide for liquid level gaging and can successfully gage the liquid level in a container to $\pm \frac{1}{16}$ in. from a given base line. This type of detector has a high sensitivity and resolving power, is rugged, and has a long service life. Cadmium sulfide is often used in pulsed X-ray systems. The a-c component reaches a steady state after a few cycles, whereas the d-c component of the photocurrent tends to increase over some time. In the average CdS crystal used for X-ray detection, approximately 5 sec is required for the d-c current to reach 95 per cent of its final value.

Industrial Radiographic Practice. The purpose of this discussion is to point out what factors the operator can control in making a radiograph and what information is needed properly to adjust these factors. It is obvious that the radiographer has no control over the material of the specimen or its thickness. He does have control, however, over the operational characteristics of the X-ray tube and the nature of the

TABLE 6-9. TYPICAL X-RAY MACHINES AND THEIR APPLICATIONS

Maximum voltage, kv peak	Intensifying screens	Application and approximate practical thickness limits
50	None	Extremely thin metallic sections. Wood, plastics, biological specimens, etc.
150	None or lead foil	Light alloys, 4½-in. aluminum or equivalent. 1-in. steel, or equivalent
	Calcium tungstate	1½-in. steel, or equivalent
250	Lead foil	2-in. steel, or equivalent
	Calcium tungstate	3-in. steel, or equivalent
400	Lead foil	3-in. steel, or equivalent
	Calcium tungstate	4-in. steel, or equivalent
1,000	Lead foil	5-in. steel, or equivalent
2,000	Lead foil	8-in. steel, or equivalent

TABLE 6-10. APPROXIMATE RADIOGRAPHIC EQUIVALENCE FACTORS OF
SEVERAL METALS

Metal	50 kv peak	100 kv peak	150 kv peak	200 kv peak	1,000 kv peak	Gamma rays
Aluminum..	1.0	1.0				
Magnesium..........	0.6	0.6				
24 ST alloy..........	1.4	1.2				
Steel	...	2.2	1.0	1.0	1.0	1.0
18-8 stainless steel.....	...	....	1.0	1.0		
Copper..............	...	....	1.5	1.4		
Zinc................	...	....	1.4	1.3		
Brass*..............	...	....	1.4*	1.3*	1.2*	1.1*
Lead................	...	....	12	12	5	2.3

* Tin or lead alloyed in the brass will increase these factors.

recording device employed. Furthermore, he can adjust the relative
positions of the X-ray source and the recording device over a considerable range. The problem, therefore, is to correlate these variable factors
so that the film will adequately record the nonhomogeneous nature of the
specimen. The radiographer must know the size of the smallest defect
that can be recorded under the conditions which he finally selects because
this often determines the applicability of the method. Table 6-9 gives
practical thickness limits of materials which can be inspected by conventional industrial X-ray sources.

Radiographic equivalence factors for several materials are given in
Table 6-10. These factors may be used to determine exposure factors
for one metal from technique charts compiled for other materials. In
Table 6-10 aluminum is the standard metal at 50 and 100 kv peak,
whereas steel is used as the standard metal at higher voltages. To obtain
the approximate equivalent thickness of the standard metal, the thickness of the metal being radiographed is multiplied by the corresponding
factor in Table 6-10. For example, to radiograph 0.5 in. of copper at
200 kv peak, multiply 0.5 in. by the factor 1.4, obtaining an equivalent
thickness of 0.7 in. The exposure needed is that required for 0.7 in. of
steel.

The radiographic contrast is an important characteristic of a radiograph. Contrast is a measure of the difference in density between an
image and its immediate surroundings and enables the radiographer to
see the image. The contrast is the result of different intensities in the
X-ray image. The ability of the film to differentiate between two
intensities is given by the slope of the characteristic curve at some particular density. This slope is called the "gradient." If the curve has a
considerable straight-line portion, such as curves in Fig. 6-27, the slope
of this part is called the "gamma" of the film. The average gamma is the
slope of a straight line joining two points of specified density on the

characteristic curve. The characteristic curve of a particular brand of film provides a measure of its speed. A shift of the characteristic curve to the left means an increase in speed. The same density can be obtained with lower intensity or shorter exposure time. Speed is sometimes indicated by the inertia, as shown in Fig. 6-27. A more practical method is to compare speeds in terms of the relative exposure time required to obtain a given density. The lower the voltage used, the greater the radiographic contrast. This leads to the general rule of choosing the lowest voltage which will give a proper film density in a reasonable length of time.

Closely associated with contrast is the latitude of the film. Latitude is defined as the interval on the log intensity axis of the characteristic curve corresponding to a specified density range. The latitude is important in considering the range of thickness of the specimen which can be successfully recorded with a single exposure.

Technique Charts. It is helpful in practical radiography to construct technique charts to guide the operator in choosing the proper operating conditions. These charts consist of a series of curves in which the exposure necessary to produce a given film density is plotted against the thickness of the specimen. Each chart is prepared for a specific material, a definite source-to-film distance, and a particular film-screen combination. The three variable factors are the tube voltage, material thickness, and exposure. By choosing two of these variables, the third can be found from the exposure chart. The exposures may be expressed in milliampere-seconds, or the chart may be made for a specific tube current and the exposure expressed in seconds. The relationship between thickness and exposure time is nearly linear if the latter is plotted on a logarithmic scale. The charts are frequently provided by the manufacturer of the X-ray equipment or may be readily prepared in the laboratory. A set of exposure charts covering the usual range of materials and of film-screen combinations should be readily available to the radiographer. Using the technique charts, the problem of properly exposed radiographs become very simple. For example, a welded section 1 in. thick is to be radiographed using lead screens. It is desired to use a 10-ma tube current with an exposure time of 300 sec. From Fig. 6-38b the required voltage would be 180 kv. A given exposure chart can be used for conditions other than those specifically given by the use of the proper correction factors. A correction in source-to-film distance can be made by applying the inverse square law equation (6-26). If a different material is to be used, the equivalent thickness factor, Table 6-10, can be used. If a different type of film is to be used, the film manufacturer will usually supply the necessary information to correct for film speed.

The sharpness or detail of the image of the defect as it appears on the radiograph is also a matter of several factors. It is important because the eye detects an image of low contrast much more readily if its outline is sharply defined. The sharpness of the X-ray image is primarily

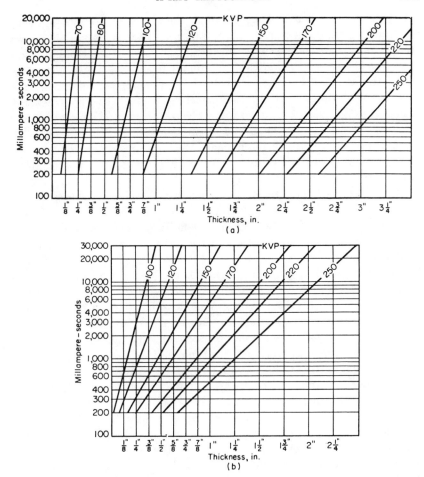

Fig. 6-38. Exposure chart. (a) Technique chart for rolled steel using calcium tungstate screens, source-to-film distance 36-in. high-speed film. (b) Technique chart for rolled steel using lead filter screens. Screens 0.005 in. lead, density 1.5, film Eastman Type A, Ansco Superay A, or Du Pont 506. (*Courtesy General Electric Company, X-Ray Division.*)

dependent on the geometrical factors of size of focal spot and distance from target to defect and from defect to film. It has been shown that the width of the unsharp region or penumbra surrounding the true shadow of the defect is equal to $F \times t/d$, where F is the effective diameter of the focal spot, t the distance from defect to film, and d the distance from target to defect. The sharpness of the image is improved by using a tube with a smaller focal spot, by moving the film closer to the defect, or by moving the X-ray tube farther away. Any distortion in a radiograph is usually undesirable. Distortion can sometimes be useful: when magnification is desired; when one part is superimposed on another, and it is desired to eliminate superimposition; and when one wishes to determine the depth of a flaw below the surface by triangulation.

Object contrast is a limiting factor in the sensitivity obtainable with materials of low density and atomic number. For example, aluminum has a lower absorption per unit thickness than steel. Therefore, it takes a greater change in thickness of aluminum to cause a given change in X-ray transmission than with steel. Consequently, aluminum is said to have less object contrast than steel. The fact the lower energy radiation can be used to radiograph a given thickness of aluminum than the same thickness of steel partially compensates for the difference in object contrast. A 1 per cent change in thickness in steel can be detected on a film. But with aluminum and the lighter metals, it is difficult to record 2 per cent thickness changes.

Penetrameters. The term "radiographic definition" refers to the sharpness with which the radiograph shows any discontinuities in the specimen. To check the radiographic definition a penetrameter is used. This consists of a strip of the same material as the specimen and having a thickness equal to some definite percentage of the specimen thickness. When making a radiograph the penetrameter is usually placed on the surface facing the source of radiation and adjacent to the area being radiographed. If it is difficult or impossible to position the penetrameter on the specimen, the penetrameter may be placed on the film. If the outline of the penetrameter can be seen on the radiograph and the thickness of the penetrameter is, for example, 2 per cent of the specimen thickness, it indicates that the radiographic sensitivity is at least 2 per cent. Holes of various diameters are usually drilled in the penetrameter. The image of the holes provides an indication of the clarity with which a defect will be revealed. The penetrameter may be thought of as a superimposed flaw. Since it is placed on the source side of the object, it is in the worst possible position with respect to the film. The penetrameter constitutes an ideal defect since its edges and its holes are sharply defined. Such holes give an abrupt change in metal thickness, whereas a "real" flaw usually has rounded edges giving a gradual change in thickness. In addition, real flaws are often diffuse and irregular in shape. Consequently, a cavity of the same diameter and thickness as one of the holes may be invisible while the hole is visible.

Four types of penetrameters are used: flat, step, wire, and bead. Figure 6-39 shows various penetrameters. In practice the various inspection codes specify the nature of the penetrameter and the degree of visibility that is acceptable. It should be remembered that the penetrameter is not a measure of the size of the smallest detectable defect but is merely an index of a certain standard of contrast and sharpness. The particular standard chosen is one which experience has indicated as capable of revealing defects which are considered harmful. A penetrameter sensitivity of 1 to 2 per cent can be obtained with present techniques over a wide range of thickness of steel. In radiography the word sensitivity is sometimes referred to in another sense; in this text it will be

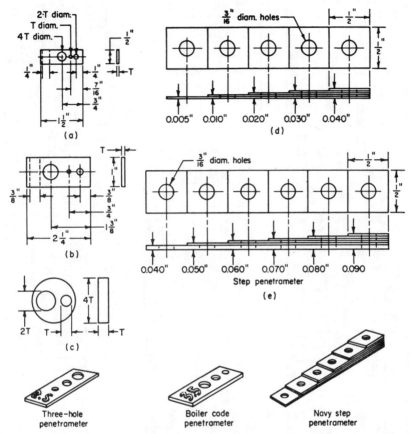

FIG. 6-39. Penetrameters. (a) For metals up to 2½ in. in thickness—penetrameter thickness from 0.005 to 0.050 in.; (b) for metals up to 8 in. in thickness—penetrameter thickness from 0.060 to 0.160 in.; (c) for metals over 8 in. in thickness—penetrameter thickness of 0.180 in. and over.

referred to as radiographic sensitivity in contrast to penetrameter sensitivity discussed above. It has been shown[7] that the difference in film density due to a small change in thickness in a homogeneous specimen is given by

$$\Delta D = 0.434\gamma\mu\Delta T \qquad (6\text{-}28)$$

where ΔD = change in density

γ = film gradient at density D

μ = linear absorption coefficient for radiation

ΔT = change in thickness

The radiographic sensitivity

$$\frac{\Delta T}{T} = \frac{2.3\Delta D}{\gamma\mu T} \qquad (6\text{-}29)$$

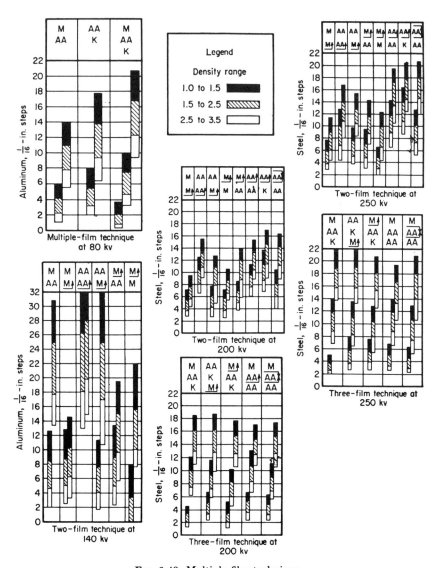

Fig. 6-40. Multiple film technique.

If ΔD is taken as the smallest preceptible change in density (values given for ΔD range from 0.005 to 0.02) and assuming an average value of 0.01,[8] then the radiographic sensitivity is given by

$$\frac{\Delta T}{T} = \frac{0.023}{\gamma \mu T} \tag{6-30}$$

This equation is strictly true only for monochromatic radiation. It should be pointed out that there is a difference between the sensitivity given by this equation and the penetrameter sensitivity.

Multiple-film Technique. In the multiple film technique the exposure is made with two or more films in the same cassette. This technique eliminates retakes due to film defects, since any defects not present on both films are not defects in the specimen. The multiple-film technique is also used to cover a wide range of thickness in one exposure. Two or more films of unequal speed are placed in the same cassette. For example, by using Kodak No Screen and Ansco Super-ray A film with an exposure factor of 100, steel thicknesses up to 3.0 in. can be inspected. With an exposure factor of 200, steel thicknesses of 0.5 to greater than 4.0 in. can be seen. Turner[8] discusses the use of multiple-film techniques for examination of multiple-thickness specimens. Figure 6-40 shows results obtained by Turner for various materials and energies.

The use of intensifying screens and filters was discussed earlier. Intensifying screens and films also reduce the influence of scattered radiation from within the specimen. Another way to reduce the scatter is to confine the incident primary X-ray beam only to the specimen being radiographed. The primary beam may be confined by diaphragms or by blocking with copper shot, lead, high-absorption liquids, or barium clay around the specimen. Floors, table tops, or any material behind a film holder will generate sufficient scattered radiation to fog a film. A sheet of at least 1/8-in. lead should be used between the film holder and any material behind it.

To summarize, the radiographic contrast is influenced by the tube voltage, the amount of scattered radiation, and the gradient of the characteristic curve of the film-screen combination at the particular density employed. Any modification of the technique which will permit the use of lower voltages, of reduced scattered radiation, or of an increased gradient will result in improved radiographic contrast.

Weld Radiography. Flaws which may be found by radiography in welds include the following: porosity, slag inclusions, inadequate penetration, incomplete fusion, undercutting, cracking, and pinholes.

The term "porosity" is used to describe globular voids free of any solid material, a gaseous inclusion. The gases forming the voids are released during the welding process and are entrapped in the molten metal. These voids may be scattered more or less uniformly through-

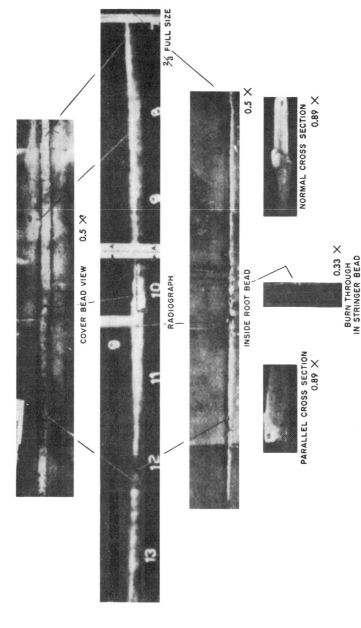

FIG. 6-41a. Weld defect porosity. The radiograph reveals the roughness of the stringer bead but shows no serious defects. One point near the longitudinal seam shows indications of a crack. The longitudinal section reveals clean metal with the roughness above the specimen surface. Several small gas pockets may be observed in this area. The section across the suspected crack shows a slight burnthrough with a shrinkage crack extending through several successive passes. (*Courtesy Armand Barkow.*)

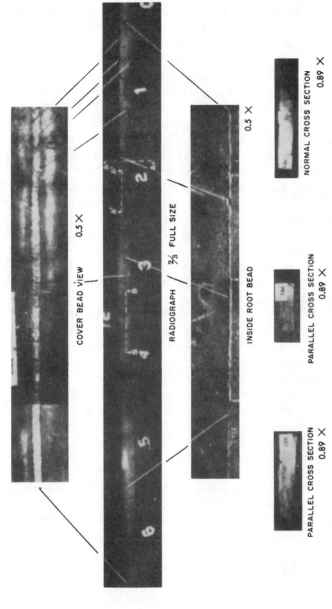

COVER BEAD VIEW 0.5 ×

RADIOGRAPH ⅔ FULL SIZE

INSIDE ROOT BEAD 0.5 ×

NORMAL CROSS SECTION 0.89 ×

PARALLEL CROSS SECTION 0.89 ×

PARALLEL CROSS SECTION 0.89 ×

Fig. 6-41b. Weld defect porosity. The cover bead in this section is rather narrow and some undercutting may be observed. The radiograph indicates some porosity. The cross section shows both the depth of the window with a shrinkage crack in the second bead and lack of fusion at the edges. On the cover bead the longitudinal section shows gas pockets both in the root pass and in later passes. (*Courtesy Armand Barkow.*)

out the weld or may occur in groups or clusters. Figure 6-41a and b
shows radiographs of porosity in welds. The term "slag" or "nonmetallic
inclusions" is used to describe the oxides and other solids which are
found as elongated or round inclusions in the weld. Figure 6-42 is a
radiograph of a weld having slag inclusions. "Undercutting" is the
term used to describe the reduction in the base metal thickness where
the last bead is fused to the surface. This appears as a dark line on the
radiograph. Figure 6-43 is a radiograph of a weld which shows undercutting. The first or stringer weld bead which initially ties the two parts
together is referred to as the "root bead." The term "incomplete" or
"inadequate penetration" refers to the incomplete filling of the bottom of
the weld groove with weld metal. Figure 6-44 shows radiographs of butt
and fillet welds having inadequate penetration. The term "incomplete
fusion" designates the lack of bond between beads or between the weld
metal and the base metal. Figure 6-44 shows a radiograph of a weld
having incomplete fusion. Several types of cracking may occur in weld
joints. Good radiographic inspection will show almost all types of weld
metal cracks and many base metal cracks. Cracking of welded joints results from the presence of localized stress which exceeds the ultimate
strength of the material. Three different types of cracks can occur in
weld metal: transverse cracks, longitudinal cracks, and crater cracks.
Figure 6-42 shows these three types of cracks. Base metal cracking may
occur in the heat-affected zone of the metal being welded. Hardness and
brittleness in the heat-affected zone are the major causes which tend to
promote base metal cracking.

Casting Radiography. Practically all casting defects can be put into
two categories: defects which are due to improper foundry technique
and control and defects which result from the natural shrinkage that
takes place when molten metal cools and solidifies. Into the first category fall such defects as gas holes, cold shuts, various inclusions, and
misruns. The second category includes such defects as hot cracks, cold
cracks, and various types of shrinkage cavities. Misruns result from low
pouring temperatures which cause the molten metal to solidify before
completely filling the mold cavity. The resulting casting is missing
certain parts, especially thin sections. Generalized gas porosity, which
is characteristic of aluminum and magnesium alloy castings, is caused by
precipitation of hydrogen from the molten metal. This can be reduced
or eliminated by taking proper precautions to prevent the solution of the
gas in the metal or by degassing the melt before pouring. Lowering the
pouring temperature will speed up solidification, and the gas will remain
in solid solution rather than forming gas bubbles.

One of the most common casting defects due to molding practice are
gas holes. These result either from moisture in the mold or from air
entrapped in the mold cavity. Sand molds are of two types, dry sand

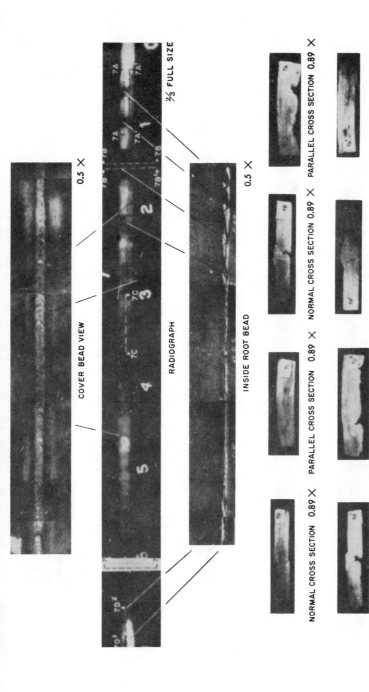

COVER BEAD VIEW   0.5 ×

⅔ FULL SIZE

RADIOGRAPH

INSIDE ROOT BEAD   0.5 ×

NORMAL CROSS SECTION 0.89 ×   PARALLEL CROSS SECTION 0.89 ×   NORMAL CROSS SECTION 0.89 ×   PARALLEL CROSS SECTION 0.89 ×

NORMAL CROSS SECTION 0.89 ×   PARALLEL CROSS SECTION 0.89 ×   NORMAL CROSS SECTION 0.89 ×   PARALLEL CROSS SECTION 0.89 ×

FIG. 6-42. Weld defect, slag inclusion. An arc burn next to the weld can be found on the radiograph. The radiograph shows a number of very dark spots indicating voids of considerable magnitude. The irregular dark areas indicate slag inclusions. The longitudinal sections show the depths of the voids in the slag infested region. The cross section shows the depth of the void at the center of the window and the shrinkage crack at the second pass nearer the edge of the window. There is a large slag pocket in the third pass at the edge of the weld with some cold lap in the next pass.

The radiograph shows the partial crack of the root bead but no indications that exist. The slag inclusions and slag string which occur as a dark line in the radiograph can be seen indirectly above the dark crack. The longitudinal section shows one of the slag pockets and dispersal of gas pockets. (*Courtesy Armand Barkow.*)

137

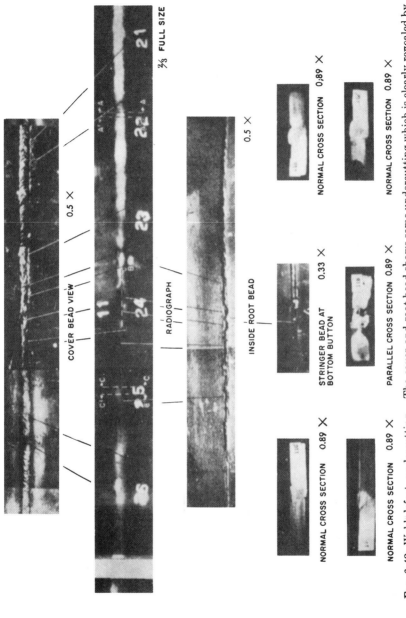

COVER BEAD VIEW 0.5 ×

⅔ FULL SIZE

RADIOGRAPH

INSIDE ROOT BEAD

0.5 ×

NORMAL CROSS SECTION 0.89 ×

NORMAL CROSS SECTION 0.89 ×

STRINGER BEAD AT BOTTOM BUTTON 0.33 ×

NORMAL CROSS SECTION 0.89 ×

PARALLEL CROSS SECTION 0.89 ×

NORMAL CROSS SECTION 0.89 ×

Fig. 6-43. Weld defect, undercutting. The cover and root bead shows some undercutting which is clearly revealed by the radiograph. The longitudinal section shows the burnthrough voids almost penetrating the weld and the general dirty condition of the material. The radiograph shows a fairly sound weld with the root bead near one side instead of the center of the window pass and considerable undercutting on some depth and width. The cross section shows the slagged beads and the lack of root bead in fusing both ends of the stock. Some minor gas pockets may also be indicated. (*Courtesy Armand Barkow.*)

138

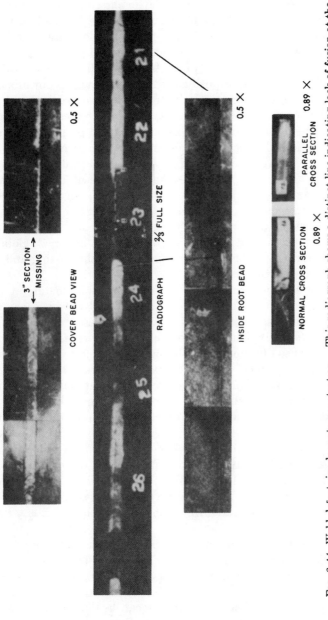

0.5 ✕

← 3" SECTION →
MISSING

COVER BEAD VIEW

RADIOGRAPH | ⅔ FULL SIZE

0.5 ✕

INSIDE ROOT BEAD

0.89 ✕

NORMAL CROSS SECTION

PARALLEL
CROSS SECTION

0.89 ✕

Fig. 6-44. Weld defect, inadequate penetration. This radiograph shows a distinct line indicating lack of fusion at the right. The distinct blur at one point indicates lack of fusion of some magnitude. The longitudinal section shows the general lack of penetrant indicated on the radiograph with one straight line, with a blur actually showing at the end of the section. The cross section shows the full extent of the blur. The affected weld is less than 50 per cent of the weld thickness. (*Courtesy Armand Barkow.*)

139

molds and green sand molds. Green sand molds contain considerable amounts of moisture which turns to steam when the hot molten metal strikes the mold. This steam can be trapped and held in the solidifying casting, producing voids. Dry sand molds are not so prone to gas holes from this cause. However, all types of molds will produce castings containing gas holes from entrapped air if the mold cavity is not properly vented. Die castings are particularly troubled by entrapped air, for the skin forms very quickly and prevents the air from escaping through the vents. Any mold design which has not provided proper gating can cause swirling actions during the pouring of the casting which will either draw air into the mold or cause a skin to form. Cold shuts form during the pouring of a casting. In most cases they are the result of two streams of molten metal failing to unite at the interface because of surface oxidation of the molten metal. Frequently, however, molten metal may splash onto the mold wall and solidify; subsequently molten metal will not bond to this, producing a cold shut. Cold shuts are small globules of metal embedded in, but not entirely fused with, the casting and are generally associated with a gas hole.

A cold crack is a definite rupture or fracture of the solid metal, producing a distinct separation. Cold cracks always start at the surface of the casting. They are a result of internal or external stresses or a combination of both. Radiographically a cold crack appears as a relatively straight dark line, usually continuous throughout its length, and generally exists singly. A hot crack consists of one or several parallel fissures caused by the internal rupture or fracture of the material while in the hot semiplastic state. On the film, hot cracks appear as irregular dark lines of variable width and density and often with numerous branches tending to follow dendritic outlines. Inclusions are metallic or nonmetallic foreign material which is mechanically embedded in the metal. In iron and copper alloys, the entrapment of oxides of the metal constituents in the metal is commonly called slag inclusion. In aluminum and magnesium alloys, oxide inclusions are referred to as dross inclusions. When sand or refractory material from the mold is trapped, it is termed sand inclusion.

Microshrinkage is a special condition existing in a magnesium alloy casting in which shrinkage voids occur at grain boundaries. This results in a film appearance of dark feathery streaks or peculiarly mottled dark patches.

Pipe Radiography. Radiography on pipe can be done in three ways, as shown in Fig. 6-45. There is the single-wall technique, in which the source is inserted inside the pipe and film is placed around the welded seam. The practicability of this technique depends not only upon the wall thickness but also on the possibility of introducing the source into the pipe itself. This method while completely radiographing the walls

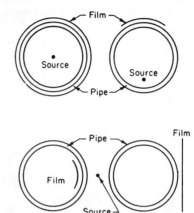

FIG. 6-45. Radiography of pipe.

has severe limitations and is not always practical with all diameter to wall thickness ratios. A practical ratio of source-to-film distance is $7t$, where t is the wall thickness. The second technique is the double-wall technique. The wall is examined in sections, the film being placed on only a portion of the wall at a time. The new film and source are then moved around to the next appropriate position. The third technique is a variation of the double-wall technique. The source is positioned some distance from the pipe so that the image of both the top and bottom of the wall is produced on one film. Moving the source away from the pipe, the exposure time is lengthened, so that while only one exposure is required the advantage is readily offset by the increased exposure time.

Codes. There are many existing codes that go into detail on X-ray and gamma-ray radiography. These codes may specify such items as source-to-film distance, penetrameters, and other pertinent information. Some of the existing radiographic codes include those of the American Society of Mechanical Engineers, Power Boiler and Pressure Vessel Code, United States defense establishments, American Welding Society, American Petroleum Institute, and boiler codes of various states. In addition, many company specifications are written to cover specific items produced or procured by the individual company.

Reference Radiographs. A set of standard radiographs covering steel castings for high-pressure steam and other severe service was prepared by the New York Naval Shipyard. These have now been issued in a revised form to contractors and naval inspectors by the Bureau of Ships, Navy Department. The ASTM has made this information available by issuing it as ASTM Industrial Radiographic Standards on Steel Castings (E71-52). The ASTM has also issued comparison radiographs for steel welds. These radiographs were made from sample steel welds containing all types of defects usually found in production welding. After obtaining the weld specimens, the specimens were radiographed using an opti-

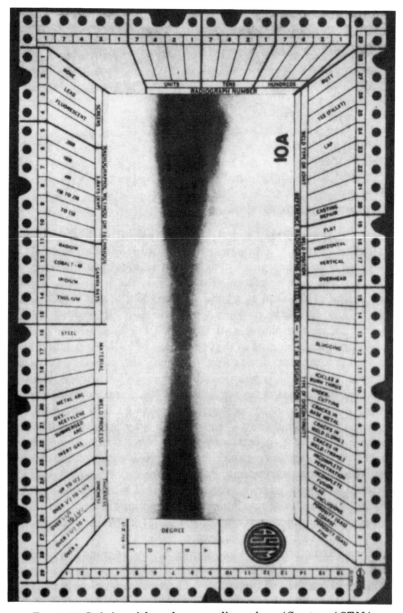

FIG. 6-46. Coded card for reference radiographs. (*Courtesy ASTM.*)

mum X-ray technique to reveal all details present in the weld and the adjacent base metals. Positive prints were then made of all radiographs, and these were mounted on specially designed sorting cards for ease of examination and to facilitate sorting during evaluation. Figure 6-46 shows the coded side of the sorting card which identifies:

1. Test plate source
2. Weld process
3. Weld type
4. Radiographic method
5. Degree and type of discontinuity

The discontinuities are identified as:

1. Gas porosity (small)
2. Gas porosity (large)
3. Slag inclusions
4. Lack of fusion
5. Lack of penetration
6. Transverse cracks in weld metal
7. Longitudinal cracks in weld metal
8. Cracks in base metal
9. Undercutting
10. Icicles (tear drops)
11. Burn-through
12. Stubbing or slugging

The Bureau of Aeronautics developed the Reference Radiographs for the Inspection of Aluminum and Magnesium Castings. These radiographs were ultimately adopted by the ASTM and issued as ASTM Standard E98-53T.

There has been a large number of papers in the literature dealing with the application of radiography to various industrial problems. Special attention is called to four papers.[9-12] The paper by Kolm discusses the use of precision radiography and the mass production of radiographic images with detail normally expected only under laboratory conditions. The paper by Kahn and associates illustrates the appreciation and value of radiography in the bronze foundry. The papers by Mooradian and Fadikov and Samokhvalov discuss and illustrate the application of scintillation counters in radiography. The paper by Barkow[13] discusses in detail the interpretation of pipeline welds.

Correlation. As yet there has not been too much work done on correlating the relationship between the radiograph and the mechanical properties of the specimen. However, the reader is referred to papers by Masi,[14] Carr,[15] Wallmann,[16] and Feinberg.[17]

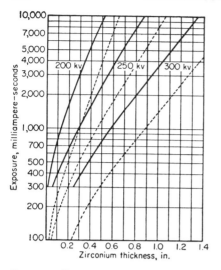

FIG. 6-47. Exposure curve for zirconium. Solid lines are filtered exposures; dashed lines are unfiltered exposures. Kodak M film, developed 8 min at 68°F with Kodak liquid developer; source-to-film distance, 36 in.; 5-mil lead front screen; 10-mil lead back screen; density = 2.0; 10-mil lead filter. (*Courtesy Westinghouse Electric Corporation, Bettis Atomic Power Division.*)

In the field of nuclear energy the quality of the radiographs very often has to be better than that accepted in routine industrial radiography. Various new techniques make such improvement possible. Vacuum cassettes are used in order to obtain an intimate contact between an intensifying screen and film. Using increased object-to-film distance in order to obtain an enlarged radiograph improves the accuracy of measurements on the film. Single-coated films are used in order to obtain sharp photographic enlargements. The technique of radiographing a radiograph is used in order to increase the latitude. The Los Alamos Scientific Laboratory has used this latter technique in radiographing zirconium coated with graphite. A conventional radiograph is prepared, but because of the difference in atomic number of the materials, the graphite will not be delineated. However, the silver in the emulsion has been affected. By making a low-kilovoltage exposure of the developed radiograph, the graphite can then be resolved.

In radiography applied to the nuclear field one complicating factor is the high density of the materials involved. Another factor is the necessity for better than normal radiographic sensitivities. Table 6-11[18] shows the steel equivalent thickness for zirconium, uranium, and hafnium.

TABLE 6-11. STEEL EQUIVALENT THICKNESS FOR ZIRCONIUM, HAFNIUM, AND URANIUM

Kilovoltage	Thickness of steel equivalent to		
	1 in. of zirconium	1 in. of hafnium	1 in. of uranium
150	~2.30		
200	~2.05		
250	~1.95	~14	~18
300	~1.67	~12	~16
1,000	~1.0	~3	~14

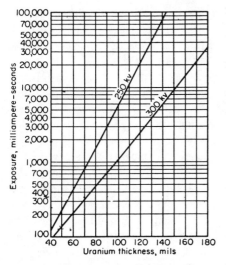

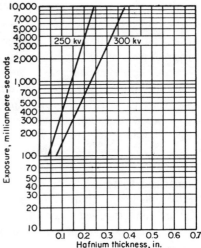

FIG. 6-48. Exposure curve for uranium. Kodak No-Screen film, developed 8 min at 68°F, with Kodak liquid developer; source-to-film distance, 36 in.; 5-mil lead front screen; 10-mil lead back screen; density = 2.0. (*Courtesy Westinghouse Electric Corporation, Bettis Atomic Power Division.*)

FIG. 6-49. Exposure curve for hafnium. Kodak No-Screen film, developed 8 min at 68°F, with Kodak liquid developer; source-to-film distance, 36 in.; 5-mil lead front screen; 10-mil lead back screen; density = 2.0. (*Courtesy Westinghouse Electric Corporation, Bettis Atomic Power Division.*)

Figures[19] 6-47 to 6-49 show exposure curves for zirconium, uranium, and hafnium.

When radiographing radioactive materials the radiation from the specimen may fog the film or completely obliterate the radiographic image. The problem is one of discriminating between the natural radiation and the radiation that causes the radiographic image. When the energy emitted by the specimen is lower than the radiographic energy, a lead filter can be used between the specimen and film. When the emitted energy is similar to or greater than that of the radiographic radiation reaching a film, both a lead filter and a large specimen-to-film distance can be used.

Fluoroscopy. Among the properties of X rays listed earlier in this chapter was their ability to cause certain materials to fluoresce. This property is utilized in fluoroscopy in which the X-ray energy is converted to visible light. The advantages of fluoroscopy are that it is fast and economical. Unfortunately, fluoroscopy has inherent limitations which have hitherto confined its application to a fairly narrow field. There is no permanent record of the X-ray image seen by the observer. The image can be photographed; then the technique becomes known as photofluorography. A good radiograph is a valuable record of the internal soundness of the article. A weakness of the technique is that sensitivity, definition, and contrast of the image are below those obtained by conventional radiography. Since the fluoroscopic screens are coarse grained, the screen lacks the resolving power of film. Recent developments in the design of screens and fine focus tubes have reduced the lack of sensitivity between the two techniques. Because a higher tube voltage must always be used in fluoroscopy than would be needed in radiography, the technical inferiority of the fluoroscopic image still persists. Because the fluoroscopic screen does not accumulate the effect of X rays as a film emulsion does, fluoroscopy can be applied only to comparatively thin metallic specimens or thicker objects of lower density. During the early part of World War II, the limiting sensitivity of industrial fluoroscopy was considered to be 0.050 in. or 7 per cent, whichever was greater. Generally speaking, fluoroscopy of steel with wall thickness greater than 1.4 in. had proved unsatisfactory. Consequently, fluoroscopy was looked upon as a supplement to radiography, a technique to be used for culling out specimens containing gross defects. During the past few years, improvements have been made in fluoroscopic equipment and techniques, so that the sensitivity approaches that of film radiography, which allows moderate thickness of light alloy metals to be examined. With fluoroscopy, as with many other methods of nondestructive testing, the judgment of the observer is all-important.

In fluoroscopy, the eye is the registering medium. In the discussion of visual testing, it was pointed out that the eye was not a highly accurate registering device. In fluoroscopy the eye is required to differentiate variations in light intensities, a task for which the eye is especially unreliable. The operator of a fluoroscopic unit should spend at least 20 min in total darkness before beginning to interpret images. Figure 6-50 shows how the sensitivity of the eye changes with the time in total darkness. The sensitivity increases slowly during the first 10 min, then begins to increase in the period between 10 and 15 min. In the period between 15 and 20 min, the sensitivity increases by a factor of 5. The process slows down, but there is still a definite increase in sensitivity. The brightness of a radiographic image using an illuminator is controllable and usually has for good viewing a value in the range of 10 to 1,000 mL, as shown in Fig. 6-51. In addition, there is the factor of eye fatigue, which limits the operator to short periods of viewing. The discomfort of viewing in a darkened room may influence the operator's judgment.

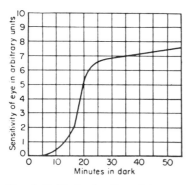

FIG. 6-50. Dark adaption of the eye.

A fluoroscopic unit consists of an

X-ray source, a fluoroscopic screen, and protective barriers. The X-ray source is usually a conventional low-voltage X-ray source having a smaller focal spot.

One of the major problems in fluoroscopy has been to find a suitable substance for coating the fluorescent screen. Roentgen's original screen was coated with barium platinocyanide, and this substance was in use until about 1910. When subjected to X rays it fluoresced with a brilliant green light, but it had the disadvantages of being costly and deteriorating rather rapidly with use. The next coating was zinc orthosilicate, and this was in general use for only a few years.

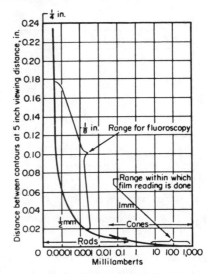

FIG. 6-51. Brightness and visual acuity

These screens also gave a green light, but they produced afterglow and a distinct lack of fluorescence with high-voltage X rays. In 1914, C. V. S. Patterson introduced cadmium tungstate screens, and these were in use for almost 20 years. Cadmium tungstate gives no afterglow and can be used satisfactorily for long periods without marked deterioration. It fluoresces with a bluish-white light but is lacking in brilliance. In 1933, L. Levy and D. W. West[20] developed zinc-cadmium screens which are incorporated in most present-day fluoroscopic units. These screens fluoresce with a yellowish-green light ten times as bright as that given by the older cadmium tungstate screens. The fluorescent materials are held together by a suitable binder and coated on cardboard. The cardboard side is toward the X-ray source, and the inspector sees the coated side. The performance of a fluoroscopic screen is a function of many parameters such as thickness, chemicals used, chemical purity, crystal size, and radiation response. The screen brightness is a function of the X-ray intensity at a given voltage. Figure 6-52 shows the relative brightness as a function of effective kilovoltage. This curve shows that, above

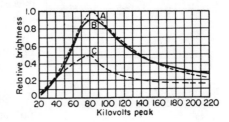

FIG. 6-52. Relative brightness vs. kilovoltage for various screens. (A) Radelin screen Fb; (B) Patterson screen B2; (C) Radelin screen F.

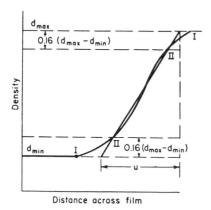

Fig. 6-53. Unsharpness across screen.

approximately 160 kv peak, fluoroscopy becomes less and less useful. The curves also indicate the difference in light output of screens due to the grain size, the larger the grain size, the greater the light output and the poorer the resolving power.

A fluorescent screen or X-ray film is not perfect in delineating edges. An attempt to obtain "knife-edge" changes in brightness or density on any viewing surface can never be wholly successful because of the finite size of the sensitive particles of which the surface is composed. The geometric unsharpness is a major factor in determining the ability of the eye to pick up discontinuities. The method of H. A. Klassens[21] and D. T. O'Connor[22] has been applied in Fig. 6-53 to illustrate the unsharpness of a knife-edge image formed under poor geometric conditions. In Fig. 6-53 the knife-edge shadow blur equals $\frac{1}{50}$ in. The most serious obstacle to high-sensitivity fluoroscopy is the large grain size of the fluoroscopic screens.

O'Connor and Polansky[23] have developed an expression for the minimum size defect visible in terms of the magnification, the focal spot size, and the unsharpness of the fluorescent screen. Figure 6-54 shows the minimum defect visible using various size focal spot tubes, and a Patterson B-2 screen. Table 6-12 shows the brightness of four selected screens. Figure 6-55 shows a typical response curve for brightness as a function of X-ray intensity.

The protective barriers must perform two tasks simultaneously: protect the operators from radiation and permit a clear undistorted view of the screen. If glass is used as a barrier, special glass such as Hi-D lead silicate must be used because the common glasses such as Pyrex and lead and barium oxide discolor under continued exposure to radiation. A liquid barrier such as lead perchlorate, potassium iodide, or zinc iodide has

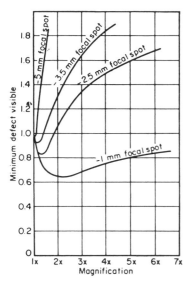

Fig. 6-54. Minimum defect vs. magnification for various size focal spots.

TABLE 6-12. BRIGHTNESS OF SELECTED SCREENS

Screen type	Brightness, ft-c	
	80 kv peak, 120 r/min	100 kv peak, 170 r/min
B-2	3.5	6.2
PFG	3.4	6.2
FG	4.0	6.8
F	2.1	2.6

been used. Lead perchlorate performs satisfactorily but is toxic and in its dry state, explosive.

Image Intensifiers. In recent years, considerable attention has been focused upon the need for brighter fluoroscopic screens. It has been pointed out that the clarity of fluoroscopic vision should be improved many times and that if possible the need for dark adaptation before fluoroscopy should be eliminated. A few years ago work was begun in at least three laboratories in the United States to develop devices by which fluoroscopic screens may be brightened or intensified. Coltman[24] at the Westinghouse Laboratories undertook investigations which culminated in the development of an electronic image tube capable of intensifying the fluoroscopic screen approximately 150 times. Comparatively little in the way of gain (about 100 times) over a conventional fluoroscopic screen is needed in order to reach the quantum limitation in the information content of the X-ray signal. The direct image converter tube developed by Coltman employs a several-stage process to convert the X-ray signal into a bright visual signal. The commercially available image amplifier is shown in Fig. 6-56. An X-ray image arriving from the left in the figure impinges on a fluorescent screen which converts the X-ray energy to light energy. The photoelectric surface converts the light energy to free slow-speed electrons. The number of electrons emitted by the photoelectric surface is determined by the amount of X radiation reaching the point on the fluoroscopic screen. An electrostatic field focuses the electrons to a sharp image on a phosphor. In addition, the electrons are accelerated by a potential of approximately 30 kv. The additional energy imparted to the electron increases the brightness on the phosphor. The image is restored to approximately full size

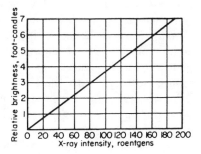

FIG. 6-55. Response curve as a function of X-ray intensity.

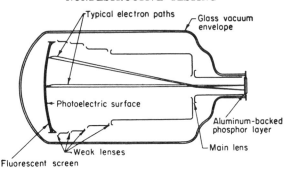

FIG. 6-56. Coltman image amplifier tube. (*Courtesy Westinghouse Electric Corporation, X-Ray Division.*)

through an easily viewed optical system. The brightness of the resultant image is approximately 100 to 300 times that of a Patterson B-2 fluoroscopic screen under the same radiation.

By increasing the light reaching the inspector's eye, the image amplifier represents an advance toward obtaining an optimum image. The increased light intensity is valuable in the following ways: the visual acuity is increased, little or no dark-adaption time is required, image contrast can be increased by employing lower kilovoltages than possible with current fluoroscopic techniques, the possibilities of photofluorography are increased, and the use of television viewing of fluoroscopic images is possible. Television projection permits further light amplification and contrast control.

The screen unsharpness of the image-forming material used in the image amplifier is approximately the same as that of screens employed in conventional fluoroscopy. The commercially available image amplifiers can resolve wire screens of 35 to 45 mesh, compared with the 70 mesh resolvable with the present high-definition techniques. With the advent of the 5,000-kvma 0.5-mm-focal-spot X-ray tubes, it is expected that the image amplifier will resolve much finer mesh. The image amplifier is not suitable for the inspection of thin sections of light-alloy materials. The filtration by the glass input window limits the inspection range to a minimum of $\frac{1}{2}$ in. of aluminum. The image amplifier's major application will be immediately below and above the upper limit of high-definition fluoroscopy with the current 900-kvma 1-mm X-ray tube. Above 1 in. of aluminum, 2 per cent penetrameter sensitivity can be obtained with the image amplifier. Below this thickness, up to 4 per cent sensitivity can be expected. If the specimen thickness varies from very thin to very thick, such as from 1.8 to 4 in., a high-definition fluoroscope should be designed to permit viewing with both the fine-grained fluorescent screen and the image amplifier.

Philips Image Intensifier. A group in Eindhoven, Holland, has developed the Philips Image Intensifier. The Norelco Industrial Image Intensifier consists of an electronic tube, shown in Fig. 6-57. In this diagram, A represents the X-ray target, of fractional focus size, and B a specimen halfway between the X-ray target and the initial phosphor. This results in an X-ray magnification of $2\times$. The glass envelope C is the evacuated container for the electronic components. Part D is a fine-grain phosphor that converts the visible radiation to electrons. Through the application of high voltage the electrons emitted from the photocathode are accelerated. The result is a brightness gain of at least $1,000\times$ on the fine-grained fluorescent screen. The electronic system results in a reduction of image size of $9\times$, but the adaptation of a suitable lens yields an image exactly comparable to the size of the image on the pickup phosphor.

Another approach to electronic amplification involves the use of a television or scanning-type presentation. One such system employs a sensitive television pickup camera to detect the image from a fluoroscopic screen. The image then is presented on a television kinescope, which, within limits, can be viewed as bright as desired. Polansky and Criscuolo[25] have investigated the use of a closed-link television inspection system.

TVX System. Still another scanning-type system is the General Electric TVX[26] system. This system employs a photoconductive X-ray camera to convert the X-ray signal directly into an electronic signal, Fig. 6-58. Contrasted with the direct image converter tube described above, the use of a relatively thick photoconductive layer eliminates the

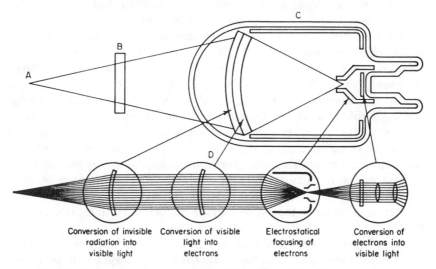

Conversion of invisible radiation into visible light   Conversion of visible light into electrons   Electrostatical focusing of electrons   Conversion of electrons into visible light

FIG. 6-57. Norelco image intensifier. (*Courtesy Philips Electronics, Inc.*)

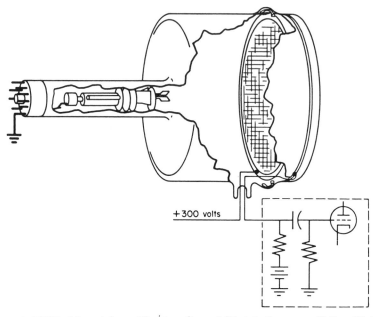

+300 volts

Fig. 6-58. TVX pickup tube. (*Courtesy General Electric Company, X-Ray Division.*)

low-efficiency step of converting X rays to light.   The photoconductive material used, lead oxide, behaves essentially as a perfect insulator in the dark and as a conductor upon exposure to radiation.   This change in conductivity due to the impinging X-ray image is used to modulate a scanning beam.   The resultant signal is amplified and presented for viewing on a kinescope.   The pickup tube itself is a relatively simple device, consisting essentially of the photoconductive layer of lead oxide and an electrostatically deflected electron gun.   The TVX unit presently being used employs the American standard of 525 lines interlaced 30 frames per second.   The pickup tubes have an outside diameter of $8\frac{1}{2}$ in. and have an effective signal plate area over 18 cm in diameter. Although this is a useful inspection area for many applications, there are many others that would require some sort of mechanical scanning device for complete inspection.   The electronic system is divided into two units, one comprising the camera or pickup unit and the other the viewing monitor.   The monitor unit which will be stationary in most installations is relatively large.   The pickup unit, which may be fitted into small areas or be moved to various inspection stations, is small and light.

One extraordinary advantage in using a highly absorbing photo-conductive detector such as lead oxide is that it responds over a very wide range of X-ray energies.   As shown in Fig. 6-59, a useful response

TABLE 6-13. CONTRAST SENSITIVITIES OF THE TVX SYSTEM

Potential, kv peak	Current	Distance	Material penetrated	Sensitivity,* per cent
100	2 ma	18 in.	2-in. Al	2
145	2 ma	18 in.	0.5-in. steel	4
1,300	0.5 ma	36 in.	1.5-in. steel	12
Co⁶⁰	9.6 r/min	1 m	1-in. steel	8
Co⁶⁰	9.6 r/min	1 m	2-in. steel	12
Co⁶⁰	9.6 r/min	1 m	3-in. steel	16
6,000	15 r/min	36 in.	1-in. steel	20

* Standard A-N penetrameters used: sensitivity taken when all holes are visible.

is obtained over a range of 40 to 1,500 kv peak. Another illustration of response is shown in Fig. 6-60. Typical contrast sensitivities obtained with the TVX unit are given in Table 6-13. These data show contrast sensitivities of the order of 2 per cent on up to 1 in. of steel when using a 6-Mev linear accelerator as a source of radiation. The loss in contrast sensitivity at these higher energies is thought to be caused by secondary scatter within the photoconductive layer.

An idea of the detail observable with the TVX system can be obtained from Fig. 6-61 which is the image of an aluminum test plate, the details of which are given in the figure caption. The smallest hole in the plate, a No. 80 drill, 0.3 mm in diameter, is clearly resolved. The actual limitation on the detail in the case of the TVX system is the spot size of the scanning beam in the pickup tube, which is of the order of 0.15 mm. At X-ray intensities of 0.01 r/min, a 0.5-mm mesh can be clearly resolved with these tubes. The speed of response of the system is such that inspection of motion within an object or of moving objects is possible.

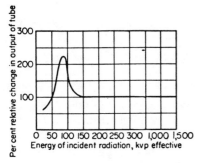

FIG. 6-59. Relative response of TVX tube to energy of X rays. (*Courtesy General Electric Company, X-Ray Division.*)

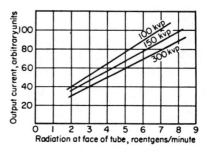

FIG. 6-60. Relative response of TVX tube to intensity of X rays. (*Courtesy General Electric Company, X-Ray Division.*)

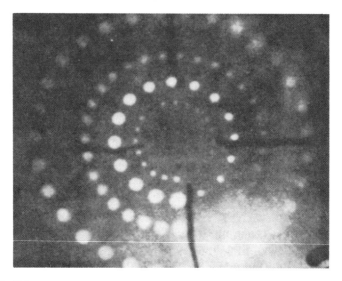

Fɪɢ. 6-61. TVX image of ¼-in. aluminum plate. The holes in the two innermost circles go through the plate and vary from No. 80 to No. 30 drill size. The third and fourth sets of holes are drilled to depths equaling the diameter of the holes in the third circle which vary from No. 60 to No. 30 drill size. (*Courtesy General Electric Company, X-Ray Division.*)

The present limitation on speed of response is imposed by the scanning rate of the system.

High-power Fluoroscopy. O'Connor and Polansky[27] describe and discuss the performance of a high-power industrial fluoroscopic tube. This tube has a focal spot of 0.5 mm. The maximum continuous rating of the high-powered X-ray tube is 4,000 kvma peak for voltages above 70 kv. Unfortunately, the power rating of the tube at the lower-voltage range is reduced because of filament limitation. As the voltage applied to the tube is reduced, more electronic emission and consequently more filament current are required to maintain constant power. The evaporation of the filament becomes severe at high currents, and therefore a practical limitation must be placed upon the maximum filament current. Table 6-14 gives the sensitivities obtainable with the 4,000 kvma peak high-power X-ray tube. The range of thickness that can be inspected is 1.8 to 3.00 in. of aluminum. From the table it can be seen that sensitivities of 1.5 per cent have been obtained. Because of tube power limitation below 70 kv, only 2 per cent sensitivity is obtainable for ¼ in. aluminum thickness.

A fluorescent screen absorbs only about 2 per cent of the energy of a 2-Mev X-ray beam. Polansky and O'Connor[28] have shown, however, that fluorescent screens are capable of a highly discriminating response at 2 Mev even though this energy is far above for their peak response.

TABLE 6-14. FLUOROSCOPIC SENSITIVITY FOR ALUMINUM
(Kvma peak-ma 4,000; magnification, 2)

Voltage, kv peak	Thickness, in.	Definition sensitivity	Current, ma
50	0.25	2	40
60	0.5	2	40
70	0.75	1.5	50
80	1.00	1.5	50
90	1.50	1.5	40
100	2.00	1.5	40
110	2.50	1.5	35
120	3.00	1.5	33

Photoradiography. In photoradiography, the image on the fluorescent screen is photographed with a conventional camera, as shown in Fig. 6-62. Although medical photoradiography is used, industrial photofluorography is not common. The results obtained using this technique are intermediate between radiography and fluoroscopy. There is a saving in film and at the same time a permanent record is made. For the same specimen thickness, somewhat higher kilovoltages are required than for radiography. In rendition of detail, photoradiography is superior to fluoroscopy largely because the film acts as an integrating device and enhances the contrast of the fluoroscopic image. Also, the film can be viewed with transmitted light, and with ample illumination the image can be enlarged. Factors which tend to diminish detail in the photoradiograph are the graininess of the fluorescent screen, the grain of the film, and the limitation of the lens system. The grain size of the film becomes important because of the reduction of the image and its magnification when viewed.

Microradiography. Microradiography is a technique used to locate and identify constituents in thin sections and to locate minute voids such as shrinkage and gas cavities. The specimen is placed in contact with a

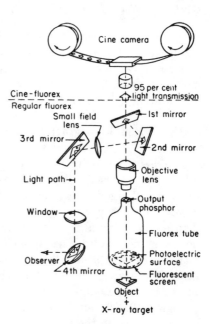

FIG. 6-62. Photoradiography. (*Courtesy Westinghouse Electric Corporation, X-Ray Division.*)

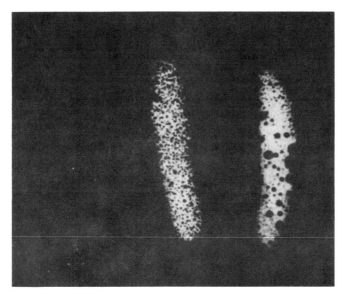

FIG. 6-63. Microradiography. These three slices from a bronze casting were taken near the cope surface. The slice at the right shows the tiny bubbles of evolved gas. The middle slice shows the slightly larger bubbles formed by coalescence. The slice at the left shows the cluster porosity at the cope surface (1.65×). (*From G. L. Clark, "Applied X-Rays," McGraw-Hill Book Company, Inc., New York*, 1955.)

very fine-grain special emulsion film. A piece of lead foil with a hole about $\frac{1}{4}$ in. in diameter is placed over the top of the specimen. Exposure to X ray produces a small spot image on the film which may be magnified and viewed through a microscope. Figure 6-63 shows a microradiograph of cast aluminum alloy containing about 8 per cent copper.

High-speed Radiography. Radiographic exposures as short as one-millionth of a second have to be used to "stop" the motion of projectiles and high-speed machinery. Special high-voltage equipment and X-ray tubes must be used. The high-voltage generators consist of capacitors which are charged in parallel and discharged in series through the X-ray tube. The X-ray tube has three electrodes: anode, cold cathode, rather than a conventional heated filament, and auxiliary cathode. In operation the high voltage is applied between the cathode and auxiliary cathode, causing a metallic arc and thus creating a source of electrons. Immediately after the initial arc, the flow of electrons is transferred to the anode and X-rays are generated. The tube current may reach values as high as 2,000 amp, but because of the extremely short exposure the heat generated in the target is not excessive.

Xeroradiography. A radiographic process that has recently come into use is xeroradiography. The name xerography, xero Greek for dry, comes from the "dry writing" thus produced. The two basic elements of xerography are the xerographic plate and the developing powder. The plate normally used has a photoconductive layer of amorphous selenium deposited in a stiff metal backing plate. In the dark, this surface can accept and hold an electrostatic charge. However, when exposed to light or penetrating radiation, the charge will decay in an amount somewhat proportional to the radiation. Thus, after exposure, a charge pattern remains on the plate surface in the form of a latent electrostatic image. The image is developed by spraying it with a finely divided powder that has been given an electrostatic charge of opposite sign by frictional electricity. The powder will be attracted to the charged portions of the plate, more powder being attracted to the areas of greater charge concentration. The image can then be seen on the plate or transferred to a second surface, such as a sheet of ordinary paper. The electrostatic charge is usually applied to the selenium surface by passing over it a fine wire assembly held at a high potential so that corona current is emitted. This technique is shown in Fig. 6-64. The charging unit is made up of three parallel wires, which are maintained at a positive electrical potential, and a screen between the wires and the plate. This assembly is driven across the plate in about 10 sec. The whole assembly is contained in a lighttight box, so that the plate is charged in the dark. The plate, thus sensitized and ready for exposure, is then covered by a dark slide.

In xeroradiography undercutting is very prevalent and is much more of a problem than with film. There are two distinct types of undercutting. The first is that type of radiographic undercutting commonly found with film. In xeroradiography lead screens cannot be put directly against the free surface of the plate to eliminate undercutting. It has been found that 0.005- to 0.010-in. lead screens placed between the specimen and the dark slide will reduce radiographic undercutting. It also has been found helpful to use a 0.005- or 0.010-in lead filter at the tube, depending upon the kilovoltage. The second undercutting phenomenon, unique to xeroradiography, has been called "xerographic undercutting." The phenomenon is due to air ionization in the air space between the dark slide and the plate surface, as shown in Fig. 6-65. In exposing a specimen of varying thickness, the plate will be discharged more at the thinner sections, forming discontinuities in the charge pattern. Across these discontin-

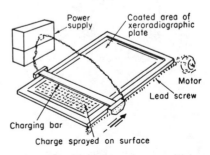

FIG. 6-64. Technique for sensitizing xeroradiographic plate. (*Courtesy The Halloid Company.*)

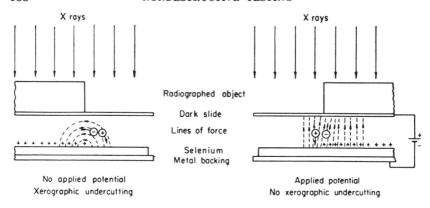

FIG. 6-65. Xerographic undercutting. (*Courtesy The Halloid Company.*)

uities strong localized fields curve sharply from one charge concentration to the other.   In addition, X-ray absorption in the air just above the plate creates positive and negative ions.   The negative ions are attracted to the plate and tend to discharge the more highly charged side of the boundaries.   As the exposure continues, such edges become discharged and the ions will follow, causing more and more undercutting. To eliminate this effect, a d-c voltage is applied between the dark slide and the metal backing of the plate.

To develop the image, the plate is suspended in a large box, and a cloud of finely divided white powder is sprayed into the box.   The powder is charged by turbulent action in the nozzle of the spray gun. Figure 6-66 is a sketch showing development taking place.   Development time is in the order of 20 to 30 sec.   The powder will be attracted to the plate, areas of higher charge attracting more powder than those of lesser charge.   This can be seen in Fig. 6-67.   However, at sharp

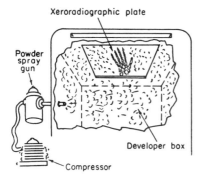

FIG. 6-66. Developing xeroradiographic image. (*Courtesy The Halloid Company.*)

discontinuities in the charge pattern on the plate, such as those formed by voids in the specimen, fringing fields will be created.   These fringing fields will attract powder to the edge of the area of higher charge, leaving the edge of the area of lesser charge devoid of powder.   At the bottom of Fig. 6-67, a profile of the powder deposition for the object is shown.   The image can be viewed immediately after the completion of development.   A total time of only a minute is needed to get a readable image, exclusive of the setup

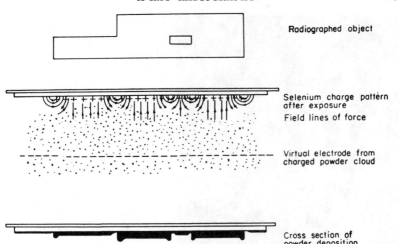

Radiographed object

Selenium charge pattern after exposure

Field lines of force

Virtual electrode from charged powder cloud

Cross section of powder deposition

Fig. 6-67. Principle of development of xeroradiographic plate. (*Courtesy The Halloid Company.*)

and exposure time. After the plate has been read, the image can be removed completely by passing it once or twice under a rotating fur brush. The plate is then ready to be used again after it has been rested.

At low kilovoltages and short exposures, plates can be used again within a few minutes. In the range above 100 kv peak, where longer exposure times are normally used, fatigue of the plate is likely to occur. This results in lowered contrast sensitivity, ghost images, or both. This fatigue phenomenon might persist for hours, depending upon the intensity and kilovoltage of the radiation. Fortunately there is a technique for eliminating this effect. After the plate is cleaned, and a dark slide put in place, the back of the plate is exposed to infrared heating lamps for a few seconds to raise the temperature of the plate to 110 to 120°F. It is then rapidly cooled to room temperature by a forced air blast. This completely restores the plate to a fully recovered condition, and it can be used immediately. Two of the serious problems with xeroradiographic plates are the inhomogeneities and "defects" in the selenium plates. A permanent record of a xeroradiograph must be made in many instances. One method of getting a permanent record is to photograph the image on the plate. In Fig. 6-68 a camera can be seen built into the viewer for just this purpose.

Over the entire X-ray spectrum, the xeroradiographic plate has a speed about the same as type A radiographic film. At the low kilovoltages, it is somewhat faster. As the kilovoltage is raised, because lead intensifiers are used with film and not with xeroradiographic plates, the plate becomes somewhat slower than film. The photoconductive

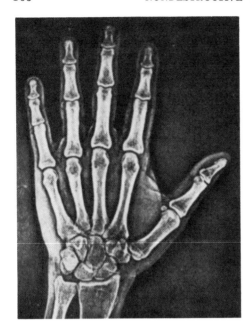

FIG. 6-68. Photograph of a xeroradiograph of a hand. (*Courtesy The Halloid Company.*)

layer of the xeroradiographic plate will deteriorate if left in strong artificial light or sunlight. Deterioration of photoconductive material
destroys the usefulness of the plate. The life of the plate is dependent
upon mechanical handling and the deterioration of the photoconductive
material. Consequently, it is essential that the plate remain covered
except when being exposed. Plates should not be allowed to lay around
without the dark slide in position. The sensitive surface of the plate is a
relatively soft material which is quite easily damaged. Any unnecessary
contact with the surface, including touching by fingers, should be avoided.

Taylor and Tenney have made a field evaluation of industrial xeroradiography.[29] Their experiments furnished data on sensitivity and resolution which are listed in Tables 6-15 to 6-17. The tabulated comparison
of the xeroradiographic data and those for type A film shows that the
results are identical for aluminum absorber thickness up to 2 in. At
3 in. of aluminum, the radiographic films furnish slightly better sensitivity and resolution. Table 6-16 contains the data on sensitivity and

TABLE 6-15. RELATIVE SPEED OF XERORADIOGRAPHIC PLATES

Kilovoltage	Sensitivity
150 kv	2.67
250 kv	2
1,000 kv	2
22 mev	1.4

TABLE 6-16. PER CENT RESOLUTION AND SENSITIVITY FOR XERORADIOGRAPHIC
PLATES AND TYPE A FILMS AT OPTIMUM KILOVOLTAGE

Thickness, in.	Xeroradiography		Type A film	
	Sensitivity	Resolution	Sensitivity	Resolution
0.5	. . .	4	. . .	4
1	1	2	1	2
1.5	1	2	1	2
2	1	1.5	1	1.5
3	1.5	2	1.3	1.3

TABLE 6-17. PER CENT RESOLUTION AND SENSITIVITY FOR XERORADIOGRAPHIC
PLATES AND TYPE A FILMS AT 250 KV PEAK

Steel thickness, in.	Xerox plates		Type A film	
	Sensitivity	Resolution	Sensitivity	Resolution
¼	. . .	4	2	4
½	2	4	2	4
1	2	4	1	2
1½	2	4	0.8	2

resolution, with similar data for type A film. As the thickness was increased the per cent sensitivity and the resolution were constant for xeroradiography, but both decreased for radiographic film. The xeroradiographic plates are considerably faster at low kilovoltages than type A film, but their relative speed decreases with increasing energy of the radiation, as shown in Table 6-15.

Betatron. The betatron developed by Kerst[30] in 1940 is used for the production of high-energy electrons. These electrons can be used for the production of hard X rays. Figure 6-69 shows a cross-sectional view of a betatron. Figure 6-70 shows a top of the doughnut. To produce high-speed electrons the betatron employs the magnetic inductive effect used in a transformer. In a transformer the primary winding connected to an a-c voltage source establishes a varying flux in an iron

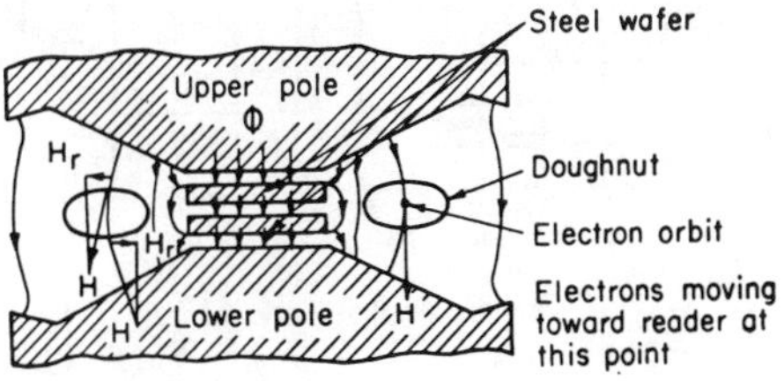

FIG. 6-69. Cross section of betatron.

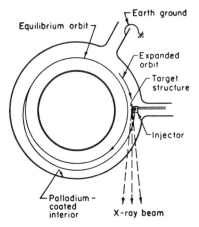

FIG. 6-70. Top view of betatron doughnut.

core. The secondary winding wound on this core has induced in it a voltage. The voltage induced is equal to the product of the number of turns in the secondary winding and the time rate of change of flux. In the transformer the electric current which flows is made up of the free electrons present in the wire. The betatron is essentially a simple transformer, except that instead of wire, the secondary is a circular hollow tube. This tube, called a "doughnut," is used to contain the electrons during their travel of many thousand revolutions. The tube is usually made of porcelain and is coated on the inside with a conductive layer of palladium connected to ground. The doughnut is placed between the poles of an electromagnet that produces a pulsating field. Electrons injected into the tube as the magnetic field increases will be accelerated in a circular path. The force acting on the particles is proportional to the rate of change of flux and the magnitude of the field. The energy gain per revolution is equal to the time rate of change of flux linking the orbit. Since the electrons circle the orbit a great many times before coming to rest, the energy gain is tremendous. For example, in a 24-million-volt betatron, the electrons circle the orbit about 350,000 times, traveling a distance of 260 miles. All this happens in one-fourth of a cycle, or less than 1.4 msec, since the excitation frequency of the magnet core is 180 cps. The average voltage gain per turn at the orbit is about 70 volts, which give approximately 24 million volts. As the electrons reach maximum energy, they are deflected by an electrical pulse and caused to spiral outward until they strike the "target." The X rays produced have an energy of 24 million volts. The complete mathematical theory of the betatron is quite complex; it can be found in standard nuclear physics textbooks.[31]

Figure 6-71 shows the linear absorption coefficient of iron as a function of betatron energy. The transmission is maximum at approximately 22 million volts. The absorption coefficient for iron has a minimum·

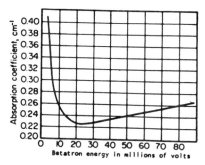

FIG. 6-71. Linear absorption coefficient of X rays of iron.

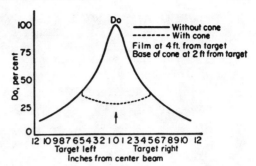

FIG. 6-72. Energy distribution about target center of 22-Mev betatron. (*Courtesy Los Alamos Scientific Laboratory.*)

value between 6 and 8 million volts. The spectrum of the X-ray energies from a 22-million-electron-volt betatron is such that the average energy is in this minimum absorption range. Thus optimum penetration is achieved when operating a betatron in the energy range of 20-30 Mev. A further advantage of this energy range is the characteristic of extremely wide latitude. It is possible to see a flaw in a steel forging or casting throughout a section varying from 12 to 18 in. with a single exposure at a latitude of 6 in. on a single film. Flaws $1/32$ in. deep can be seen in steel sections 3 to 12 in. thick. Flaws $1/16$ in. deep can be seen in steel specimens up to 20 in. thick. These flaws may be as narrow as 0.005 in. Where minute flaws exist, the small focal spot, 0.005 by 0.10 in., makes possible the enlargement of a minute flaw by simply increasing the specimen-to-film distance. A gap as narrow as 0.001 in. parallel to the plane of the X-ray beam is detectable. Application of the betatron in industry is not limited to inspection of heavy sections. In one installation, transmissions and motors are being checked for internal condition and alignment of parts to determine the extent of reconditioning required. In another, internal mechanisms of military equipment are checked before shipment.

Smith at the Los Alamos Scientific Laboratory[32] made a study of the factors which are unique in betatron radiography. The X rays emitted by the target of a betatron are in the form of a small-angle cone. The intensity is greatest at the axis of the cone and falls off rapidly with increasing angle from the axis, Fig. 6-72—solid curve. This concentration of intensity can be used to advantage when radiographing solid spherical or cylindrical specimens but presents difficulties when radiographing flat specimens. A compensating filter in the form of a lead cone can be used to flatten the intensity of the beam. Although the cone tends to decrease the beam intensity, the elimination of the hot spot permits using a shorter target-to-film distance. With the beam flattening cone, the necessity for multiple film is eliminated. Experiments indicate that the minimum front screen necessary to give maximum protection against scattered radiation is 0.120 in. of lead. The heavy

Fig. 6-73. Stroboradio-graph of paint spraying pump. (*Courtesy U.S. Ordnance Corps—Detroit Arsenal.*)

front screens required for protection against scattered radiation have no intensifying effect. Back screens up to a thickness of 0.020 in. are important as a density producing component, but decrease the resolution.

Stroboradiography. Betatrons have been used for stroboradiography. For stroboradiography a synchronizing device is used to control the betatron. The primary function of this device is to produce a demand for an X-ray pulse every time the specimen is in the desired position. The accumulation of many weak images in exactly the same spot results in a sharp radiograph of required film density. The synchronizer may be activated by the specimen to be radiographed. Figure 6-73 shows a stroboradiograph of a small paint spray air pump. The speed of the pump was 1,400 rpm, and the exposure time at 4 ft from target to film was approximately 15 min.

Magna-Scanning. Magna-Scanning is a combination of two regular industrial radiographic techniques, magnification and scanning. Figure 6-74 shows a radiograph produced using Magna-Scanning. The magnification is accomplished by direct geometrical enlargement, with the film placed at some distance from the specimen. The requirements for this technique are a fine focal spot and a minimum of secondary scattering. Scattering of secondary radiation normally obscures details on a film placed remotely from the specimen. In Magna-Scanning the betatron is rotated during the exposure. It is necessary that the focal spot remain essentially fixed in space while the betatron rotates. The beta-tron was raised 16 ft above the floor and rotated so that the beam was directed downward. The two rifles were placed on a stand, 12 ft from the floor. The film was placed on a sheet of masonite 4 ft above the floor.

Picker-Polaroid Process. The Picker-Polaroid Self Developing Radiograph has been applied to industrial radiography. This process produces a radiographic image within a minute or two after the exposure, without the use of conventional processing. It was developed originally for medical radiography. The characteristic curve of any photographic material gives quantitative data about its principal properties. These are: the maximum black, the contrast or gradient, the total exposure scale, and the density range. Figures 6-75 and 6-76 show experimentally determined exposure curves for Polaroid paper for 200-kv peak X rays. The slopes of these curves are negative, whereas the slopes of the characteristic curves for conventional photographic materials are positive. Zero exposure results in maximum black. A radiographic image is usually more easily and more accurately interpreted if the image densities

Fig. 6-74. Arrangement for exposing the world's largest radiograph produced by Magna-Scanning with the 22-million-volt betatron.

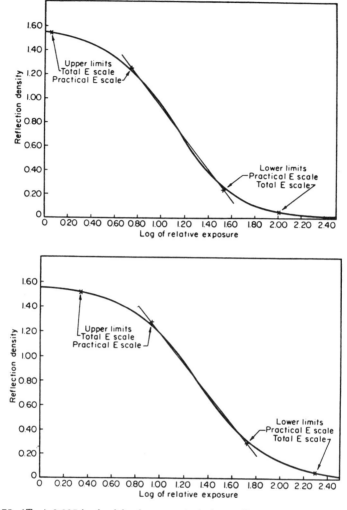

FIG. 6-75. (*Top*) 0.005-in.-lead back-screen technique; (*bottom*) no-screen technique.

are in a straight portion of the characteristic curve. The Polaroid curves have little or no straight portion. But a straight portion can be approximated by a line that does not deviate from the characteristic curve by more than 0.02 density units.

Table 6-18 gives calculated and experimentally determined minimum discernible thickness changes. Tables 6-19 and 6-20 show that the Polaroid Radiographic Process as now commercially available is best suited to low-energy radiography using the salt-screen technique. The short density range of the Polaroid paper is most applicable to radiography of high-contrast specimens. It also can be considered suitable for

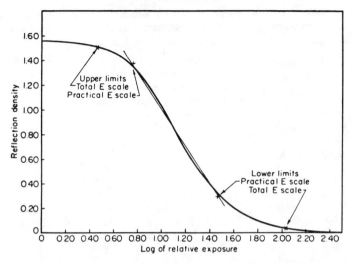

Fig. 6-76. Characteristic curve of Polaroid paper for 200-kv peak X rays (Patterson screen). (*Courtesy Los Alamos Scientific Laboratory.*)

detecting large cavity defects or inclusions in an object. The mottled pattern, generally experienced to various degrees in the Polaroid radiographs, seriously impairs visualization of detail and density changes. The over-all speed of the process can be of value in certain instances when a quick on-the-spot radiograph is needed.

TABLE 6-18. CHARACTERISTICS OF POLAROID PAPER (No. 1001) FOR 200-KV PEAK X RAYS

Characteristic	Salt screen	No screen	Lead screen	Units
Maximum black	1.6	1.6	1.6	Density
Maximum slope	1.75	1.43	1.58	$\frac{\Delta D}{\Delta \log E}$
Practical E scale	0.71	0.79	0.79	log E
Practical D range	1.08	0.99	1.02	Density
Total E scale	1.56	1.95	1.95	log E
Total D range	1.47	1.47	1.47	Density
Working D	0.7	0.7	0.7	Density
Slope	1.67	1.4	1.5	$\frac{\Delta D}{\Delta \log E}$
$\Delta E/E$	2.8	3.4	3.1	Per cent
Calculated sensitivity at 1-in. steel	0.007	0.009	0.008	In.
Observed sensitivity at 1-in. steel	0.020	0.030	0.020	In.
Claimed radiographic sensitivity at 1-in. steel	0.030	0.040	0.030	In.

TABLE 6-19. PER CENT SENSITIVITY AND RESOLUTION OF POLAROID PAPER RADIOGRAPHS OF ALUMINUM ABSORBERS AT 40 TO 150 KV PEAK

Absorber thickness, in.	Salt screen		No screen		Lead screen	
	Sensitivity	Resolution	Sensitivity	Resolution	Sensitivity	Resolution
At 40 kv peak:						
¼	2	8	8	12		
At 80 kv peak:						
1	4	6	4	6		
1½	2.7	4	5.3	6.7		
2¼	2.7	3.6				
At 120 kv peak:						
½	...	...	8	12	8	12
1½	...	...	5.3	5.3	4	5.3
2¾	2.9	3.6	4.4	5.8		
At 150 kv peak:						
1	...	...	8	10	6	8
2½	...	...	6.4	6.4	3.2	4
3	2.7	4				
4	4	4				

TABLE 6-20. PER CENT SENSITIVITY AND RESOLUTION OF POLAROID PAPER RADIOGRAPHS OF STEEL ABSORBERS AT 200 AND 1,000-KV PEAK X RAYS AND Co⁶⁰ GAMMA RAYS

Absorber thickness, in.	Salt screen		Lead screen	
	Sensitivity	Resolution	Sensitivity	Resolution
At 200-kv peak X rays:				
1	3	6	4	8
At 1,000-kv peak X rays:				
1	8	8		
1½	8	8	2.7	8
2	8	8		
2½	...	...	4	6.4
At Co⁶⁰ gamma rays:				
1	8	8	6	8
2	6	8	6	8
3	5.3	5.3	5.3	5.3

Electron Radiography. Two techniques for electron radiography will be described. They are thin-specimen radiography using fast electrons as radiation and surface electron emission upon a photographic plate. These techniques make use of secondary photoelectrons instead of X rays as a means of registering an image on the photographic emulsion. The number of emitted photoelectrons depends upon the wavelength and the intensity of the incident X rays. The number of photoelectrons emitted increases as the atomic number increases. Figure 6-77a schematically illustrates the arrangement for the transmission technique. A thin foil of heavy element such as lead is pressed in contact with the specimen and a fine-grain emulsion. The X rays incident on the foil eject electrons, some of which go through the specimen to the emulsion, exposing it in amounts depending upon the energy and number of the electrons which have penetrated the specimen. The experimental arrangement for the back-emission technique is shown in Fig. 6-77b. In back-emission electron radiography, a polished surface is placed in direct contact with a fine grain emulsion. The X rays pass through the emulsion and eject electrons from the specimen surface. The emulsion is exposed in amounts depending mainly on the atomic numbers of the surface elements in contact with it. In order to obtain the desired results it is necessary that the emulsion be in closest possible contact with the surface to be investigated. It is necessary that the photographic emulsion have a low absorption coefficient for X rays and a high absorption coefficient for photoelectrons.

One advantage of transmission electron radiography over X radiography is that a standard high-voltage X-ray unit may be used for thin-specimen radiography. Electron transmission radiographs are influenced essentially by physical density changes, while low-voltage X-ray transmission is highly sensitive to atomic number differences. The advantage of back-emission electron radiography over photographic

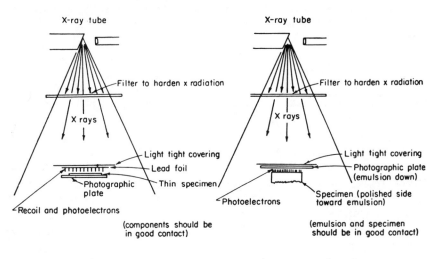

Transmission electron radiography

(a)

Back-emission electron radiography

(b)

Fig. 6-77. Electron radiography.

analysis is that surface components which may be indistinguishable by visible light might easily be differentiated without the necessary etching.

Soft X rays must be prevented from reaching the sensitive surface as they contribute only to the background density, producing negligible photoemission. Because of the dependence of X-ray absorption on wavelength in all elements, hard X rays are reduced only slightly in intensity, whereas soft X rays are absorbed almost entirely. For electron radiographs taken with the X-ray tube operating at 250 kv or below, a filter consisting of $\frac{3}{16}$ in. of copper plus $\frac{3}{16}$ in. of aluminum is satisfactory. The aluminum is added to absorb characteristic fluorescent radiation and Compton scatter from the copper. As is obvious, decreasing the amount of filtration shortens the exposure time but may introduce excessive background density. It has been found that ultra-high-resolution spectroscopic plates are the most satisfactory for electron radiography.[33] Lead is the most satisfactory electron-emitting material because of its high atomic number and high electron density. Lead can be molded about the specimen in tight contact without damaging it. A lead thickness of 25 μ is satisfactory. Increasing the thickness of lead only increases the exposure.

Neutron Radiography. The absorption of neutrons by matter is comparable with that of X rays or gamma rays having energies of some hundreds or thousands of kilovolts. However, the relative absorption by the various elements is quite different. Neutrons can be absorbed by a nucleus to form a different nucleus, or they may be scattered by the nucleus. For some elements such as iron and carbon, scattering accounts for most of the total absorption. In Table 6-21 are given the mass and linear absorption coefficients of the elements for thermal neutrons. There is no obvious relationship between the neutron absorption and scattering coefficients and the atomic number as there is for X rays. The variation of absorption from element to element is quite random. The largely random distribution of neutron absorption coefficients among the elements permits radiography on the same film of a greater range of materials than is possible with X rays or gamma rays. This also permits examining thicker sections or reducing exposure times. In certain circumstances, materials that cannot be examined satisfactorily by X rays or gamma rays can be examined by neutron radiography. It is possible to differentiate by neutron radiography between elements which have similar absorption coefficients for X rays or gamma rays but quite different ones for neutrons, such as boron and carbon, cadmium and barium.

Unfortunately the direct photographic effect of neutrons is negligible. The most usual detector for thermal neutrons is the boron trifluoride proportional counter, which uses the ionization produced by the alpha-

TABLE 6-21. ABSORPTION COEFFICIENTS FOR NEUTRONS

Element	Atomic no.	Neutrons (λ = 1.08 A)			
		(μ/ρ) true	(μ/ρ) scattering	(μ/ρ) total	μ total
H	1	0.11	48.4	48.5	
Li	3	3.5	0.17	3.7	2.0
Be	4	0.0003	0.50	0.50	0.92
B	5	24	......	24*	60*
C	6	0.00015	0.26	0.26	0.60
N	7	0.048	0.43	0.48	
O	8	<0.00002	0.15	0.15	
F	9	<0.0003	0.11	0.11	
Ne	10	0.006	......	0.006*	
Na	11	0.007	0.092	0.099	0.097
Mg	12	0.001	0.092	0.093	0.16
Al	13	0.003	0.033	0.036	0.97
Si	14	0.001	0.043†	0.044†	0.10†
P	15	0.002	0.060†	0.62†	0.12†
S	16	0.0055	0.023	0.029	0.058
Cl	17	0.33	0.255	0.59	
A	18	0.0060	......	0.006*	
K	19	0.018	0.031	0 049	0.042
Ca	20	0.0037	0.053	0.057	0.088
Sc	21	0.09	0.175†	0.27†	0.68†
Ti	22	0.044	0.075	0.119	0.54
V	23	0.033	0.060	0.093	0.56
Cr	24	0 021	0.044	0.065	0.46
Mn	25	0.083	0.024	0.107	0.79
Fe	26	0.015	0.126	0.141	1.1
Co	27	0.21	0.051	0.26	2.2
Ni	28	0.028	0.185	0.213	1.9
Cu	29	0.021	0.074	0.095	0.85
Zn	30	0.0055	0.039	0.045	0.32
Ga	31	0.015	......	0.015*	0.089*
Ge	32	0.011	0.071	0.082	0.45
As	33	0.020	0.056	0.076	0.44
Se	34	0.056	0.076	0.132	0.59
Br	35	0.029	0.045	0.074	
Kr	36	0.0002	......	0.0002*	
Rb	37	0.0029	0.059	0.042	0.064
Sr	38	0.0048	0.065	0.070	0.18
Y	39	0.0056	......	0.0056*	0.021*
Zr	40	0.0006	0.046	0.047	0.31
Nb	41	0.0041	0.040	0.044	0.37
Mo	42	0.009	0.046	0.055	0.55
Ru	44	0.009	......	0.009*	0.11
Rh	45	0.53	......	0.53*	6.6*
Pd	46	0.023	0.027	0.050	5.7

TABLE 6-21. ABSORPTION COEFFICIENTS FOR NEUTRONS (*Continued*)

Element	Atomic no.	Neutrons ($\lambda = 1.08$ A)			
		(μ/ρ) true	(μ/ρ) scattering	(μ/ρ) total	μ total
Ag	47	0.20	0.039	0.24	2.5
Cd	48	11.2	. . . .	11.2*	97†
In	49	0.60	. . . .	0.60*	4.4*
Sn	50	0.002	0.025	0.027	0.20
Sb	51	0.016	0.021	0.037	0.25
Te	52	0.013	0.018	0.031	0.19
I	53	0.018	0.018	0.036	0.18
Xe	54	0.083	. . . .	0.083*	
Cs	55	0.077	0.032	0.109	0.20
Ba	56	0.0027	0.015†	0.018†	0.068†
La	57	0.023	0.040	0.063	0.39
Ce	58	0.0021	0.012	0.014	0.097
Pr	59	0.029	0.017	0.046	0.30
Nd	60	0.11	0.10	0.21	1.5
Sm	62	25	. . . .	25*	195*
Eu	63	10	. . . .	10*	52*
Gd	64	84	. . . .	84*	497*
Tb	65	0.09	. . . .	0.009*	0.75*
Dy	66	2.0	. . . .	2.0*	17.2*
Ho	67	0.015	. . . .	0.015*	1.3*
Er	68	0.36	0.054	0.41	2.0
Tm	69	0.25	. . . .	0.25*	2.3*
Yb	70	0.076	. . . .	0.076*	0.42*
Lu	71	0.22	. . . .	0.22*	2.1*
Hf	72	0.20	. . . .	0.20*	2.3*
Ta	73	0.044	0.023	0.067	1.1
W	74	0.036	0.022	0.058	1.1
Re	75	0.16	. . . .	0.16*	3.4
Os	76	0.028	. . . .	0.28*	0.63*
Ir	77	0.80	. . . .	0.80*	18*
Pt	78	0.05	0.035	0.050	11
Au	79	0.17	0.027	0.20	3.9
Hg	80	0.63	0.080	0.71	9.6
Tl	81	0.006	0.021†	0.027†	0.32†
Pb	82	0.0003	0.034	0.034	0.39
Bi	83	<0.00003	0.029	0.029	0.28
Th	90	. . . .	0.033	0.033‡	0.37‡
U	92	0.005	0.023†	0.028‡	0.52

* Scattering not included.
† Incoherent scattering not included.
‡ Scattering only.

particles emitted in the reaction

$$\underset{5}{\overset{10}{B}} + \underset{0}{\overset{1}{n}} \rightarrow \underset{3}{\overset{7}{Li}} + \underset{2}{\overset{4}{He}} \qquad (6\text{-}31)$$

and is relatively insensitive to gamma rays. Such a counter, of suitable size, might be used in neutron radiography in combination with a scanning technique. The "loading" of a photographic emulsion with a heavily absorbing material such as boron or cadmium should, however, make the direct photographic registration of neutrons possible. Neutron scintillation crystals are now available and may prove of use in this field. Other methods involve the use of an intermediate foil of some material which emits alpha, beta, or gamma rays during or after exposure to neutrons.

The most useful of these methods have been summarized by Kallmann.[34] A thin layer of boron or lithium is used as a "converter" of neutrons to alpha particles. The alpha particles impinge on a fluorescent screen, the light from which affects the photographic film. A thin sheet of aluminum is placed between the boron or lithium and the fluorescent screen. The purpose of the aluminum sheet, which is less than 0.5 μ in thickness, is to reflect the light from the fluorescent screen. The sheet is sufficiently thin not to slow down appreciably the alpha particles of highest energy. This method records not only neutrons indirectly but directly any gamma rays that are present in the neutron beam. Moreover, the lack of contact between the photographic film and the alpha emitting layers leads to loss of definition. In another technique cadmium foil is placed in direct contact with the photographic film. Cadmium emits gamma rays on exposure to neutrons. Any gamma rays accompanying the primary neutron beam represent a source of unwanted blackening. In still another technique a foil of some material is used which becomes radioactive on exposure to neutrons and emits beta or gamma rays. Such a foil made of indium, silver, or gold can be irradiated in the absence of a photographic film. The radioactive image formed on the foil is subsequently transferred to the film. In an alternate technique the film and intermediate foil are exposed to the neutron beam together and then removed. The film and foil are left in contact for a period to permit the transfer of the image to take place. This enhances the neutron blackening and does not lead to a prohibitive amount of unwanted gamma-ray blackening. This latter technique has been used at Harwell[35] by Thewlis and coworkers. A combination of the latter two techniques in which layers of silver and cadmium are employed has been used by Peter.[36] The first neutron radiographs were published by Kallmann[34] and Peter.[36] At the present time neutron radiographs are inferior in quality to X-ray radiographs but compare favorably with those obtained by the use of gamma rays.

Beta emitting isotopes have been used in radiography. With beta emitters, the X radiation is produced by the interaction of the beta ray with other material. The radiation is composed of internal bremsstrahlung arising from the interaction of nuclear beta particles with the radiation field of the nucleus. In addition external bremsstrahlung is produced in the deceleration process of the beta particles by absorber nuclei. Also there is characteristic X-radiation produced by the interaction of the beta particles with orbital electrons either of the source or absorber material. Kereiakes and Krebs[37] have used a Tracerlab medical applicator as a source of beta radiation. This unit (300 mc of Sr^{90} in equilibrium with Y^{90}) gave an X-ray intensity of 6.4 mr/min at a distance of 7.5-cm distance from the source.

GENERAL REFERENCES

American Society for Metals: "Metals Handbook," section on Radiography of Metals, pp. 141–145, American Society for Metals, Cleveland, Ohio, 1948.

American Society for Testing Materials: Symposium on Radiography, 1943.

Clark, George L.: "Applied X-Rays," McGraw-Hill Book Company, Inc., New York, 1955.

Clauser, H. R.: "Practical Radiography for Industry," Reinhold Publishing Corporation, New York, 1952.

Crowther, J. A.: "Handbook of Industrial Radiography," Edward Arnold & Co., London, 1949.

Eastman Kodak Company: "Radiography in Modern Industry," Eastman Kodak Company, Rochester, N.Y., 1957.

Sproull, Wayne T.: "X-Rays in Practice," McGraw-Hill Book Company, Inc., New York, 1946.

St. John, Arcel, and Herbert R. Isenburger: "Industrial Radiology," John Wiley & Sons, Inc., New York, 1943.

Wiltshire, W. J.: "A Further Handbook of Industrial Radiography," Edward Arnold & Co., London, 1957.

Zmeskal, Otto: "Radiographic Inspection of Metals," Harper & Brothers, New York, 1943.

SPECIFIC REFERENCES

1. White, Gladys R.: X-Ray Attenuation Coefficients from 10 Kev to 100 Mev, *Natl. Bur. Standards Rept.* 1003, May 13, 1952.

2. Victoreen, John A.: The Calculation of X-Ray Mass Absorption Coefficients, *J. Appl. Phys.*, vol. 20, pp. 1141–1147, December, 1949.

3. Richtmyer, F. K., and E. H. Kennard: "Introduction to Modern Physics," McGraw-Hill Book Company, Inc., New York, 1942.

4. Pace, A. L.: Lead Screens Mean Better Films with Less Exposure, *Gen. Elec. Radiation Digest*, vol. 1, no. 4, p. 4, 1954.

5. Seeman, H. E.: Some Physical and Radiographic Properties of Metallic Intensifying Screens, *J. Appl. Phys.*, vol. 8, no. 12, pp. 836–845, December, 1937.

6. Frerichs, Rudolf, and John E. Jacobs: An Economical Industrial X-Ray Detector, *Gen. Elec. Rev.*, vol. 54, no. 8, pp. 42–45, August, 1951.

7. Crowther, J. A., "Handbook of Industrial Radiology," Edward Arnold & Co., London, 1949.

8. Turner, Ralph E.: The Use of Multiple-Film Technics to Speed Industrial Radiographic Inspection, *J. Soc., Nondestructive Testing*, vol. 15, no. 3, pp. 146–150, May–June, 1957.

9. Kolm, Eric A.: Radiography in Production Control and Inspection of Subminiature Tubes, *J. Soc. Nondestructive Testing*, vol. 14, no. 6, pp. 20–23, November–December, 1956.

10. Mooradian, A. J.: NRU—Some of Its Special Testing Problems, *J. Soc. Nondestructive Testing*, vol. 16, no. 2, pp. 164–170, March–April, 1958.

11. Fadikov, I. G., and A. A. Samokhvalov: Ionization Methods for Detecting Defects in Thick Sections of Metal by γ Rays, *Conf. Acad. Sci. U.S.S.R. Peaceful Uses of Atomic Energy, Moscow*, July, 1955.

12. Kahn, N. A., Solomon Goldspiel, and R. R. Waltien: Application of Radiography in the Manufacture of Bronze Castings, *American Foundrymen's Society*.

13. Barkow, A. G.: How to Interpret Radiographs of Pipe-Line Welding Defects, *Oil Gas J.*, Oct. 4, 1951.

14. Masi, Oscar: Weld Radiography: A Tentative Method for the Quantitative Evaluation of Defects, *ASTM, Spec. Tech. Pub.* 145, 1952.

15. Carr, L. R.: The Quantitative Interpretation of Radiographs in Terms of Mechanical Properties, *J. Sci. Instr.*, vol. 23, no. 10, pp. 221–227, October, 1946.

16. Wallmann, K.: Relationship between Radiograph and Properties Determined on Welded Joints, *Arch. Eisenhüttenw.*, vol. 8, no. 6, pp. 243–247, December, 1934.

17. Feinberg, Irving J.: Tensile Properties of Porosity-Graded 195 Alloy, *J. Soc. Nondestructive Testing*, vol. 15, no. 3, pp. 168–173, May–June, 1957.

18. Oaks, A. E.: Radiographic Inspection of Nuclear Core Materials and Components, *ASTM, Spec. Tech. Pub.* 223, 1958.

19. Westinghouse Electrical Corporation: *Bettis Tech. Rev.*, vol. 1, no. 2, pp. 101–103, July, 1957.

20. Levy, L., and D. W. West: Modern Applications of Luminescent Substances, *J. Soc. Chem. Ind.*, vol. 58, pp. 457–462, 1939.

21. Klassens, H. A.: Measurement and Calculation of Unsharpness Combinations in X-Ray Photography, *Philips Research Repts.*, vol. 1, no. 4, pp. 241–249, August, 1946.

22. O'Connor, D. T., and D. Polansky: Theoretical and Practical Sensitivity Limits in Fluoroscopy, *J. Soc. Nondestructive Testing*, vol. 10, no. 2, pp. 10–21, Fall, 1951.

23. O'Connor, D. T., and D. Polansky: Theoretical and Practical Sensitivity Limits in Fluoroscopy, *J. Soc. Nondestructive Testing*, vol. 10, no. 2, equations 31 and 32, Fall, 1951.

24. Coltman, J. W.: Fluoroscopic Image Brightening by Electronic Means, *Radiology*, vol. 51, pp. 359–367, 1948.

25. Polansky, D., and E. L. Criscuolo: Characteristics of a Closed-Link Television X-Ray Inspection System, *J. Soc. Nondestructive Testing*, vol. 14, no. 3, pp. 18–21, May–June, 1956.

26. Berger, Harold, and A. L. Pace: Field Performance of a Television X-Ray System, *J. Soc. Nondestructive Testing*, vol. 15, no. 1, pp. 26–29, January–February, 1957.

27. O'Connor, D. T., and D. Polansky: A High Sensitivity Fluoroscope for the Inspection of Light Alloy Castings, *NAVORD Rept.* 3791, September, 1954.

28. Polansky, D., and D. T. O'Connor, The Feasibility of Fluorography above 150 kv, *J. Soc. Nondestructive Testing*, vol. 12, no. 6, pp. 27–34, November–December, 1954.

29. Taylor, Grover M., and G. H. Tenney: Field Evaluation of Industrial Xeroradiography, *J. Soc. Nondestructive Testing*, vol. 13, no. 6, pp. 12–17, November–December, 1955.

30. Kerst, D. W.: The Acceleration of Electrons by Magnetic Induction, *Phys. Rev.*, vol. 60, pp. 47–53, July 1, 1941.

31. Holliday, David: "Introductory Nuclear Physics," John Wiley & Sons, Inc., New York, 1950.

32. Smith, Clark J.: Report on the Investigation of Betatron Radiographic Techniques, *J. Soc. Nondestructive Testing*, vol. 11, no. 7, pp. 17–22, September, 1953.

33. Berman, Arthur I.: Electron Radiography, *U.S. Atomic Energy Comm. Rept.* AECU-1853, Dec. 5, 1950.

34. Kallmann, H.: Neutron Radiography, *Research* 1, pp. 254–260, 1947.

35. Thewlis, J.: Neutron Radiography, *Brit. J. Appl. Phys.*, vol. 7, pp. 345–350, October, 1956.

36. Peter, O.: Neutron Durchleuchtung, *Z. Naturforsch.*, pp. 557–559, 1946.

37. Kereiakes, J. G., and A. T. Krebs: Technical Radiography with Beta Emitting Isotopes, *U.S. Army Med. Research Lab. Rept.* 180, Mar. 21, 1955.

ADDITIONAL REFERENCES

Buechner, W. W., R. J. van de Graaff, H. Feshbach, E. A. Burrill, A. Sperduto, and L. R. McIntosh: An Investigation of Radiography in the Range from 0.5 to 2.5 Million Volts, *ASTM Bull.*, pp. 54–64, December, 1948.

Burrill, E. A., and W. W. Buechner: Sensitometry of Radiographic Films Exposed to Two-Million-Volt X-Rays, *ASTM Bull.*, pp. 52–57, October, 1947.

Emigh, C. R., and L. R. Megill: Semi-Empirical Equations for the Spectral Energy Distribution in X-Ray Beams, *J. Soc. Nondestructive Testing*, January, 1953.

Miller, N. C., and J. D. Steely: Some Experimental Findings and Operation Practices in Betatron Radiography, *J. Soc. Nondestructive Testing*, November–December, 1953.

CHAPTER 7

GAMMA RADIOGRAPHY

Radioactive gamma-ray sources both natural and man-made are finding wide use in the field of radiography. Gamma rays are emitted during the disintegration of radioactive material and like X rays are electromagnetic radiation. However, gamma rays, as can be seen from Fig. 7-1, have a wavelength which is intermediate between the shortest and longest X-ray wavelengths. Gamma rays do not have a continuous spectrum as X rays do but consist of one or more discrete energies.

The chief merits of gamma-ray sources are (1) small size, (2) high penetration of the radiation compared with industrial X-ray sources in common use, (3) relative low cost compared with X-ray units, (4) independence from a supply of electricity and water, and (5) the lower image contrast which permits a larger range of metal thickness to be recorded at one exposure of film. However, in radiographing sections of relatively uniform thickness, low contrast may be objectional because it does not give the optimum conditions for flaw detection. The major disadvantages of such sources are that they are generally of low intensity, requiring long exposure. Some sources have a short half-life which necessitates replacement. Since radioactive sources cannot be "shut off," they must be properly stored to protect personnel at all times.

There are three major factors to be considered when selecting radioisotopes: (1) the half-life, (2) the gamma-ray energy, and (3) the material to be radiographed. Natural gamma-ray sources, primarily radium and radon, have been used for industrial radiography since the early 1930s. The use of these materials never became widespread, because of their cost and in the case of radon because of its short half-life. The development of the nuclear reactor during World War II has provided new and better gamma-ray sources useful for radiography. These sources may

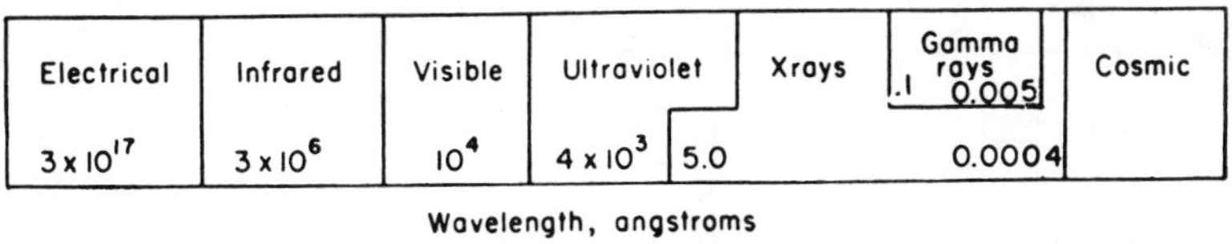

Electrical	Infrared	Visible	Ultraviolet		Xrays	Gamma rays	Cosmic
3×10^{17}	3×10^{6}	10^{4}	4×10^{3}	5.0	.1	0.005 0.0004	

Wavelength, angstroms

FIG. 7-1. Electromagnetic spectrum.

177

be prepared either by separation from fission products or by irradiation in a nuclear reactor. Co^{60}, for example, is prepared by bombarding Co^{59} with thermal neutrons. The reaction is

$$_{27}Co^{59} + {}_0n^1 \rightarrow {}_{27}Co^{60} \tag{7-1}$$

$$_{27}Co^{60} \rightarrow \text{beta} + \underset{1.17 \text{ Mev}}{\text{gamma}} + \underset{1.33 \text{ Mev}}{\text{gamma}} + {}_{28}Ni^{60} \tag{7-2}$$

Cs^{137} is an example of one of the isotopes obtained from fission by-products. One of the disadvantages of sources prepared by separation from fission products is that they may contain a mixture of isotopes having a wide variety of energies.

Fundamentals of Radioactivity. Before discussing the various sources of gamma radiations it is advisable to discuss briefly some of the fundamentals of radioactivity and associated phenomena. The radiation emitted by a radioactive material is due to the spontaneous disintegration of the atoms of the material. This disintegration with the emission of an alpha or beta particle results in the formation of new or "daughter" elements. Gamma-ray emission frequently accompanies the emission of the alpha or beta particle. All radioactive sources are characterized by the property which is known as the "half-life" of the material, the time required for half the material to decay. Figure 7-2 shows how the intensity of a radioactive source, such as Co^{60}, decreases with time. Table 7-1 lists the half-life for a number of gamma-ray emitters.

The Curie. The activity of gamma-ray sources is usually given in curies, millicuries, or microcuries. A curie is defined in terms of a disintegration rate, a disintegration rate of 3.7×10^{10} per second being 1 curie (c). A millicurie (mc) is equal to 3.7×10^7 disintegrations per second, and a microcurie (μc) is equal to 3.7×10^4 disintegrations per second. The disintegration rate is not influenced by pressure, temperature, or chemical combination.

Radioactive Decay. The amount of a radioactive material remaining after a period of time t can be calculated if the amount initially present

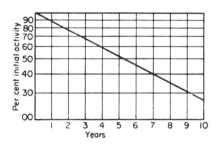

Fig. 7-2. Co^{60} decay curve.

TABLE 7-1. CLASSIFICATION OF ISOTOPES BY ENERGY OF GAMMA RADIATION

Isotope	Half-life	Isotope	Half-life
		Ba^{140}	12.8 days
	Greater than 2 Mev	Nd^{147}	11 days
Ra^{226}	1,620 years	I^{131}	8 days
Sb^{124}	60 days	Sb^{122}	2.8 days
La^{140}	40.4 hr	Mo^{99}	67 hr
As^{76}	27.6 hr	Cd^{115}	43 days: 2.3 days
Na^{24}	14.97 hr	Sm^{153}	47 hr
Ga^{72}	14.3 hr	W^{187}	24.1 hr
	1.5 to 2 Mev	Ru^{105}	4 hr
Ag^{110}	270 days		*0.25 to 0.5 Mev*
Pr^{142}	19 hr	Se^{75}	127 days
K^{42}	12.4 hr	Hf^{181}	46 days
	1 to 1.5 Mev	Hg^{203}	46.5 days
$Eu^{152, 154}$	12.4 years	Ru^{103}	42 days
Co^{60}	5.3 years	Ce^{141}	32.5 days
Cs^{134}	2.3 years	Cr^{51}	26.5 days
Zn^{65}	250 days	Ba^{131}	13 days
Ta^{182}	111 days	Hg^{197}	65 hr: 24 hr
Sc^{46}	85 days	Rh^{105}	36.5 hr
In^{114}	50 days		*Less than 0.25 Mev*
Fe^{59}	46.3 days	Pb^{210} (RaD)	22 years
Rb^{86}	19.5 days	Ac^{227}	22 years
Au^{198}	2.7 days	Eu^{155}	1.7 years
Br^{82}	35 hr	Ce^{144}	290 days
Os^{193}	32 hr	Os^{191}	15 days
Ho^{166}	27 hr	Re^{186}	3.8 days
Cu^{64}	12.8 hr	Ru^{97}	2.8 days
	0.5 to 1 Mev	Tm^{170}	127 days
Cs^{137} (Ba^{137})	33 years	W^{125}	73 days
Sb^{127}	2.7 years	Tc^{97}	90 days: 9.3 hr
Po^{210} (RaF)	138 days	Te^{127}	90 days
Ir^{192}	74 days	Cs^{131}	9.6 days
Zr^{95}	65 days	Au^{199}	3.3 days
Cd^{115}	43 days		
Cb^{95}	35 days		

is known, using the expression

$$N = N_0 \exp(-\lambda t) = N_0 \exp\left(-\frac{0.693t}{T}\right) \qquad (7\text{-}3)$$

where N = number of atoms of material remaining

N_0 = number of atoms of material initially present

λ = decay constant = $0.693/T$

t = time elapsed

T = half-life of material

In the above expression T and t must be in the same units of time.

Co^{60} has a half-life of 5.3 years. The activity of a 10-c source of Co^{60} at the end of 2 years is

$$N = N_0 \exp\left(-\frac{0.693}{5.3}\right)(2) = N_0 \exp(-0.26) = 0.771N_0 \quad (7\text{-}4)$$

The activity has decreased to 0.771 of the initial activity or to 7.71 c.

The term specific activity is used to indicate the activity of 1 g of a radioactive source. The specific activity of a given source is the number of disintegrations per unit time per gram of the radioelement. Co^{60} with a half-life of 5.3 years has

$$\text{Specific activity} = \frac{(0.693)(6 \times 10^{23})}{(5.3 \times 3.14 \times 10^7 \text{ sec})(60)} \quad (7\text{-}5)$$
$$= 4.15 \times 10^{13} \text{ dis/sec}$$
$$\text{Specific activity} = M\frac{0.693}{T} \quad (7\text{-}6)$$

where M is the number of atoms per gram of Co^{60} and T is the half-life. The number of atoms per gram of any material is found by dividing Avogadro's number 6×10^{23} by the atomic weight of the material.

Roentgen. The roentgen is defined as: "That quantity of X or gamma radiation such that the associated corpuscular emission per 0.001293 g of dry air (equal 1 cm³ at 0°C and 760 mm Hg) produces, in air, ions carrying 1 esu of quantity of electricity of either sign."

The roentgen per hour at one meter (rhm) is a physical unit of radioactive source strength. It is a way of assigning a numerical value to the total "amount" of any radioactive substance which emits gamma rays or X rays. One rhm of Co^{60} is that amount of Co^{60} whose unshielded gamma-ray emission produces 1 r/hr of ionization in air at a distance of 1 m from the source. Cobalt emits 1.17 and 1.33 Mev gamma rays per disintegration. At these energies the absorption in air is almost entirely due to Compton scattering. The absorption coefficient for air at these energies averages about 3.5×10^{-5} cm⁻¹; therefore, a 1-c source of Co^{60} would give

$$\frac{3.7 \times 10^{10} \text{ dis/sec} \times 3{,}600 \text{ sec/hr}}{4\pi(100 \text{ cm})^2} \times \frac{(1.17 + 1.33)3.5 \times 10^{-5} \text{ cm}^{-1}}{6.77 \times 10^4 \text{ Mev/cm}^3 \text{ air}}$$
$$= 1.30 \text{ rhm/c} \quad (7\text{-}7)$$

A gamma-ray source emits radiation in all directions, so that the factor $4\pi r^2$ is the total area in which the radiation is emitted. The factor 6.77×10^4 represents the energy in million electron volts to ionize 1 cm³ of air.

TABLE 7-2. COMPARATIVE GAMMA-RAY OUTPUT FROM 1-c SOURCES
(Assuming no self-absorption)

Isotope	Half-life	Approx r/hr/1 m
Au^{198}..............................	2.7 d	0.22
I^{131}..............................	8.0 d	0.24
Cs^{137}..............................	37 y	0.36
Ta^{182}..............................	117 d	0.61
Ra^{226} (0.5-mm Pt filter)........	1,620 y	0.84
Co^{60}..............................	5.3 y	1.30

Neutron Irradiation. Artificial radioactives are made by placing the material in a nuclear reactor for a period of time. The neutrons bombarding the material cause a transmutation of the material into a radioactive form. The basic equation used in calculating the specific activity of isotopes produced in a nuclear reactor is

$$S = \frac{1.64\sigma 10^{-11}F}{A}\left[1 - \exp\left(\frac{-0.693t}{T}\right)\right] \tag{7-8}$$

where S = specific activity (c/g) of isotope on removal from reactor
    σ = activation cross section of material being irradiated, barns
        (1 barn = 10^{-24} cm^2)
    F = neutron flux in reactor per cm^2 per sec
    A = atomic number of material being irradiated
    t = time in reactor
    T = half-life of isotope produced

TABLE 7-3. REACTOR IRRADIATION FACILITIES FOR RADIOISOTOPE PRODUCTION

Reactor	Location	Approx flux	Sample size
Graphite reactor	Oak Ridge National Laboratory	1×10^{11}–7×10^{11}	$\frac{3}{4}''$ diam × $2\frac{7}{8}''$ $3\frac{1}{2}'' \times 3\frac{1}{2}'' \times 10''$
Brookhaven reactor	Brookhaven National Laboratory	10^{11}–4×10^{12}	$\frac{3}{4}''$ diam × $2\frac{1}{8}''$ $12'' \times 12'' \times 24''$
Low intensity test reactor, LITR	Oak Ridge National Laboratory	10^{13}	$\frac{3}{8}''$ diam × $1\frac{3}{4}''$
Materials testing reactor, MTR	National Reactor Testing Station, Idaho	5×10^{13}	Depends on available "rabbit" or hole
Argonne research reactor, CP-5	Argonne National Laboratory	5×10^{12}	

TABLE 7-4. ACTIVATION CROSS SECTIONS

Radioisotope	Cross section, barns
Cs^{134}	26
Co^{60}	34
Ir^{194}	130
Ta^{182}	21
W^{187}	40
Eu^{152}	5,500
Eu^{154}	420

Table 7-3 gives the neutron flux available in various reactors, such as the Materials Testing Reactor, Chalk River, and Brookhaven National Laboratory which are used for producing isotopes. Table 7-4 gives the activation cross section for various materials.

If 1 g of Na^{23} is irradiated at a neutron flux of 5×10^{11} neutrons/$(cm^2)(sec)$ for a 7-day period, the specific activity of the Na^{24} produced ($\sigma = 0.6$ barn) is

$$S = \frac{(1.64 \times 10^{-11})(5 \times 10^{11} \times 0.6)}{23} \left\{ 1 - \exp \left[\frac{-0.693(7)}{15.06/24} \right] \right\}$$

$$S = 0.21 \text{ c/g}$$

The mass of radioactive material m produced per gram of material radiated can be calculated from the expression

$$m = 7.7 \times 10^{-9} SAT \qquad (7\text{-}9)$$

where m = mass in grams of radioactive material
A = atomic weight of isotope produced
S = specific activity of isotope, c/g
T = half-life in days of isotope formed

Absorption. In their interaction with matter, gamma rays behave essentially like X rays. The expression, Eq. (6-8), given for the absorption X rays also holds for the absorption gamma rays. The reader is referred to Chap. 6 for discussion of the interaction of gamma rays with matter. The half-value thickness is the thickness of absorber required to decrease the intensity of a beam to one-half the incident intensity. The half-value thickness of lead can be calculated from the equation

$$\frac{I}{I_0} = 0.5 = \exp \left[(-0.56)(cm^{-1})x \right]$$

$$x = 1.25 \text{ cm} = 0.5 \text{ in. for } E = 1.5 \text{ Mev} \qquad (7\text{-}10)$$

The half-value thicknesses for several materials and radioisotopes are given in the following table.

TABLE 7-5. HALF-VALUE THICKNESS (inches)

Material	Co[60]	Cs[137]	Ir[192]
Lead	0.49	0.25	0.19
Steel................	0.87	0.68	0.44
Aluminum...........	2.2	1.6	1.2
Concrete............	2.7	2.1	1.9
Water...............	5.3	3.6	3.2

Films for Gamma-ray Radiography. Despite the differences in quality of radiation from gamma-ray and X-ray sources, the same types of film and intensifying screens are applicable. Likewise, the processing of film follows the same procedure as that used in X-ray radiography. Gamma rays can also be detected by the same methods as described in the preceding chapter on X rays. The factors involved in making a useful radiograph with gamma rays are the same as those involved when using X rays. These factors were discussed in detail in Chap. 6. In general, gamma-ray radiographs have lower contrast than radiographs produced by low- and medium-voltage X rays. Because of this relatively low contrast, gamma-ray radiographs have wide latitude. This permits their use over a wider range of thickness than is possible with low- and medium-voltage X rays.

The intensity of gamma-ray sources is small compared with an X-ray tube. Consequently much longer exposures are required. The exposure time can be reduced by using a faster film with lower contrast. In general, gamma-ray radiographs have low contrast so that the use of faster film does not produce the desired results.

Gamma-ray Sources. Various gamma-ray sources useful in radiography are individually discussed in the following section.

Radium (Ra[226]) *and Radon* (Rn[222]). The use of radium in industrial radiography began in the late 1920s with the pioneer work of Dr. R. F. Mehl[1,*] at the Naval Research Laboratory. Radium is a natural occurring gamma emitter with a half-life of 1,590 years. Tenny and coworkers[2] list the following gamma radiation from radium: 0.24, 0.29, 0.34, 0.60, 1.12, 1.76, and 2.19 Mev. The most intense energies are at 0.6, 1.12, and 1.76 Mev. The average energy of the gamma emission from radium is about 1.7 Mev. The gamma rays from radium are approximately equivalent in penetrating power to the radiation produced by a 2-million-volt X-ray machine. A 1-c source of radium has an output of 0.84 rhm. Radium is useful for the radiography of steel sections from 2 to 6 in. The half-value layer for the radiation from radium is as follows: lead, 0.51 in.; iron, 0.90 in.; and concrete, 3 in. The gamma

* Superscript numbers indicate Specific References listed at the end of the chapter.

rays emitted by radium and radon are identical. As an alternate to radium, radon may be used, but radon has a half-life of 3.85 days. Consequently, corrections must be applied to compensate for the decrease in intensity of the source.

Cobalt (Co[60]).[2] Co[60] is prepared by the slow neutron bombardment of pure cobalt. Co[60] has a half-life of 5.3 years. Figure 7-2 shows the decay curve for Co[60] calculated using Eq. (7-3). For industrial radiography the correction to the source strength at intervals of 6 months is probably sufficiently accurate. The specific activity of Co[60] is approximately 150 c/g. The gamma rays emitted by Co[60] have energies of 1.17 and 1.33 Mev. This corresponds closely to the effective energy of a 2-million-volt X-ray machine. One curie of Co[60] may be considered equivalent to 1.5 of radium. Co[60] can be used to radiograph steel in the range from 2 to 6 in.

Iridium (Ir[192]).[3] Ir[192] is produced by irradiation of iridium metal with neutrons in a nuclear reactor. The high absorption of thermal neutrons by iridium facilitates the rapid preparation of sources with high specific activity. It is commercially available in specific activities of over 300 c/g. This permits strong sources with very small effective focal spot sizes. The usual size of the source is $\frac{1}{8}$ in. diameter by $\frac{1}{8}$ in. long. The half-life of iridium is 74.4 days. Figure 7-3 shows the decay curve for iridium. Ir[192] decays with the emission of some 20 known gamma rays ranging in energy from 136 to 1,157 kv. The principal energies are 310, 470, and 600 kv. The radiation from Ir[192] is roughly equivalent to an X-ray tube operating at 800 to 900 kv peak.

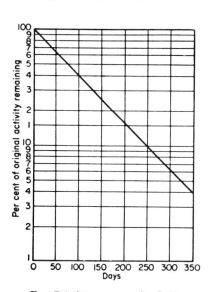

FIG. 7-3. Decay curve for Ir[192].

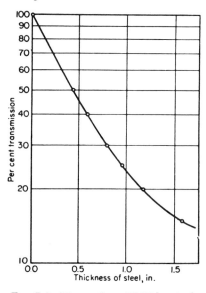

FIG. 7-4. Attenuation of Ir[192] by steel.

TABLE 7-6. THICKNESS LIMITS FOR Ir[192]

Material	Lower range, in.	Upper range, in.
Aluminum	$\frac{3}{8}$	10
Brass	$\frac{1}{8}$	5
Bronze	$\frac{1}{8}$	5
Monel	$\frac{1}{8}$	5
Steel	$\frac{1}{2}$	$2\frac{1}{2}$
Titanium	$\frac{1}{4}$	8
Zirconium	$\frac{1}{4}$	8

Figure 7-4 shows an attenuation curve for Ir[192]. This curve changes slope with increasing thickness of steel because of the gradual absorption of the less energetic components of the radiation. The upper and lower range of thickness limits for various materials with 2 per cent sensitivity is given in Table 7-6. Garrett[3] has discussed the use of Ir[192] for fluoroscopy.

For thin sections better radiographic sensitivity can be obtained with Ir[192] than with Co[60], but for thickness greater than 1 in. the sensitivity should be about the same. Figure 7-5 gives an exposure curve for Ir[192]. Table 7-7 compares the sensitivity of Ir[192] and other radiations for various steel thicknesses.

Cesium (Cs[137]). Cs[137] is one of the by-products of the fission process. A mixture containing approximately equal parts of Cs[133], Cs[135], and Cs[137] is obtained. Cs[133] is not radioactive, and Cs[137] emits only beta particles. Cs[137] decays with a half-life of 37 years to an isomeric state of Ba[137]. The gamma radiation from a cesium source originates from the decay of this barium isotope, and a gamma ray of 0.667-Mev energy is emitted. By enclosing the cesium in a container of sufficient thickness to stop the beta particles, a monochromatic source of 0.667-Mev gamma radiation is obtained. A Cs[137] source is roughly equivalent to a million-volt X-ray machine. The output of a Cs[137] source has been calculated to be 0.36 rhm/c. Cs[137] is a suitable source for the radiography of steel $1\frac{1}{2}$ to $2\frac{1}{2}$ in. thick.

Table 7-8 lists the resolution obtained by Dutli and Taylor[4] using Cs[137] for aluminum and steel. Resolution is

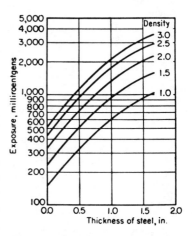

FIG. 7-5. Exposure curve for Ir[192]. Kodak no-screen film, developed 8 min at 68°F, with Kodak rapid X-ray developer; 10-mil lead front screen; 15-mil lead back screen.

TABLE 7-7. SENSITIVITY OF Ir[192]

Radiation	Steel thickness, in.	Sensitivity (observed)
Ir[192]	0.5	1.6
Ir[192]	1.0	0.8
Ir[192]	1.5	0.5
Ir[192]	1.75	0.9
Ir[192]	2.5	0.6
200 kv	1.0	0.5
200 kv	1.75	0.6
400 kv	2.0	0.7

TABLE 7-8. RESOLUTION USING Cs[137]

Aluminum		Steel	
Thickness, in.	Resolution, %	Thickness, in.	Resolution, %
1.6	3.7	0.5	8
3.0	2.6	1.0	4
4.6	2.6	1.5	3.3
		2.0	3
		2.5	2.4

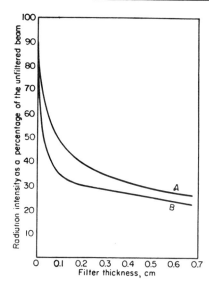

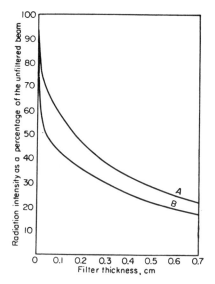

FIG. 7-6. Attenuation of thulium radiation by copper. *A* source, Tm, 4 by 4 mm; *B* source, Tm[81], 2 by 2 mm.

FIG. 7-7. Attenuation of thulium radiation by aluminum. *A* source, Tm, 4 by 4 mm; *B* source, Tm[81], 2 by 2 mm.

defined here as the visualization of the radiographic image of a cylindrical hole with equal diameter and height.

Thulium (Tm^{170}).[11,12] Tm^{170} decays by two beta emissions accompanied by a 0.084-Mev gamma ray. A 52.2-kev, Yb X ray is also emitted. Thulium has a half-life of 129 days. The output of a thulium source is calculated to be 45×10^{-4} rhm for 1 c. Figures 7-6 and 7-7 show the attenuation of the radiation from thulium by aluminum and copper.

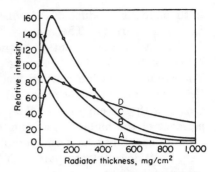

FIG. 7-8. Effect of source size on bremsstrahlung from thulium. *A*, Tm^{170}, 84 kev; *B*, YbK X rays; *C*, Tm K X rays + bremsstrahlung 30 to 85 kev; *D*, brehmsstrahlung 85 kev.

In view of the low penetrating power of beta particles it might appear that the only effective radiation were the 84-kev gamma ray and the Yb X ray. This is not the case, because the beta particles interact with the thulium atoms to produce X rays. This radiation known as "Bremsstrahlung" is present in Tm^{170} sources with energies of 500 kev. The relations in intensities between the 84-kev gamma radiation and the bremsstrahlung are influenced by the source size. The graph in Fig. 7-8, although only schematic, illustrates variations due to source size.

On steel specimens less than ½ in. thick Tm^{170} can be used to obtain better-quality radiographs than with any other isotope available at present. For thicknesses greater than ½ in. the sensitivity available becomes comparable to that obtained with Ir^{192}. Table 7-9 shows the sensitivity obtained for various thicknesses of steel. Tm^{170} can be used for the radiography of aluminum in the range from ⅛ to 2 in.

Europium (Eu^{155}). Eu^{155} decays with the emission of a number of soft gamma rays. The highest energy reported is 136.8 kv. The two principal gamma-ray energies are 100 (40 per cent) and 85 (60 per cent) kv, which gives an average energy of approximately 100 kv. The half-life of Eu^{155} is 1.7 years. The two principal gamma rays are emitted in cascade, which gives a value of the total disintegration energy of approximately 200 kev. Two methods can be used to procure europium: neutron bombardment of samarium and recovery from fission products.

TABLE 7-9. SENSITIVITY USING Tm^{170}

Steel thickness, in.	Calculated sensitivity
0.1	1.9
0.2	1.3
0.4	1.1
0.5	1.2

Both methods produce Eu^{154} which has a 16-year half-life and emits hard gamma rays. The energy of the Eu^{155} is comparable to that of Tm^{170}, but has the advantage of the longer half-life.

Cerium (Ce^{144}). Ce^{144} is a fission product with a 290-day half-life. The gamma radiation emitted has an energy of approximately 135 kv. Ce^{144} gives an output of about 60 rhm/c. Ce^{144} decays to Pr^{144} which decays with the emission of 2.2-, 1.5-, and 0.7-Mev gamma rays. The presence of the high-energy gamma rays from Pr^{144} seriously interferes with its capability as a radiographic source. From the standpoint of cost, Ce^{144} does not compare with Tm^{170} for use with thin sections and light metals.

Americium (Am^{241}). Am^{241} which has a half-life of 470 years is formed by the neutron irradiation of plutonium. A 59-kev gamma ray is emitted with some internal conversion X rays of energies down to 14 kev. Attenuation measurements give a linear absorption coefficient of $\mu = 24.8$ in.$^{-1}$ for steel up to 0.15 in. in thickness and $\mu = 1.92$ in.$^{-1}$ for aluminum up to 1.6 in. in thickness.

Xenon (Xe^{133}). Xe^{133} has a half-life of 5.3 days and emits a gamma ray of 0.081 Mev.

Tantalum (Ta^{182}). Ta^{182} is used as a substitute for Co^{60}. It is produced by neutron capture in a nuclear reactor. It has a half-life of 120 days and a specific activity of 48 c/g. The gamma-ray spectrum is quite complex, with over 40 components present. The absorption characteristics of Ta^{182} are very similar to those of Co^{60}. The half-value thickness for lead is 0.05 in.

Antimony (Sb^{124}). Sb^{124} has a half-life of 60 days. The spectrum includes energies of 2.0 and 1.7 Mev, so that it has advantages over Co^{60} and radium for the radiography of thick steel sections. The low specific activity of 5 c/g is a serious disadvantage.

A comparison of the various gamma-ray sources discussed here is given in Table 7-10.

Within the United States the procurement, delivery, possession, use, transfer, and disposal of radioactive isotopes are governed by Federal regulations except for certain very low-level sources. To obtain radioactive material it is first necessary for the prospective user to obtain the proper license from the United States Atomic Energy Commission. The customer must provide the radioisotope supplier with a purchase order on the approved Atomic Energy Commission Purchase Order Form. The Atomic Energy Commission will make periodic checks to determine if proper safety precautions and safety regulations are being observed. Radioactive materials may be shipped to destinations outside the United States without prior approval from the United States Atomic Energy Commission provided that the shipment is made directly to the destination.

TABLE 7-10. HARD GAMMA EMITTERS

Radioisotope	Half-life*	Energy,† mev
Na²⁴	15.06 h	2.754
Ga⁷²	14.3 h	2.51
La¹⁴⁰	40 h	2.50
Ir¹⁹⁴	19 h	2.1
Sb¹²⁴	60 d	2.04
As⁷⁶	26.8 h	1.7
Pr¹⁴²	19.2 h	1.59
Ag¹¹⁰ᵐ	270 d	1.516
K⁴²	12.44 h	1.51
Eu¹⁵²,¹⁵⁴	13 y; 16 y	1.40
Co⁶⁰	5.27 y	1.33
Br⁸²	35.87 h	1.312
Fe⁵⁹	45.1 d	1.289
Ta¹⁸²	115 d	1.223
Sc⁴⁶	85 d	1.12
Zn⁶⁵	250 d	1.12
Rb⁸⁶	19.5 d	1.08
Rh¹⁰⁶ (Ru¹⁰⁶)	30 s	1.045
Cs¹³⁴	2.3 y	0.794
W¹⁸⁷	24.1 h	0.78
Zr⁹⁵	65 d	0.754
Cb⁹⁵	35 d	0.745
Ba¹³⁷ᵐ (Cs¹³⁷)	2.6 m	0.662

* h, hour; d, day; y, year; m, minute; s, second.

† The energy given in each case is the maximum gamma energy which occurs to an extent of 5 per cent or more.

Personnel Protection. There is considerable interest regarding the possible danger from the use of X rays and gamma emitting isotopes. Actually, X rays and radioactive materials involve no worse hazards than do some other industrial processes. If proper precautions are exercised, X-ray equipment and radioactive materials can be used for industrial radiography without danger to personnel. Radiation need not be feared when properly understood.

Radiation protection has been the object of much study in recent years, and protection against external sources is now fairly well understood. The problem of protection against internal radiation sources is much less understood. Fortunately, the radiographer is concerned entirely with external radiation sources. Early workers with X rays noticed sunburnlike changes on the skin which had been exposed repeatedly to X rays. These X-ray burns frequently did not heal and often led to severe and painful malignant changes of the skin and, eventually, to death. In addition to the superficial skin injuries, damaging effects upon the blood and blood-forming organs were observed which often led to severe anemias and leukemias. Damage to the reproductive organs was reported which led to impaired fertility and often sterility.

Damage from exposure to X rays and gamma rays is of two types, local and general. Local damage, which usually takes the form of burns, is due to excessive exposure to radiation. Nothing is felt or noticed at the time, but serious results may appear in a short time. To prevent such burns and their consequent effects, the hands and all other parts of the body must be kept well away from unshielded or improperly shielded sources. Irradiation of the body, if of too great an amount, causes anemia due to damage to the bone marrow in which the blood corpuscles are formed. This anemia gives rise to a general state of ill health and difficulty in the healing of wounds, and may prove fatal.

X rays and gamma rays are called ionizing radiations because they produce ion pairs in the air and tissue which they traverse. As far as is known today, all biological effects in body tissues are a direct consequence of the ionization produced. There is a continuous repair activity going on in irradiated cells. Permanent harm is done only if the cells are destroyed at a greater rate than they are repaired.

In evaluating possible dangers from radiation, other factors in addition to the specific ionization of the radiation must be considered. The rate of application of radiation is important. The shorter the time of application of a given dose of radiation, the more severe the biological effect. A dose of 400 r of X rays or gamma rays applied within a few minutes over the total body would kill one-half of the population. But 400 r applied over a period of 25 years would produce no detectable damage.

Another factor is the volume of tissue which is irradiated. For therapeutic purposes, many times 400 roentgens are applied through a volume of about 8 by 8 by 8 in. without producing permanent changes. Radiation effects are cumulative, and the body tissues vary widely in their radiation sensitivity. Gonads, lymphocytes, bone marrow, and the gastrointestinal tract are more sensitive than muscle and nerve tissue. Today, despite the vast quantities of radioactive materials handled, the *loss* of working hours due to radiation damage is practically zero.

Permissible Exposure. The permissible exposure is the total amount of radiation that a person may receive continuously without suffering any detectable damage. The permissible dosage rate for X rays and gamma rays in controlled areas[5] is now set at 0.3 roentgen (r) or 300 milliroentgens (mr) per week. This is *not* a tolerance dose since it is not absolutely known that the body is so constructed as to tolerate any radiation. However, up to the present, such levels of radiation have produced no ill effects in large numbers of persons. The permissible dosage rate is based on the total body irradiation. The exposure may

be distributed uniformly over the working time in a
week but may be obtained in larger short time exposures provided that the total does not exceed 0.1 r in
any one week.   The value of 0.1 r/week corresponds to
0.05 r/day (48-hr week), or 6.25 mr/hr, or 0.1 mr/min.

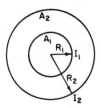

Factors in Protection.   Distance and shielding are
the two most important factors in the protection of
personnel from X-ray and gamma radiation.   In any
radiation shielding problem, there are six basic
principles which should be considered:

FIG. 7-9. Radiation
field surrounding
point source.

1. The major portion of the protection should be as near as possible to
the tube or gamma-ray source.

2. Wherever feasible, the direction of the useful beam should be
restricted.

3. Advantage should be taken of the inverse square law to reduce the
intensity of the radiation.

4. Advantage should be taken of any architectural features of the
building.

5. Radiation is reduced in penetration and in intensity upon each
scattering.

6. Radiation is attenuated in ordinary air.

When radioactive material is confined to a point source there is equal
probability that equal amounts of radiation will go in all directions.   It
follows that, as the radiations proceed from the source, the amount per
unit area will progressively decrease.   With a point source of radiation
as shown in Fig. 7-9 the intensity I at any point on the periphery of
sphere A_1 can be determined as follows:

$$I_1 = \frac{D}{4\pi(R_1)^2} \tag{7-11}$$

where R_1 is the radius of the sphere A_1 and D is the number of disintegrations per second of the source.   Considering the larger sphere A_2, the
intensity I_2 at any point on the periphery can be determined as above;
thus

$$I_2 = \frac{D}{4\pi(R_2)^2} \tag{7-12}$$

where R_2 is the radius of the sphere A_2.   These two expressions can be
combined to give

$$\frac{I_1}{I_2} = \frac{R_2^2}{R_1^2} \tag{7-13}$$

the inverse square law mentioned in Chap. 6.

Dosage Rate. The following expression[6] can be used to calculate the dosage rates for point sources emitting gamma rays. It gives a good approximation in the range from 0.5 to 2.0 Mev and a fair approximation in the range from 0.3 to 3.0 Mev.

$$R_f = 5.60CE \qquad (7\text{-}14)$$

where R_f = roentgens per hour at 1 ft
C = activity of source, c
E = total energy of gamma radiation
For a 1-c source of Co⁶⁰ this expression gives (Co⁶⁰ emits two gamma rays having energies of 1.17 and 1.33 Mev, respectively)

$$R_f = 5.60(1.17 + 1.33)(1) = 14.0$$
$$14r/c, \text{ hr at 1 ft}$$

If shielding and scattering are negligible, the dosage rate R at any other distance D is given by

$$R = \frac{R_f}{D^2} \qquad (7\text{-}15)$$

For determining the dosage rate from radium filtered through 0.5 mm of platinum, the following formula is more suitable:

$$S = \frac{8.4M}{d^2} \qquad (7\text{-}16)$$

where S = source strength, r/hr
M = mass of radium, mg
d = distance in centimeters from source to point of measurement
Table 7-11 gives the dosage rate at 1 ft/c for a number of gamma energies.

TABLE 7-11. DOSAGE RATE

E, Mev	R_f per curie (r/hr at 1 ft)
0.1	0.5
0.2	1.1
0.3	1.8
0.5	3.1
0.8	5.0
1.0	6.0
1.5	8.2
2.0	10.0
2.5	11.6
3.0	13.0
4.0	15.0

The National Bureau of Standards Handbook 54[7] gives the following expression for the danger range, the distance from an unshielded source of radioactive material at which the gamma radiation is 6.25 mr/hr.

$$\text{Danger range (cm)} = \sqrt{\frac{R \times M}{0.00625}} \qquad (7\text{-}17)$$

where R = roentgens per millicurie-hour at 1 cm from a source
$\qquad M$ = source strength, mc

Given a 100-mc source of Co^{60}, and using the inverse square law, and the value of 1.30 rhm calculated previously for 1 c of Co^{60}, the danger range is found to be

$$\frac{1.30 \times 10^{-3}}{R} = \frac{(1)^2}{(100)^2}$$

$$R = 13$$

$$\text{Danger range} = \sqrt{\frac{13 \times 100}{0.00625}}$$

$$= 432.5 \text{ cm} = 14.4 \text{ ft}$$

Using Eq. (7-14) and the maximum permissible dose per week as 0.3 r, the time which a person can be permitted to remain at different distances from cobalt and iridium sources of various strengths is given by

$$T = \frac{1.3D^2}{SC} \qquad \text{for cobalt}$$

$$T = \frac{6.3D^2}{SC} \qquad \text{for iridium}$$

where T = time, min/week
$\qquad D$ = distance, ft
$\qquad S$ = fraction of radiation transmitted by the shielding, from Figs. 7-10 to 7-12
$\qquad C$ = strength of source, c

For the present discussion, radiation will be divided into three categories: primary, scattered, and secondary radiation. The primary radiation is the radiation coming directly from the source and includes the useful beam. The useful radiation is that radiation which passes through the aperture or collimator of the source enclosure. The leakage radiation is that coming from the source, excluding the useful beam. The scattered radiation is the radiation that has been deviated in direction and usually has its energy diminished. The secondary radiation is the radiation which originates in the irradiated material.

Primary Radiation. Equation (6-8) for the attenuation of radiation by an absorber is often referred to as narrow beam absorption. The shielding problems actually encountered usually fail to meet the ideal conditions of narrow beam attenuation. The sources of radiation may not approximate points, and broad beam, wide, thick absorbers

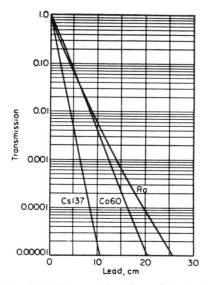

FIG. 7-10. Transmission through lead of gamma rays from radium, Co[60], and Cs[137]. (*National Bureau of Standards Handbook No. 54.*)

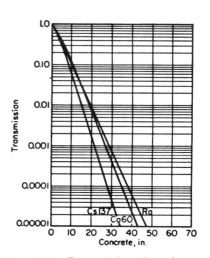

FIG. 7-11. Transmission through concrete (specific gravity 147 lb/ft³) of gamma rays from radium, Co[60], and Cs[137]. (*National Bureau of Standards Handbook No. 54.*)

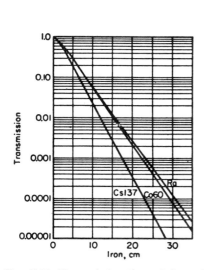

FIG. 7-12. Transmission through iron of gamma rays from radium, Co[60], and Cs[137]. (*National Bureau of Standards Handbook No. 54.*)

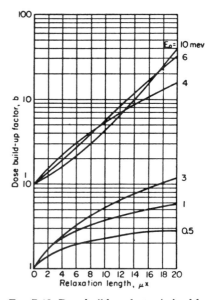

FIG. 7-13. Dose build-up factor in lead for a point isotropic source.

TABLE 7-12. COMPARATIVE THICKNESS TO GIVE EQUIVALENT ABSORPTION

E, Mev	Pb	Fe	Al	H_2O
0.2	1	4.7	15	36
0.5	1	2.7	7.4	19
1.0	1	1.75	4.8	11.5
1.5	1	1.4	4.1	10
2.0	1	1.5	4.3	10
2.5	1	1.5	4.8	11
3.0	1	1.6	5.0	12
4.0	1	1.8	5.8	15
5.0	1	2.0	6.5	16

are usually involved. In such cases X rays or gamma rays may be scattered into the detector as well as away from it. A correction must be made to allow for the effect of the scattered radiation. Equation (6-8) now becomes $I = bI_0 \exp(-\mu x)$, where b is the build-up factor. The value of the build-up factor depends upon the absorber material, the energy of the radiation, and the arrangement of the source, absorber, and detector. Figure 7-13 shows the build-up factor for lead.

Shielding calculations for a geometrically small source are based on shields that surround the source. These shields have the same shielding value when placed at any point between the detector or receptor and the source. The amount of primary radiation escaping in a given direction will depend on the effective shielding thickness in that direction. However, the amount of secondary radiation will increase, especially for large walls. This may introduce errors in the calculated value. If the shield is flat and relatively close to the receptor, an additional thickness of about one-half layer may be required to reduce the dosage rate to the desired level. In practice, it is desirable to add a safety factor of 10 to the calculated radiation value, particularly for stationary shields when weight is of little importance. For example, one would base calculations on a tolerance of 0.625 mr/hr, rather than 6.25 mr/hr or, for simplification, 1 mr/hr may be used. This safety factor will require very little additional shielding material. Figures 7-10 to 7-12 are graphs showing the reduction factor for gamma radiations from Co^{60}, Ra^{226}, and Cs^{137} plotted against shield thickness for various shielding materials. The reduction factor is the dosage rate of the gamma rays entering the shield divided by the dosage rate of the gamma rays

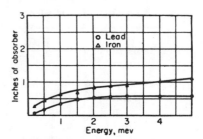

FIG. 7-14. Half-value layers for various photon energies.

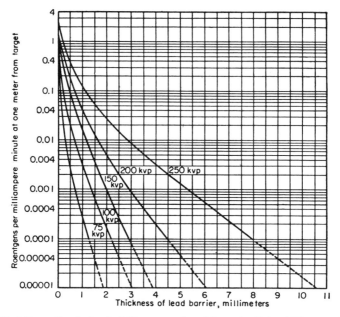

FIG. 7-15. Attenuation in lead of X rays produced by potentials of 75 to 250 kv peak. The curves were obtained with a half-wave generator and with a 90° angle between the electron beam and the axis of the X-ray beam. The filter was 3 mm of aluminum for the 150-, 200-, and 250-kv peak curves and 0.5 mm of aluminum for the 75- and 100-kv peak curves. Direct-current potentials require 10 per cent thicker barriers than for the pulsating potentials given above.

leaving the shield. These shielding data are considered as broad beam data since they include absorption of scattered radiation. Bennett[8] gives a monogram for calculating shielding for Co^{60}.

The comparative thickness of various materials to give equivalent absorption is given in Table 7-12.

The number of half-value thicknesses or layers N to reduce the original intensity I_0 of a radiation source to an intensity I is given by the expression

$$I = I_0 \exp{[-N(0.693)]} \qquad (7\text{-}18)$$

TABLE 7-13. TENTH AND HALF-VALUE THICKNESSES (inches)

Radiographic source	Lead		Iron		Concrete*	
	$\frac{1}{10}$	$\frac{1}{2}$	$\frac{1}{10}$	$\frac{1}{2}$	$\frac{1}{10}$	$\frac{1}{2}$
Co^{60}	1.62	0.49	2.90	0.87	9.0	2.7
Ra^{226}	1.85	0.56	3.03	0.91	9.6	2.9
Cs^{137}	0.84	0.25	2.25	0.68	7.1	2.1
Ir^{192}	0.64	0.19	1.5	0.44	6.2	1.9

* Density of concrete assumed to be 147 lb/ft³.

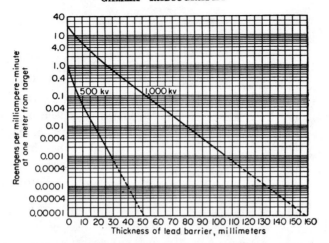

FIG. 7-16. Attenuation in lead of X rays produced by potentials of 500 and 1,000 kv. The curves were obtained with a d-c generator and with an angle of 0°. The inherent filtration was 2.8 mm of tungsten, 2.8 mm of copper, 2.1 mm of brass, and 18.7 mm of water. (*National Bureau of Standards Handbook No. 50.*)

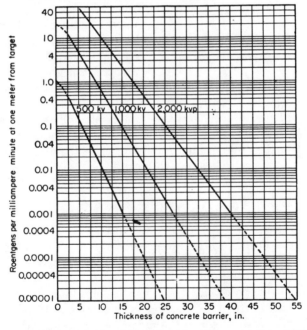

FIG. 7-17. Attenuation in concrete of X rays produced by potentials of 500, 1,000, and 2,000 kv peak. The 500- and 1,000-kv peak curves were obtained with a d-c generator; the 2,000 kv peak was obtained with a resonance generator. All data were obtained with an angle of 0° between the electron beam and the axis of the X-ray beam. The filter at 500 and 1,000 kv was 2.8 mm of tungsten, 2.8 mm of copper, 2.1 mm of brass, and 18.7 mm of water. At 2,000 kvp the filter was 1.6 mm of tungsten, 5.1 mm of copper, and 6.8 mm of water. (*National Bureau of Standards Handbook No. 50.*)

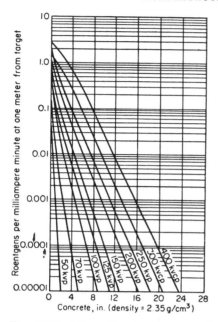

FIG. 7-18. Attenuation in concrete of X rays produced by potentials of 50 to 400 kv. (*National Bureau of Standards Handbook No. 60.*)

which can be written in the more usable form

$$N = -1.44 \log_e I/I_0 = 1.44 \log_e (I_0/I) = 3.32 \log_{10} (I_0/I) \quad (7\text{-}19)$$

Figure 7-14 shows the half-value thicknesses for various materials as a function of photon energy. Table 7-13 gives the tenth and half-value thicknesses for lead, iron, and concrete for various gamma-ray sources.

The ordinate of Figs. 7-15 to 7-18[9] is the dosage rate in roentgens per milliampere tube current at a point 1 m from the target. The abscissa is the thickness and material of the barrier. From these figures the barrier thickness required to reduce the dosage rate to the permissible value of 0.4 mr/min can be calculated. By simple calculations the graphs may be used for other conditions. If the distance is doubled, the distance factor reduces the radiation intensity by 4. If the tube current is increased the dosage rate will be increased by the same factor. The required lead equivalent of the barrier may be obtained from the appropriate curve by using as the ordinate the dosage rate X_u given by the following expression:[9]

$$X_u = 9.7 \frac{D^2}{i} \times 10^{-6} \quad (7\text{-}20)$$

where D = distance in feet between source of radiation and nearest position to be occupied by personnel during exposure

i = tube current, ma

For example, suppose that the closest personnel can get to a target is 4 m. The radiation then will be reduced by a factor of 16, and the barrier needs therefore only to reduce the radiation to 1.6 mr/ma-min at 1 m. From Fig. 7-15 the lead thickness need be only 1.2 mm of lead for 100-kv peak X rays. If the tube current is increased to 10 ma, the barrier must reduce the radiation to 0.01 mr/min which would require 2.5 mm of lead for 100-kv peak rays. Suppose that it is necessary to reduce the radiation to the permissible level at a point 10 ft from a target with a tube current of 10 ma at 150 kv. The value of X_u from Eq. (7-20) is

$$X_u = \frac{(9.7)(10)^2}{10} \times 10^{-6} = 97 \times 10^{-6} \qquad (7\text{-}21)$$

Using as the ordinate $97 \times 10^{-6}(0.000097)$ on the 150-kv curve, the lead barrier thickness is found to be 2.9 mm of lead.

Tables 7-14 to 7-18 show the barrier requirements obtained using Eq. (7-17) and Figs. 7-15 to 7-18.

TABLE 7-14. PRIMARY PROTECTIVE-BARRIER REQUIREMENTS FOR 10 MA AT THE
PULSATING POTENTIALS* AND DISTANCES INDICATED

Target distance, ft	Lead thickness (mm) with peak of				
	75 kv	100 kv	150 kv	200 kv	250 kv
2	2.2	3.4	4.3	6.7	11.8
3	2.0	3.1	4.0	6.2	10.9
5	1.7	2.7	3.6	5.5	9.6
8	1.5	2.4	3.2	4.8	8.5
10	1.3	2.2	3.0	4.5	8.1
15	1.1	1.9	2.6	4.0	7.1
20	1.0	1.7	2.4	3.6	6.4
50	0.5	1.1	1.7	2.4	4.3
Approximate half-value layer measured at high filtration...........	0.18	0.24	0.3	0.5	0.8
Radiation filter (mm Al).........	0.5	0.5	3.0	3.0	3.0

* X-rays excited by d-c potentials require the order of 10 per cent greater thickness than those given here for pulsating potential.

TABLE 7-15. PRIMARY PROTECTIVE-BARRIER REQUIREMENTS FOR 400-KV PEAK
PULSATING POTENTIAL WITH REFLECTION TARGET*

Target distance, ft	Lead thickness (mm) with target current of		
	1 ma	3 ma	5 ma
5	16.5	20	22
8	14.0	17.0	18.5
10	12.5	15.5	17.0
15	11.0	13.5	14.5
20	9.5	11.5	13.0
50	5.5	8.0	9.0
Approximate half-value layer measured at high filtration.........	2.0		

* Radiation filter = 0.4 mm tin + 0.75 mm copper + 2.0 mm aluminum.

TABLE 7-16. PRIMARY PROTECTIVE-BARRIER REQUIREMENTS FOR 500-KV
CONSTANT POTENTIAL WITH TRANSMISSION TARGET*

Target distance, ft	Barrier thicknesses with target current of					
	1 ma		3 ma		5 ma	
	Lead, mm	Concrete, in.	Lead, mm	Concrete, in.	Lead, mm	Concrete, in.
5	36	18.0	42	20.5	44	21.5
8	31	16.0	37	18.5	39	19.5
10	29	15.0	35	17.5	37	18.5
15	25	13.5	31	16.0	33	17.0
20	22	12.5	28	14.5	30	16.0
50	14	8.5	19	11.0	21	12.0
100	8	6.0	13	8.0	15	9.5
Approximate half-value layer measured at high filtration......	...	....	3.0	1.5		

* Radiation filter = 2.8 mm tungsten + 2.8 mm copper + 2.1 mm brass + 18.7 mm water.

The density of this concrete is 147 lb/ft³.

TABLE 7-17. PRIMARY PROTECTIVE-BARRIER REQUIREMENTS FOR 1,000-KV
CONSTANT POTENTIAL WITH TRANSMISSION TARGET*

Target distance, ft	Barrier thicknesses with target current of					
	1 ma		2 ma		3 ma	
	Lead, mm	Concrete, in.	Lead, mm	Concrete, in.	Lead, mm	Concrete, in.
5	123	30.5	131	32.5	136	33.5
8	113	28.0	120	29.5	125	30.5
10	107	27.0	115	28.5	120	29.5
15	97	24.5	105	26.5	110	27.5
20	91	23.0	99	25.0	103	26.0
50	69	18.5	77	20.5	82	21.0
100	53	15.0	61	17.0	66	18.0
Approximate half-value layer measured at high filtration......	...	....	8	1.8		

* Radiation filter = 2.8 mm tungsten + 2.8 mm copper + 2.1 mm brass + 18.7 mm water.

The density of this concrete is 147 lb/ft³.

TABLE 7-18. PRIMARY PROTECTIVE-BARRIER REQUIREMENTS FOR 2,000-KV
PEAK PULSATING POTENTIAL WITH TRANSMISSION TARGET*

Target distance, ft	Concrete thickness (in.) with target current of		
	0.5 ma	1.0 ma	1.5 ma
5	42.5	45.0	46.5
8	39.5	42.0	43.5
10	38.5	40.5	42.0
15	35.5	38.0	39.5
20	34.0	36.0	37.5
50	28.0	30.0	31.5
100	23.5	25.5	27.0
Approximate half-value layer measured at high filtration..........	....	2.3	

* Radiation filter = 1.6 mm tungsten + 5.1 mm copper + 6.8 mm water.
The density of this concrete is 147 lb/ft^3.

Scattered Radiation. Unfortunately, there are little data available on the attenuation of scattered radiation. The intensity and energy of the scattered radiation depend upon a number of factors. These include strength or intensity of the source, distance between source and scattering medium, composition and geometry of scattering medium, angle between scattering medium and central beam, angle of scattering, size of field at scattering medium, and distance between scattering medium and point of measurement. Table 7-19 shows that the 90° scattered radiation does not in general exceed 0.1 per cent of the incident radiation. For scattering angles greater than 90°, the amount scattered is less than 0.1 per cent. For scattering angles less than 90°, the amount scattered is considerably larger than 0.1 per cent. However, most scattering is 90° or larger so that

TABLE 7-19. RADIATION SCATTERED AT 90° TO THE USEFUL BEAM

X-ray tube potential	Field size	Percentage of incident beam scattered
75 kv peak	25 × 18 cm	0.1
80 kv peak......	8-cm diam	0.002
	25-cm diam	0.028
	35-cm diam	0.073
200 kv peak............	15-cm diam	0.04
200 kv peak............	6 × 8 cm	0.034
	10 × 15 cm	0.09
	20 × 20 cm	0.22
1 my peak.............	20 × 20 cm	0.076

for practical conditions the scattered radiation does not exceed 0.1 per cent of the incident radiation. The energy scattered in the forward direction (same direction as the primary radiation) has higher energies than that scattered in the opposite direction, Eq. (6-15). Absorption curves are not available for scattered radiation. However, in practical cases it is customary to use the same absorption curve for the scattered radiation as for the primary beam when the X-ray potential is less than 500 kv and the 500-kv curve for all higher X-ray tube voltages. The dosage rate must be reduced to at least 0.1 mr/min at positions occupied by personnel. The relationship

$$S = \frac{9d^2D^2}{i} \times 10^{-4} \tag{7-22}$$

where S = radiation at 1 m from target, r/(ma)(min)

d = distance between scatter and position occupied by personnel

D = distance between source and scatter, ft

The above relationship applies for X-ray tube potentials of 500 kv or below. For 1,000- and 2,000-kv potentials Eq. (7-14) should be divided by 10 and 60, respectively. The abscissa corresponding to the ordinate value calculated above on the appropriate curve gives the barrier thickness required for protection against scattering.

It is difficult to determine the amount of protection required for the scattered radiation. The amount and energy of the scattered radiation are dependent upon several factors pertinent only to a given installation. The only way to be sure that a radiation hazard does not exist is continually to monitor with radiation detection instruments.

Film Protection. Undeveloped photographic film requires even more protection than do personnel. A total exposure of approximately 0.15 mr of 100-kv peak radiation over a portion of the film may produce undesirable shadows. If film is placed in radiation having a dosage rate of 0.1 mr/min, the film will receive its maximum permissible exposure of 0.15 mr in 1.5 min. Table 7-20, which gives the barrier requirements for

TABLE 7-20. BARRIER REQUIREMENTS FOR PHOTOGRAPHIC FILM PROTECTION
AGAINST 100-KV-PEAK RADIATION

Target distance, ft	Exposure time and lead barrier thickness, mm					
	15 min	30 min	60 min	120 min	8 hr	32 hr
5	2.3	3.6	2.8	3.0	3.5	4.0
10	1.9	2.1	2.3	2.5	3.0	3.5
15	1.6	1.8	2.0	2.2	2.7	3.2
20	1.4	1.6	1.8	2.0	2.5	3.0

Target distance, ft	Barrier, mm	
	10 ma	⅓ ma
5	2.7	1.5
8	2.4	1.2
10	2.2	1.1
15	1.9	0.8
20	1.7	0.7
50	1.1	0.2

personnel protection, may be used to obtain the barrier thickness required to protect films from the useful beam radiation emitted by X-ray equipment. Table 7-21 indicates the values so obtained for an average current of 10 ma and for ⅓ ma. These values are valid for a high-work-load radiographic unit when the useful beam can be pointed directly at the film. Approximately one-half this thickness is required if the beam can never be pointed at the stored film.

Radiation Surveys. In working with radioisotopes, radiation surveys should be made to evaluate protective measures and to establish safe exposure periods. Since the data from such surveys do not indicate cumulative exposures to individual workers, it is necessary that the radiographer wear an integrating radiation meter. Such integrating devices indicate the effectiveness of protective facilities. These radiation detection instruments will be considered in two groups, the rate instruments usually used in radiation surveys and the integrating or cumulative-dose types suitable for personnel monitoring.

Rate Devices. The rate instruments are those which indicate the exposure received per unit time. The Gieger-Müller (GM) and other ionization-type survey instruments are examples of this group. The usual radiation detection range of GM survey instruments is from a fraction of a milliroentgen per hour to 20 mr/hr. Ionization-type survey instruments are routinely available with ranges from approximately 1 mr/hr to many roentgens per hour. An instrument capable of measuring exposure rates from 1 or 2 mr/hr to approximately 5 r/hr is desirable for use in industrial radiography. Survey instruments should be used routinely around radiographic operations to determine that the exposure rate, in occupied or frequented areas, is not excessive. One of the most important precautions in maintaining an adequate health-safety program is a periodic check of survey instrument calibration.

A simple procedure is indicated below for calibrating a gamma-ray survey meter. Instruments should be calibrated in an area free of metal objects and furniture. All radioactive sources except the source used for calibration should be removed from the calibration area. The instrument to be calibrated should be supported so that its sensitive volume can be properly oriented with respect to the source. The distance between source and instrument should be large compared with the dimensions of the instrument chamber. Primary standard sources are available from the National Bureau of Standards. Secondary standards are sources which have been calibrated in terms of the primary standards. Secondary standards are more readily available and are satisfactory for most calibration purposes.

Integrating Devices. Integrating instruments indicate the total radiation dose received in any given period of time. For work with gamma-ray emitters the most commonly used integrating devices for personnel monitoring are: (1) ionization pocket meters, (2) pocket electroscopes (dosimeters), and (3) film badges. Each type of device has its own advantages and limitations. However, in most field radiographic operations and in unusual circumstances sometimes encountered in plant radiography, it is advisable that the radiographer wear both the pocket chamber or dosimeter and the film badge. The film badge is useful for recording the total exposure received in a week, whereas the dosimeter will enable the radiographer to determine his exposure at any time during radiographic operations. This will enable him to be certain he does not exceed the maximum permissible daily exposure. Film badges are also frequently used for measuring the dose of radiation received by personnel. The developed film shows various degrees of blackening, depending upon the amount of radiation absorbed. Film badges are capable of measuring a wide range of exposures and provide an excellent record of such exposures. Since films are usually worn for a week before being developed and read, it is difficult to relate exposure from any one particular operation to the total reading. They are not, therefore, suited for indicating the need for immediate corrective measures. Film badges will show total dosage received within the limits of approximately 30 mr to many roentgens.

Handling and Storage. Special devices in which the sources are permanently fixed are a practical means of handling radiographic sources either in plant or field

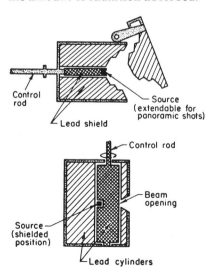

FIG. 7-19. Isotope cameras.

radiography. Such radiography units, Fig. 7-19, are designed so that the gamma radiation may be directed in the form of a beam. Directional beam devices capable of safely housing multicurie sources are available commercially. It is recommended that such devices be used when the activity of the source is greater than 1 c of Co^{60} or its equivalent.

Figure 7-20 shows the Isoscope developed by the Babcock & Wilcox Company. The Isoscope has a 1,008-c source of Co^{60}. This particular unit makes use of a lead rotor turning 180° inside a lead cylinder. Figure 7-21 is a top and side view of the Isoscope which shows the Hevimet which gives minimum shielding of 8¾ in. lead equivalent. Hevimet, made by alloying tungsten (90 per cent) with nickel (6 per cent) and copper (4 per cent), has a density of about 17.0 g/cm³. Malloy 1000 metal is a similar high-density homogeneous alloy. The source in the "on" position is allowing a beam of radiation to escape. By rotating the inner rotor 180° the source would be in the "off" position, where it would have a minimum of 8¾ in. of lead shielding it. This source is comprised of 22 disks of cobalt 1 cm in diameter by 2 mm in thickness; the total activity of the source is 1,008 c. Because of self-absorption there is only 464 effective curies or an emission at 1 m of 660 r/hr. The lead sphere which is encased in a steel jacket is mounted on a modified electric platform lift truck for ease of movement; it also has an angulating mechanism as well as a turn-

Fig. 7-20. Isoscope. (*Courtesy Babcock & Wilcox Company.*)

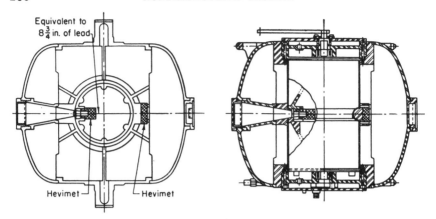

Equivalent to
$8\frac{3}{4}$ in. of lead

Hevimet   Hevimet

Fig. 7-21. Isoscope.   Top and side view.   (*Courtesy Babcock & Wilcox Company.*)

table base which permits easy positioning of the unit so as to beam the radiation in the desired direction.

The special containers used for shipping isotopes are designed to reduce the radiation intensities at their surfaces to safe predetermined levels. Beta-ray emitters are shipped in concrete containers. Gamma-ray emitters are shipped in heavy lead containers. It is suggested that gamma-ray emitters be stored in lead (or cast-iron) pots patterned after the shipping containers used for the respected isotopes. The pots may be stored on the floor or on a bench in an unused portion of the laboratory. An additional shielding barricade will not be required.

Transportation. Transportation of radiographic sources must be in accordance with the shipping regulations pertaining to the carrier involved.[10] Sources shipped by rail or commercial motor vehicle are subject to Interstate Commerce Commission regulations. These regulations are enforced by the Bureau of Explosives of the Association of American Railroads. Shipment by air is controlled by Civil Air Regulations, whereas transportation by water comes under the jurisdiction of the United States Coast Guard. The United States Post Office Department has regulations covering shipment of radioactive materials by the postal service. After the radioactive material has been placed in a shipping container, the container must be checked for outside radiation. The ICC regulations specify that there must be no more than 200 mr/hr of radiation on the surface of the container and no more than 10 mr/hr at 1 m from the center of the package.

Recommendations for the Safe Use of Radioactive Material in Radiography. Close observance of the following rules is recommended.

1. Only trained personnel should be permitted to use radioactive sources.

2. Radioactive sources should be used, stored, and transported in a manner so that the total radiation exposure to any part of a person's body does not exceed 0.1 r/week.

3. All areas where sources are used should be roped off unless other barriers exist. Roped-off areas should be of such dimensions that the radiation intensity at the boundaries thereof does not exceed 6.25 mr/hr.

4. In order to reduce personnel exposure to a minimum, all radiographic procedures, including placement of the film, should be completed before the source is put into position for the film exposure.

5. Long tongs or forceps should be used in handling unshielded radioactive sources.

6. Periodic radiation survey measurements should be made of all areas where persons might be exposed to radiation during use or storage of radioactive material.

7. Employees working with radioactive material should be required to wear film badges or pocket ionization chambers, and adequate records of personnel exposure should be maintained.

8. Radiation monitoring instruments should always be maintained in proper calibration.

9. Sources stored outside buildings or in unattended areas should be kept in locked and adequately shielded containers.

10. Radiation warning signs should be displayed prominently in all areas where radioactive material is stored or used.

GENERAL REFERENCES

American Society for Metals: "Metals Handbook," section on Radiography of Metals, pp. 141–145, American Society for Metals, Cleveland, Ohio, 1948.

American Society of Testing Materials: Symposium on Radiography, 1943.

Clark, George L.: "Applied X-Rays," McGraw-Hill Book Company, Inc., New York, 1955.

Clauser, H. R.: "Practical Radiography for Industry," Reinhold Publishing Corporation, 1952.

Crowther, J. A.: "Handbook of Industrial Radiography," Edward Arnold & Co. London, 1949.

Eastman Kodak Company: "Radiography in Modern Industry," Eastman Kodak Company, Rochester, N.Y., 1957.

Eldorado Mining and Refining Ltd., "Handbook on Radiography," Eldorado Mining and Refining Ltd., Ottawa, Canada, 1944.

Sproull, Wayne T.: "X-Rays in Practice," McGraw-Hill Book Company, Inc., New York, 1946.

St. John, Arcel, and Herbert R. Isenburger: "Industrial Radiology," John Wiley & Sons, Inc., New York, 1943.

Wiltshire, W. J.: "A Further Handbook of Industrial Radiography," Edward Arnold & Co., London, 1957.

Zmeskal, Otto: "Radiographic Inspection of Metals," Harper & Brothers, New York, 1943.

"National Bureau of Standards Handbook"

No. 42—Safe Handling of Radioactive Isotopes, 1949.

No. 59—Permissible Dose from External Sources of Ionizing Radiation, 1954.

No. 50—X-Ray Protection Design, 1952.

No. 60—X-Ray Protection, 1955.

No. 51—Radiological Monitoring Methods and Instruments, 1952.

No. 55—Protection against Betatron-Synchrotron Radiations up to 100 Million Electron Volts, 1954.

SPECIFIC REFERENCES

1. Mehl, R. F.: Nondestructive Testing by Gamma Rays, *J. Am. Soc. Naval Engrs.*, vol. 43, pp. 371–395, 1931.
2. Tenney, G. H., J. W. Dutli, and J. E. Withrow: Radium, Tantalum 182 and Cobalt 60 in Industrial Radiography, *J. Soc. Nondestructive Testing*, vol. 8, no. 3, Winter, 1949–1950.
3. Garrett C., A. Morrison, and. G. Rice: Fluoroscopy and Radiography with Iridium 192, *ASTM, Spec. Tech. Pub.* 145, pp. 9–20, 1952.
4. Dutli, James W., and Grover M. Taylor: Application of Cesium 137 to Industrial Radiography, *J. Soc. Nondestructive Testing*, vol. 12, no. 2, pp. 35–38, March–April, 1954.
5. "National Bureau of Standards Handbook" No. 59—Permissible Dose from External Sources of Ionizing Radiation, 1954.
6. Morgan, G. W.: Some Practical Considerations in Radiation Shielding, *U.S. Atomic Energy Comm., Isotopes, Div. Circ.* B-4, November, 1948.
7. "National Bureau of Standards Handbook" No. 54—Protection against Radiations from Radium, Cobalt 60 and Cesium 137, 1954.
8. Bennett, George A.: Monogram for Calculating Shielding for Co^{60}, *Nucleonics*, vol. 8, no. 4, pp. 55–58, April, 1951.
9. Glasser, O., E. Quimby, L. Taylor, and J. Weatherwax: "Physical Foundations of Radiology," Paul B. Hoeber, Inc., Medical Department of Harper & Brothers, New York, 1944.
10. Evans, Robley D.: Problems Associated with the Transportation of Radioactive Substances, *Natl. Acad. Sci., Nuclear Science Ser. Prelim. Rept.* 11, 1951.
11. Dutli, James W., and Dana E. Elliott: An Evaluation of the Application of Thulium-170 to Industrial Radiography, *J. Soc. Nondestructive Testing*, vol. 15, no. 2, pp. 112–114, March–April, 1957.
12. Sidhu, S. S., F. P. Compos, and D. D. Zauberis: Radiography with Thulium Sources, *ASTM, Spec. Tech. Pub.* 223, 1958.

CHAPTER 8

ULTRASONICS

Ultrasonic techniques are finding increased uses and importance in the field of nondestructive testing. Early attempts to use ultrasonics for testing were not too successful because the available instrumentation was not adequate. In addition the existing equipment was too complicated to be operated by anyone but technically trained personnel on a laboratory scale. The radar type of electronic circuitry developed during World War II has been incorporated into ultrasonic testing equipment. Such instrumentation has been developed and improved so that ultrasonic testing is now becoming an approved and accepted method. Commercial instruments are now available which can be used successfully by non-technically trained personnel of ordinary skill.

Striking a specimen and listening for the characteristic "ring" has been used as a means of detecting flaws. The ringing note emitted by a steel specimen containing a crack is dull and harsh compared with the note emitted by an identical "good" specimen. This "ringing" technique will detect only gross defects. The wavelength of audible sound waves is generally large in comparison with the size of the defect, and the sound travels around the defect. Many attempts have been made to improve this technique and to increase the sensitivity of such a test. Electro-magnetically driven hammers have been made to strike the specimen with rapid as well as even blows. Such devices as stethoscopes, telephone receivers, microphones, and electronic amplifiers have been used to listen for changes in the audible sound emitted. The "Welding Handbook"[1,*] mentions the use of the stethoscope in searching for defective welds. In order to prevent extraneous sounds from reaching the observer a binaural stethoscope has been used. The stethoscope is placed near the weld and the weld tapped lightly and continuously with a hammer. Variations in the sounds emitted by the weld as the stethoscope is moved along indicate irregularities in the weld. For example, a high-pitched reedy sound indicates a lack of fusion in the weld. A badly cracked weld can be found, but there is considerable doubt as to the ability of an inspector to detect porosity. The value of such a test is questionable, especially when the specimen has a complicated shape.

* Superscript numbers indicate Specific References listed at the end of the chapter.

209

With the development of reliable methods for generating and detecting ultrasonic waves, small defects can now be found.   This is due to the fact that the wavelength of ultrasonic waves is approximately equal to the size of the defects to be found.   Fortunately, most metals because of their good elastic properties readily transmit ultrasonic vibrations.   If discontinuities exist, a measurable scattering or reflection will occur because of the acoustic mismatch.

Vibrational waves which have a frequency above the hearing range of the normal ear are called "ultrasonic" waves.   This term generally includes all waves having a frequency greater than about 20,000 cps. The physical laws of acoustics which apply at audible frequencies also apply at these frequencies.   However, certain phenomena which occur at ultrasonic frequencies are not usually observed in the audible range.

Types of Waves.   There are several types of ultrasonic waves, namely, longitudinal, transverse, and surface.   The transmission of ultrasonic energy depends on particle vibration.   The particles of the medium are displaced as the wave travels through the medium.   Longitudinal waves are waves in which the particles of the transmitting medium move in the same direction as the wave is being propagated.   These waves are sometimes known as compressional, dilational, or irrotational waves.   In transverse or shear waves the particles of the transmitting medium vibrate at right angles to the direction of wave propagation.   These waves are sometimes referred to as shear waves or distortional waves.   Figure 8-1 illustrates the difference between longitudinal and shear waves.   An isotropic solid is capable of transmitting both longitudinal and transverse waves; thus either of these types of waves can be and are used for nondestructive testing.

Under the proper conditions ultrasonic waves of considerable amplitude can be propagated on the surface of material.   Surface waves can be divided into three classes: Rayleigh waves, Love waves, and Stonely waves. Waves which are propagated over the surface of a solid whose thickness perpendicular to the surface is large compared to the wavelength of the wave are known as Rayleigh waves.   These waves are roughly analogous to water waves, and the motion of the particles is both transverse and longitudinal.   The vibrations occur in the plane containing the direction of propagation and the normal to the surface of the body.   The oscillations normal to the surface of the body are highly damped, and there is no particle displacement on the surface perpendicular to the direction of propagation.   The velocity of propagation of Rayleigh waves is less than that of body waves; for metals it is approximately 0.9

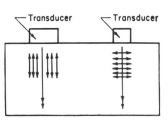

FIG. 8-1. Longitudinal and shear wave propagation.

the shear wave velocity. Waves which are propagated in the body of a solid whose thickness is comparable to the wavelength of the waves are known as Lamb waves. A thin plate is capable of transmitting an infinite number of Lamb waves. The velocity of propagation is determined by the product of the thickness of the plate and the frequency. Love waves travel on the surface without any vertical component. Such waves require that a thin layer of some material of different density be present on the bulk material below. Love waves propagate with velocities which are dependent upon frequency, the velocity decreasing with increasing frequency. Waves analogous to Rayleigh and Love waves can be propagated along the interface of two materials differing in elastic properties and density.[2]

Wave Velocity. The velocities of longitudinal waves, transverse waves, and Rayleigh waves are given by the following expressions.

Longitudinal Wave Velocity:

1. The thin rod velocity—velocity in a rod whose diameter is much less than a wavelength V

$$V = \left(\frac{Y}{\rho}\right)^{\frac{1}{2}} \tag{8-1}$$

2. The bulk velocity—velocity in a medium whose dimensions are much greater than a wavelength V_L

$$V_L = \left[\frac{Y(1 - \sigma)}{\rho(1 + \sigma)(1 - 2\sigma)}\right]^{\frac{1}{2}} \tag{8-2}$$

Transverse Wave Velocity:

$$V_T = \left(\frac{G}{\rho}\right)^{\frac{1}{2}} = \left[\frac{Y}{\rho}\frac{1}{2(1 + \sigma)}\right]^{\frac{1}{2}} \tag{8-3}$$

Rayleigh Surface Wave Velocity V_R:

$$V_R = KV_T \sim 0.9V_T \tag{8-4}$$

where Y = Young's modulus
ρ = density
σ = Poisson's ratio
G = modulus of rigidity
K = root of following equation

$$K^6 - 8K^4 + 8(3 - 2\alpha^2)K^2 + 8(\alpha^2 - 2) = 0 \tag{8-5}$$

where α is a constant of the material and is given by the expression $\alpha = [(1 - 2\sigma)/(2 - 2\sigma)]^{\frac{1}{2}}$. σ is Poisson's ratio.

TABLE 8-1. ULTRASONIC VELOCITY AND IMPEDANCE

Material	Specific gravity*	Poisson's ratio	Moduli, lb/in² × 10⁶†		Velocity, ips × 10⁵‡			Relative acoustic impedance§	
			Young's Y	Rigidity G	V_L	V_T	V_S	V_L	V_T
Aluminum (17ST)	2.699	0.355	10.4	3.90	2.46	1.22	1.10	6.64	3.29
Beryllium	1.82	0.05	43.0	20.4	5.04	3.43	3.10	9.18	6.24
Brass (cast yellow)	8.44	0.374	15.0	5.51	1.85	0.84	0.76	15.6	7.09
Copper	8.89	0.37	16.0	5.87	1.82	0.84	0.76	16.2	7.46
Hafnium	11.3		14.0		1.52	0.82	0.76	17.2	9.27
Lead	11.34	0.43	2.4	0.83	0.77	0.25	0.23	8.73	2.84
Magnesium	1.74	0.31	6.62	2.33	2.27	1.20	1.08	3.95	2.1
Monel	8.90	0.327	26.0	9.50	2.1	1.07	0.97	9.55	8.63
Mercury	13.55				0.57			7.72	
Nickel	8.90	0.34	30.0	11.5	2.37	1.18	1.06	21.1	10.5
Platinum	21.45	0.303	26.2	9.28	1.24	0.68	0.61	26.6	14.6
Silver	10.49	0.38	10.85	3.92	1.43	0.64	0.57	15.00	6.7
Stainless steel (347)	7.91	0.30	28.43	10.98	2.26	1.22	1.1	17.9	9.65
Tin	7.30	0.34	7.97	3.01	1.33	0.658	0.592	9.71	4.80
Tungsten	19.25	0.28	58.9	23.10	2.04	1.13	1.04	39.3	21.75
Uranium	18.5–19	0.25	25.5	10.2	1.33	0.76	0.69	25.1	14.3
Zirconium	6.5	0.35	14.5	4.76	1.79	1.02	0.96	11.6	6.63
Air	1.293×10^{-3}				0.13			1.7×10^{-4}	
Carbon tetrachloride	1.596				0.37			0.59	
Glass	2.32				2.22	1.29	1.16	5.15	2.99
Glycerin	1.26				0.781			0.98	
Lucite	1.182	0.40	0.48	0.20	1.05	0.43	0.39	1.24	0.51
Nylon (6-6)	1.11	0.40	0.52	0.177	1.03	0.42	0.38	1.14	0.47

TABLE 8-1. ULTRASONIC VELOCITY AND IMPEDANCE (Continued)

Material	Specific gravity*	Poisson's ratio	Moduli, lb/in² × 10⁵†		Velocity, ips × 10⁵‡			Relative acoustic impedance§	
			Young's Y	Rigidity G	V_L	V_T	V_S	V_L	V_T
Oil transformer........	0.92	.....	.....	.....	0.55	.....	.....	0.51	
Polyethylene........	0.90	0.458	0.11	0.038	0.77	0.21	0.19	0.69	0.19
Polystyrene........	1.056	0.405	0.766	0.17	0.92	0.44	0.40	0.97	0.46
Quartz........	2.65	.....	11.6	.....	2.26	.....	.....	5.99	.....
Water........	1.00	.....	.....	.....	0.59	.....	.....	0.59	.....

The values given in this table have been gathered from a number of sources and are in general sufficiently accurate for most practical applications.

  * The density of the material can be calculated by multiplying the specific gravity by 1 g/cm³ for density in the metric system or 62.4 lb/ft³ for density in the English system.

  † 1 dyne/cm² is equal to 1.450 × 10⁻⁵ lb/in².

  ‡ Velocity in cm/sec can be obtained by multiplying velocity in in./sec by 2.54.

  § The acoustic impedance values are "relative," being the product of the specific gravity and velocity.

Since most engineering handbooks and reference books give the values of the various parameters in Eqs. (8-1) to (8-3) in engineering units, these equations can be written in a more convenient form.

$$V' = 103.5 \sqrt{\frac{Y'}{S}} \tag{8-1'}$$

$$V'_L = \frac{103.5 \sqrt{Y'(1 - \sigma)}}{\sqrt{S(1 + \sigma)(1 - 2\sigma)}} \tag{8-2'}$$

$$V'_T = 103.5 \sqrt{\frac{G}{S}} = \sqrt{\frac{Y'}{S} \frac{1}{2(1 + \sigma)}} \tag{8-3'}$$

where V' = velocity of longitudinal waves, ips
Y' = Young's modulus, psi
S = specific gravity
V'_L = velocity of longitudinal waves, ips
V'_T = velocity of transverse waves, ips

Table 8-1 gives the values for some of the properties as well as the velocities for various engineering materials.

The relationship between frequency f, wavelength λ, and velocity of propagation V for all types of ultrasonic waves except Lamb waves is given by the expression

$$f\lambda = V \tag{8-6}$$

It should be remembered that the velocity of propagation depends only on the medium; and when a wave of frequency f travels from one medium to another, a change in wavelength occurs.

Beam Spreading. Because of their short wavelength, ultrasonic waves travel essentially in a straight line. It is this property which makes ultrasonic waves so useful for locating defects. As the wavelength becomes shorter, the waves more closely approach the ideal condition of rectilinear propagation. However, there is always some spreading of the beam as the waves travel from the source of ultrasonic waves. The angle of spread Θ is given by the relation

$$\sin (\Theta/2) = \frac{1.2\lambda}{d} \tag{8-7}$$

where λ = wavelength of emitted sound waves
d = diameter of source

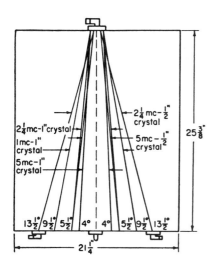

FIG. 8-2. Beam spread.

This expression shows that the higher frequencies are more directive. A 1-in.-diameter crystal at any frequency is more directive than a ½-in.-diameter crystal. Figure 8-2 shows the beam spread for various frequencies and source sizes. If the frequency is decreased until λ approaches the dimensions of the source, the waves are sent out in all directions.

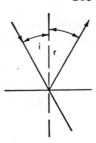

FIG. 8-3. Reflection at interface.

When an ultrasonic wave strikes an interface between two media, part of the energy is reflected and part is transmitted, Fig. 8-3. The amount of energy reflected is determined by the specific acoustic impedance of each of the two media. The specific acoustic impedance is defined as the product of the velocity V of the ultrasonic wave in the medium and the density of the medium ρ. The specific acoustic impedance for various engineering materials is given in Table 8-1. The ratio of the reflected to incident ultrasonic energy at an interface is given by the following expression:

$$\frac{E_r}{E_i} = \left(\frac{\rho_1 V_1 - \rho_2 V_2}{\rho_1 V_1 + \rho_2 V_2}\right)^2 \tag{8-8}$$

where E_r = reflected energy
E_i = incident energy
ρ_1 = density of first medium
ρ_2 = density of second medium
V_1 = velocity of propagation in first medium
V_2 = velocity of propagation in second medium

Figure 8-4 shows the ratio of the reflected to incident energy for various materials. In the case of a water-to-steel boundary, 88 per cent of the

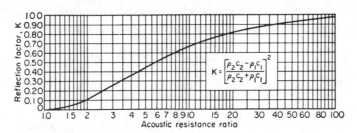

FIG. 8-4. Ratio of reflected to incident energy. Approximate acoustic resistance in grams per cubic centimeter per second: air, 42; transformer oil, 1.25×10^5; water, 1.484×10^5; glycerin 2.5×10^5; glass, 14.3×10^5; mercury, 19.0×10^5; steel, 3.90×10^5.

$$K = \left(\frac{\rho_2 C_2 - \rho_1 C_1}{\rho_2 C_2 + \rho_1 C_1}\right)^2$$

$\rho($ ' = acoustic resistance of medium

where ρ = density, g/cm³, and C = velocity of sound in medium, cm/sec.

incident energy is reflected and 12 per cent is transmitted. In the case of an air-solid interface, 100 per cent of the incident energy is reflected.

Expression (8-8) applies when the thickness of the second medium is much larger than the wavelength of the ultrasonic waves. When the medium through which the wave travels has a thickness of a wavelength or less, the following expression must be used:

$$\frac{E_r}{E_i} = \frac{(\rho_1 V_1/\rho_2 V_2 - \rho_2 V_2/\rho_1 V_1)^2}{4 \cot^2 (2\pi t/\lambda) + (\rho_1 V_1/\rho_2 V_2 + \rho_2 V_2/\rho_1 V_1)^2} \qquad (8\text{-}9)$$

where t is the thickness of the medium. Maximum transmission occurs when t is an integral number of half wavelengths. Minimum transmission will occur when t is an odd number of quarter wavelengths. When the incident wave strikes the reflecting surface at an angle, the reflected wave will make a similar angle with the surface.

One of the practical problems in ultrasonic testing is transmitting the ultrasonic energy from the source into the specimen. This is done by interposing a "couplant" between source and test specimen. The coupling medium may be oil, water, or mercury. It can be shown that the best condition exists when the couplant medium has an impedance which is the mean of the impedances of the source and test specimen. For optimum transmission the couplant should have a thickness equal to one-half wavelength or a multiple thereof, Eq. (8-9). Ernst[3] points out that losses due to impedance mismatch can be reduced by inserting plates with parallel faces of materials matching the two impedances. These plates are called "transmission" plates.

Refraction. Ultrasonic waves crossing obliquely the boundary separating two media undergo abrupt changes in direction if the velocity of propagation is different in the two media. This phenomenon is called "refraction." Figure 8-5 shows what takes place at such an interface.

The angle of refraction is given by the following expression:

$$\frac{\sin i}{\sin r} = \frac{V_1}{V_2} \qquad (8\text{-}10)$$

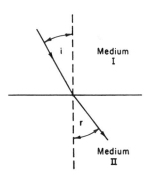

Fig. 8-5. Refraction of an interface.

where i = angle of incidence
r = angle of refraction
V_1 = velocity of propagation in first medium
V_2 = velocity of propagation in second medium

This expression holds for both longitudinal and transverse waves. The velocities V_1 and V_2 refer to either shear or transverse waves. When the velocity of propagation is greater in

the second medium than in the first,
it is possible to have an angle of incidence which makes the angle of refraction 90°.   The angle of incidence
for which the angle of refraction is
90° is referred to as the "critical"
angle.   For angles greater than the
critical angle, the wave is totally
reflected and no energy enters the
second medium.   In the case of a
water to steel interface, the critical
angle for a longitudinal wave is

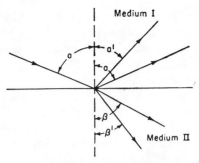

Fig. 8-6. Mode conversion of interface.

approximately 15°.   When the angle of incidence is greater than 15°,
no longitudinal waves enter the steel, all the ultrasonic energy being
reflected back into the water.

Mode Conversion.   The third phenomenon which may occur at an
interface is mode conversion.   When an ultrasonic wave of either type
impinges on an interface between two media at an angle other than at
right angles, mode conversion takes place.   Figure 8-6 shows a longitudinal wave impinging at an angle α on an interface.   In the most
general case, four separate waves are produced, a wave of each type is
reflected, and a wave of each type is refracted.   The relationship between
the various angles of reflection and refraction is given by the following
expression:

$$\frac{\sin \alpha}{V_L} = \frac{\sin \alpha'}{V_T} = \frac{\sin \beta}{V_L'} = \frac{\sin \beta'}{V_T'} \qquad (8\text{-}11)$$

where Θ = angle of incidence and reflection of longitudinal wave
$\qquad \Theta'$ = angle of reflection of transverse wave
$\qquad r$ = angle of refraction of longitudinal wave
$\qquad r'$ = angle of refraction of transverse wave
$\qquad V_L$ = velocity of longitudinal wave, medium I
$\qquad V_T$ = velocity of transverse wave, medium I
$\qquad V_L'$ = velocity of longitudinal wave, medium II
$\qquad V_T'$ = velocity of transverse wave, medium II

The velocity of the two types of waves being different, they have
different angles of reflection and refraction.   The different types of waves
also have different critical angles.   This makes it possible totally to
reflect one type of wave and have only one refracted wave entering the
second medium.   This permits the use of either L or T waves for ultrasonic testing.

Knott[4] has worked out the general case of reflection and refraction at a
plane boundary between two media.   Equations showing the relationship

between the amplitudes of the incident wave and the resulting waves are given in reference 4.

Attenuation. The loss of energy in an ultrasonic wave propagated through material can be attributed to four different mechanisms: heat conduction, viscous friction, elastic hysteresis, and scattering. It is difficult to determine by simple experimentation which of these mechanisms is of major importance. The observed losses depend greatly on the type and structure of the material and its pretreatment. These mechanisms have some frequency dependence and obey different laws with regard to their frequency dependence. Only scattering will be discussed here.

According to the classical theory, sound absorption is caused only by thermal conduction and internal or viscous friction. Sound waves cause adiabatic compressions and rarefactions as they are propagated through material. There is a momentary temperature rise in the compressions and a momentary temperature fall in the rarefactions. Heat energy is taken from the compressions and restored to the rarefaction, which decreases the energy of the wave and causes attenuation. Viscous friction between the grains accounts for some of the loss of polycrystalline materials. The inelastic behavior of a solid material also accounts for some loss of energy. The discrepancy between the experiment and theory suggests that there are other mechanisms which absorb energy from the sound wave.

The attenuation of an ultrasonic wave traveling in a homogeneous medium can be given by the expression

$$A = A_0 \exp{(-\alpha x)} \qquad (8\text{-}12)$$

where A = intensity of sound after traveling a distance x
A_0 = intensity of sound
α = attenuation constant
x = distance

The attenuation in most engineering materials and alloys is small, so that ultrasonic waves can penetrate several feet of material without appreciable loss of energy.

Scattering. Energy loss occurs because of scattering of the ultrasonic energy. The ultrasonic energy is reflected or scattered from the elastic discontinuities at the grain boundaries. The amount of energy scattered depends on the relative

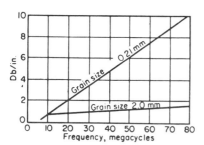

FIG. 8-7. Attenuation vs. frequency for magnesium.

TABLE 8-2. GRAIN SIZE MEASUREMENTS

Annealing temperature, °F	Average grain size by microscope, mm	Number of detectable successive reflections
890	0.01	16
1060	0.03	13
1225	0.075	6
1350	0.105	3

magnitude of the wavelength and the average grain diameter. Figure 8-7 shows the scattering as a function of grain size.

Firestone[5] suggested the use of ultrasonics for measuring the average grain size in metals. Table 8-2 gives his results for annealed brass blocks at 5 Mc. Worlton[6] and Grossman[7] have used ultrasonics for grain size measurements. Figure 8-8 shows a sketch of the experimental apparatus used by Worlton to measure the grain size of brass. The specimens were 1½-in.-diameter cylinders. These were heat-treated to produce average grain diameters ranging from 0.025 to 0.150 mm. The technique used was to send pulsed ultrasonic waves through the water and specimen and display the transmitted pulses on an oscilloscope. Energy losses occur in the water couplant, at the brass water interface, and in the specimen. The losses in the water and at the interface were constant. Consequently, any variation in the received pulses was due to losses in the specimen. The amplitude of the received pulses was measured with a calibrated amplifier. With the specimen removed some amplifier gain G_1 was required to produce a received pulse of given amplitude on the oscilloscope. When the specimen was inserted, some increased gain G_2 was needed to produce the same pulse amplitude on the oscilloscope. The ratio of G_2/G_1 for various specimens and frequencies is given in Fig. 8-9. The attenuation can be seen to increase with grain size. At the high frequencies the attenuation appears to vary as the fourth power of the

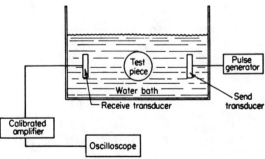
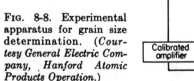

FIG. 8-8. Experimental apparatus for grain size determination. (*Courtesy General Electric Company, Hanford Atomic Products Operation.*)

frequency. The information in Fig. 8-9 is combined in a single curve, as shown in Fig. 8-10. In this curve the ratio of G_2/G_1 is plotted against the ratio of the ultrasonic wavelength and the average grain diameter. The attenuation is essentially constant for ratios of λ/D larger than 30 and increases rapidly for ratios smaller than 6. The following empirical equation can be derived for this curve:

$$D = \frac{4.2 \text{ mm}}{fN} \tag{8-13}$$

where N is the λ/D ratio and f is the frequency in megacycles. The curve given in Fig. 8-10 is applicable only for particular kind, size, and shape specimen discussed above.

Grossman used rolled alpha-brass rods of known composition and structure to determine the effects of grain size on the scattering of ultrasonic waves. Specimens with an average grain size of 0.015 to 0.150 mm were used. A Sperry Reflectoscope was used at frequencies of 2.25, 5, and 10 Mc. Both pulse-echo and transmission techniques were used. No difference in results was detected between the two techniques. Two variations of the pulse-echo technique were used by Grossman, in one case measuring the height of the first back reflection and in the other case counting the number of back reflections for a constant setting of the Reflectoscope. Figure 8-11 shows how the amplitude of the back reflection varied with grain size. Figure 8-12 shows how the number of back

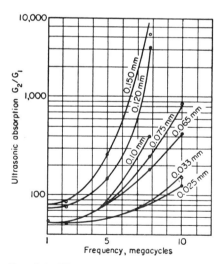

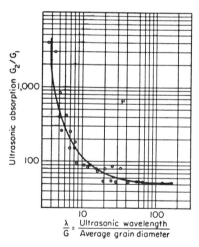

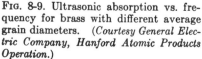

FIG. 8-9. Ultrasonic absorption vs. frequency for brass with different average grain diameters. (*Courtesy General Electric Company, Hanford Atomic Products Operation.*)

FIG. 8-10. Ultrasonic absorption curve for brass. (*Courtesy General Electric Company, Hanford Atomic Products Operation.*)

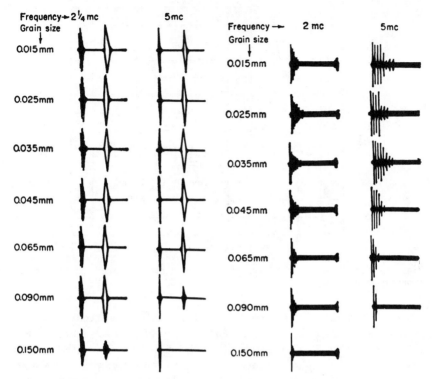

FIG. 8-11. Amplitude of back reflection vs. grain size. (*Courtesy Iron Age.*)

FIG. 8-12. Number of back reflections vs. grain size. (*Courtesy Iron Age.*)

reflections varied with grain size. Figure 8-13 is a plot of the number of back reflections as a function of the ratio of wavelength to grain size. Grossman's work can be summarized as follows: when the ratio of wavelength to grain size was greater than 30, there was little loss of energy due to scattering, but when this ratio was less than 10, the loss of energy due to scattering was rather large.

Ultrasonic Wave Production. There are a number of ways in which ultrasonic waves may be produced. However, the method of producing ultrasonic waves for nondestructive testing purposes is to utilize the piezoelectric effect. When mechanical pressure or tension is applied to certain crystals, electric charges develop in the crystal faces. The magnitude of the electric charge produced is directly proportional to the mechanical pressure or tension applied. The sign of the electric charge changes when the mechanical force changes from compression to tension. This phenomenon is known as the "piezoelectric effect." This effect has been observed in a number of crystals, such as tourmaline, quartz, and Rochelle salts. There is also a reverse piezoelectric effect in which the crystal changes dimensions if subjected to a suitable electric field.

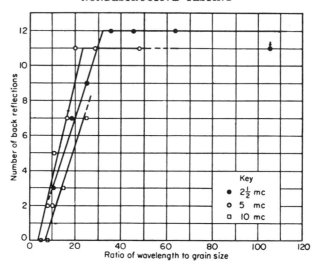

FIG. 8-13. Number of back reflections compared with ratio of wavelength to grain size. (*Courtesy Iron Age.*)

Quartz Crystals. If a quartz plate is put into an alternating electric field, the quartz will be compressed in one half of the cycle and expanded in the other half. The crystal will contract and expand with the same frequency as the applied voltage. The amplitude of vibration is a maximum when the frequency of the electrical field is the resonant frequency of the crystal. Figure 8-14 shows the amplitude of oscillation as a function of frequency for a quartz crystal.

Any device which converts one type of energy into another type is known as a "transducer." The term transducer will be used here to mean a device to convert electrical energy to mechanical energy, and vice versa. Figure 8-15 shows a quartz plate cut at right angles to the X axis which is called an X-cut crystal. Crystals cut at right angles to the Y axis are called Y cut. Two variations in oscillations are possible, longitudinal vibrations in the X direction called thickness vibrations and longitudinal vibrations in the Y direction called longitudinal vibrations. Actually such crystals do not ordinarily vibrate in one direction only, even though designed to do so. Figure 8-16 shows ways in which

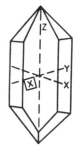

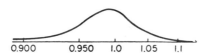

FIG. 8-14. Amplitude of oscillation of quartz crystal.

FIG. 8-15. X-cut quartz crystal.

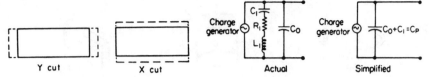

FIG. 8-16. Vibration of X- and Y-cut quartz crystals.

FIG. 8-17. Equivalent circuit of quartz crystal.

X- and Y-cut crystals vibrate. Quartz crystals can also be made to vibrate at harmonic frequencies which are multiples of their fundamental frequency. Quartz crystals have been cut to have a fundamental frequency as high as 50 Mc. However, at these frequencies the quartz is extremely thin and easily damaged.

The fundamental frequency vibrations F of a quartz transducer are

$$F = \frac{v}{2t} = \frac{115,500}{t(\text{in.})} \quad \text{cps} \tag{8-14}$$

when V is the velocity of quartz and t is the thickness of the crystal.

An analysis of the operation of a piezoelectric vibrator as a component of an electrical network is greatly simplified if the crystal and its electrodes are replaced by an equivalent electrical circuit.[8] The equivalent circuit consists of a capacitance connected in parallel with a resistance, inductance, and capacitance in series, as shown in Fig. 8-17. For testing applications the two emitting surfaces of the quartz transducer are covered with metallic electrodes. An alternating electrode potential difference from a radio-frequency generator is applied to the electrodes by special contacts.

A Y-cut quartz crystal oscillates in shear, as shown in Fig. 8-18. A Y-cut crystal on a metal radiates a beam of shear waves in a direction normal to the surface. One widely used technique for producing shear waves is to use an X-cut crystal and a plastic wedge. Using the proper wedge angle, shear waves can be produced by mode conversion. The proper wedge angle for shear waves can be calculated by using Eqs. (8-10) and (8-11).

Barium Titanate Transducers. Barium titanate, a ceramic material, is being used for transducers. Barium

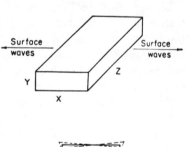

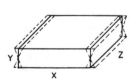

FIG. 8-18. Y-cut crystal oscillating in the shear mode.

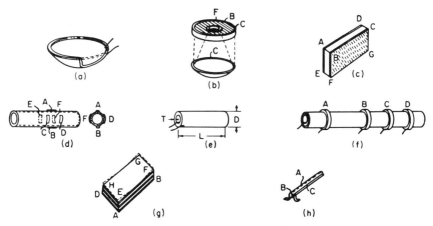

Fig. 8-19. Barium titanate transducers.

titanate is polycrystalline, and after being polarized by the application of a strong electric field, it behaves like a piezoelectric material. The shapes of transducers which can be made are practically unlimited. Figure 8-19 shows some of the shapes of the transducers which have been made from barium titanate. Such transducers are practically unaffected by moisture and withstand temperatures up to about 200°F. The impedance of barium titanate is low compared with quartz. Thus the voltage source required by exciting barium titanate transducers is lower than that required by quartz transducers.

Lithium Sulfate Transducers. Lithium sulfate is also being used for transducers. Such transducers provide improved sensitivity and greater resolution for ultrasonic nondestructive testing. It is reported that they have about fifteen times the sensitivity of quartz. Shorter acoustic pulses can be produced which in the case of pulse-echo testing result in flaw detection closer to the surface.

Rayleigh Waves. In Rayleigh waves the particles of the medium have a displacement amplitude which penetrates below the surface of the solid. The amplitude of this displacement is greatly reduced at depths of several

TABLE 8-3. VELOCITY OF RAYLEIGH WAVES

Material	Velocity, in./sec $\times 10^5$
Aluminum	1.10
Beryllium	3.10
Brass	0.76
Copper	0.76
Magnesium	1.08
Steel	1.10
Lead	0.23

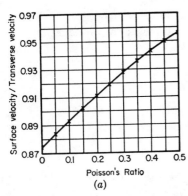
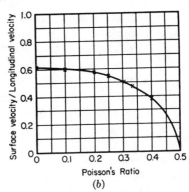

FIG. 8-20. (a) Ratio of surface wave velocity to transverse wave velocity vs. Poisson's ratio; (b) ratio of surface wave velocity to longitudinal wave velocity vs. Poisson's ratio. (*Courtesy Sperry Products, Inc.*)

wavelengths below the surface. Further, the amplitude of the wave is attenuated as it travels over the surface of the material. The velocity of propagation of Rayleigh waves depends upon the density and elastic properties of the medium. Rayleigh waves travel with a fraction of the velocity K of the velocity of shear waves. The value of K can be obtained from Eq. (8-5). Table 8-3 gives the velocity of Rayleigh waves for various materials. For other material, the Rayleigh velocity can be computed from Fig. 8-20[9] which shows the ratio of the surface to the transverse velocity for various values of Poisson's ratio. Bergmann[10] has given a useful approximation of Eq. (8-5).

$$\frac{V_R}{V_T} = \frac{0.87 + 1.12\sigma}{1 + \sigma} \tag{8-15}$$

In Rayleigh waves the particle displacement of the wave motion follows an elliptical orbit. The major axis of the ellipse is perpendicular to the surface along which the waves are traveling. The minor axis is parallel to the direction of propagation. The magnitudes of these axes are given in the following expressions:

$$u = K\left[\frac{2\left(1 - \frac{v_r^2}{v_L^2}\right)^{1/2}\left(1 - \frac{v_r^2}{v_t^2}\right)^{1/2}}{2 - v_r^2/v_t^2}\,e^{-\frac{2\pi}{\lambda}\left(1 - \frac{v_r^2}{v_t^2}\right)^{1/2}y} - e^{-\frac{2\pi}{\lambda}\left(1 - \frac{v_r^2}{v_L^2}\right)^{1/2}y}\right]$$
$$\sin(\omega t + kx) \tag{8-16}$$

$$v = -K\left[\frac{2\left(1 - \frac{v_r^2}{v_L^2}\right)^{1/2}}{2 - v_r^2/v_t^2}\,e^{-\frac{2\pi}{\lambda}\left(1 - \frac{v_r^2}{v_t^2}\right)^{1/2}y} - \left(1 - \frac{v_r^2}{v_L^2}\right)^{1/2}e^{-\frac{2\pi}{\lambda}\left(1 - \frac{v_r^2}{v_L^2}\right)^{1/2}y}\right]$$
$$\cos(\omega t + kx) \tag{8-17}$$

where u = displacement parallel to direction of propagation
    v = displacement perpendicular to direction of propagation
    V_1 = longitudinal wave velocity
    V_t = transverse wave velocity
    V_r = surface wave velocity
    x = direction of propagation
    y = depth below surface
    $\omega = 2\pi f (f = $ frequency$)$
    $k = 2\pi/\lambda$
    λ = wavelength

For example, when the value of $\sigma = \frac{1}{4}$ at a depth below the surface of 0.193 wavelengths, there is no motion parallel to the surface. For greater depths the amplitude parallel to the surface again becomes finite. However, a change of phase occurs so that the vibrations take place in the opposite phase. There is no finite depth below the surface at which the motion in a direction normal to the surface vanishes. As the distance below the surface increases, the amplitude of the vibration increases to a maximum at a depth of 0.076 wavelength and then decreases monatonically. At a depth of 1 wavelength the amplitude has decreased to 0.19 of the value at the surface. From equations 8-16 and 8-17 the displacement functions can be computed. The ratio of the surface wave velocity to the transverse wave velocity and longitudinal wave velocity for various valves of Poisson's ratio can be found from Figure 8-20.

Attenuation of Rayleigh Waves. Cook and Van Valkenburg[11] made an experimental study of the attenuation of surface waves as a function of distance below the surface. Their study was made on aluminum at a frequency of 2.25 Mc. The specimen was made with seven 0.018-in.-diameter holes drilled to a depth of $\frac{1}{4}$ in., as shown in Fig. 8-21. The holes were drilled at seven steps below the surface ranging from 0.019 to 0.089 in. Their results using a Reflectoscope are shown in Fig. 8-22. Figure 8-23 shows the results of some experiments using a frequency of 2.25 mc on a polished aluminum surface. Since a reflection technique was used to obtain the data, the distances indicated on the abscissa should be multiplied by 2.

The path traversed by a surface wave can be found by placing the

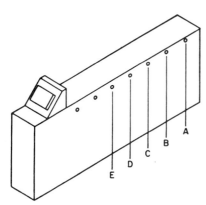

Fig. 8-21. Test block for study of surface waves. (*Courtesy Sperry Products, Inc.*)

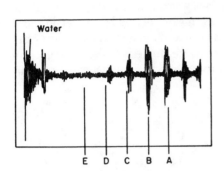

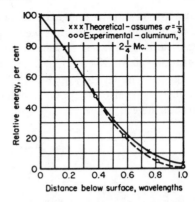

FIG. 8-22. Results using test block of Fig. 8-21. (*Courtesy Sperry Products, Inc.*)

FIG. 8-23. Attenuation of surface waves on aluminum. (*Courtesy Sperry Products, Inc.*)

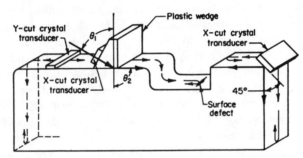

FIG. 8-24. Methods of producing surface waves. (*Courtesy Sperry Products, Inc.*)

finger on the surface of the material. A reduction in the height of the received signal indicates that some of the sound energy is being absorbed. Any surface on which surface waves are to be transmitted should be as free as possible of any dirt, grease, etc. Such substances damp or absorb the waves appreciably and in some instances produce reflections of the beam.

Production of Rayleigh Waves. Firestone[12] suggested the use of a Y-cut quartz crystal to generate Rayleigh waves. A Y-cut crystal whose width to thickness ratio is approximately 7:1, Fig. 8-24, will produce a low-amplitude surface wave. The surface wave produced will be propagated in both X directions of width equal to the Z dimension. An X-cut quartz crystal placed against the sharp corner of a plate at an angle of about 45° with the surface will produce surface waves.

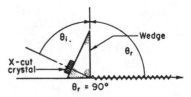

FIG. 8-25. Production of surface waves by means of wedge.

These waves are propagated in both directions from the corner. Both of these techniques produce weak surface waves. A third type of surface wave unit generates surface waves by means of mode conversion and refraction. An X-cut quartz crystal is placed against the surface of the wedge, as shown in Fig. 8-25. The wedge angle is chosen so that the angle of refraction Θ_r is 90°. The proper wedge angle Θ_i can be calculated using Eq. (8-18).

$$\frac{\sin \Theta_i}{\sin \Theta_r} = \frac{V_L}{V_R} \tag{8-18}$$

or

$$(\Theta_i) = \sin^{-1}\frac{V_L}{V_R}$$

Using a plastic wedge of the proper angle, surface waves will be propagated in the forward direction.

Lamb Waves. When ultrasonic waves travel in a specimen whose thickness is comparable to the wavelength, Lamb waves are produced. The theoretical and mathematical formula for these waves was worked out by Prof. Horace Lamb in 1917.[13] Lamb's analysis showed that a plate is capable of transmitting an infinite number and kinds of waves. Lamb classified these waves into two main types, namely, symmetrical and asymmetrical. The type is determined by whether the particle motion is symmetrical or antisymmetrical with respect to the medial plane of the specimen. Each of these is further divided into first mode, second mode, third mode, etc. Figure 8-26 shows various types and modes of these waves. In the above figures the arrow indicates the direction of propagation of the waves. The circular and elliptical figures represent the general directions of the displacements in one wavelength. The figures show the displacements in sections of plates carrying Lamb waves, each section being cut out by a plane at right angles to the principal face of the plates; this plane includes the direction of propagation.

Figures 8-27 and 8-28 show computed phase velocities of different modes for both types of Lamb waves in aluminum. The abscissa is the product of frequency and sheet thickness in inches per second. The ordinate is the phase velocity in inches per second. The velocity with

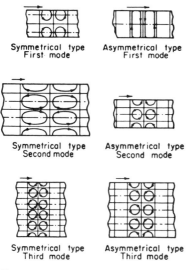

Symmetrical type
First mode

Asymmetrical type
First mode

Symmetrical type
Second mode

Asymmetrical type
Second mode

Symmetrical type
Third mode

Asymmetrical type
Third mode

FIG. 8-26. Types and modes of lamb waves.

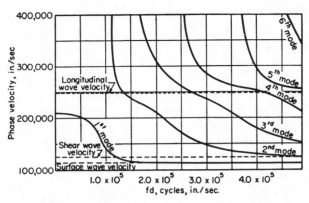

FIG. 8-27. Phase velocity vs. product of frequency and plate thickness, symmetrical type.

which individual Lamb waves travel along the sheet is called the "phase" velocity. The relationship between the phase and group velocity can be found by drawing a tangent to the curve at a given ratio of wavelength to thickness. The point at which the tangent intersects the ordinate gives the group velocity. For short wavelength high-frequency waves, the phase and group velocities become equal. The first mode of both symmetrical and asymmetrical types approaches the velocity of a surface wave. For higher modes, both types approach the velocity of shear waves. The velocity of propagation of Lamb waves depends upon the ratio of the wavelength to the thickness of the material. Figures 8-27 and 8-28 given by Firestone[12] show the wave phase velocity plotted as a function of the product of frequency and plate thickness for symmetrical and asymmetrical types of Lamb waves, respectively.

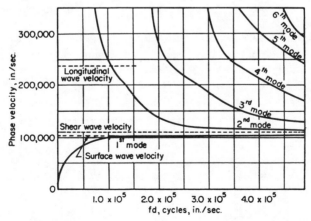

FIG. 8-28. Phase velocity vs. product of frequency and plate thickness, asymmetrical type.

Production of Lamb Waves. Lamb waves can be generated in thin sheets by using a Y-cut quartz crystal whose X dimension is seven times its thickness. Such a crystal will produce a strip of waves in both X directions, having a width equal to the Z dimension of the crystal. The Lamb waves so produced will consist of a mixture of types and modes. Firestone[12] describes a means of generating Lamb waves by using longitudinal waves. Longitudinal waves of a predetermined velocity and frequency impinge on a surface of a sheet at a given angle of incidence. The proper angle of incidence θ_i can be computed from the expression

$$\sin \theta_i = \frac{V_L}{V_P} \qquad (8\text{-}19)$$

where V_P is the phase velocity of the desired Lamb wave and V_L is the velocity of propagation of the incident wave. For the first-mode symmetrical-type Lamb wave with an fd product of 1×10^5 in water ($V_L = 59,000$ ips), from Fig. 8-27, $V_P = 140,000$ ips.

$$\sin \theta_i = \frac{59 \times 10^3}{140 \times 10^3} = 0.42$$
$$\theta_i = 25°$$

Figure 8-29 shows the transducer devised by Firestone.

Table 8-4 shows the type and mode of waves for various angles of incidence.

Echoes may be observed with Lamb waves in thin material. Firestone[12] observed the following Lamb waves: symmetrical first mode, asymmetrical first mode, and symmetrical third mode having group velocities of about 110,000 110,000, and 156,000 ips. When Lamb waves are reflected at an edge, they can change part of their energy into another mode or type.

Technique. Several different techniques have been used in ultrasonic testing such as: (1) pulse echo, (2) transmission, (3) resonance, (4) frequency modulation, and (5) acoustic image.

TABLE 8-4. ANGLE OF INCIDENCE FOR LAMB WAVES

Type	Mode	V_P, ips $\times 10^3$	θ_i
Asymmetrical	1st	105	33
Symmetrical	1st	110	31
Asymmetrical	2d	135	25
Symmetrical	2d	210	16
Symmetrical	3d	310	10.6
Asymmetrical	3d	335	9.8

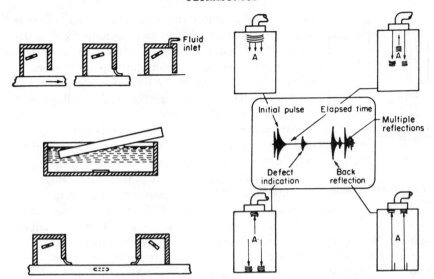

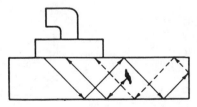

Fig. 8-29. Firestone transducer for lamb waves.

Fig. 8-30. Principle of pulse echo technique.

Pulse Echo. In the pulse echo technique a pulsed ultrasonic beam is sent through the couplant into the specimen. At the opposite face, the beam is reflected and the echo picked up by a transducer, as shown in Fig. 8-30. The transmitting transducer may serve as the receiving transducer, or a separate transducer may be used. A discontinuity or flaw in the specimen will also send back an echo. The time intervals that elapse between the initial pulse and the arrival of the echoes are measured with a cathode ray oscilloscope. In the echo pattern a flaw can be recognized by the relative position and amplitude of its echo. There may also be a number of multiple reflections displayed in the echo pattern. The resolution of this technique depends on the duration of the ultrasonic pulses. The shorter the duration of the pulse, the thinner the specimen that can be successfully tested. A variation of this technique is the angle pulse echo or shear wave technique shown in Fig. 8-31. The shear wave technique is used primarily for testing thin materials. The wave enters the specimen at an angle and travels in a zigzag path between the top and bottom of the sheet. As in the preceding case, echoes are produced by flaws and by the edge of the sheet.

Transmission. The transmission technique requires a separate sending and receiving transducer. A

Fig. 8-31. Principle of angle pulse echo technique.

FIG. 8-32. Principle of transmission technique.

pulsed continuous or modulated ultrasonic beam is sent through the specimen and the amplitude of the transmitted beam measured. Inhomogeneities in the sample decrease the amplitude of the transmitted beam because of the reflection and scattering that take place. The principle of this technique is shown in Fig. 8-32.

Resonance. In the resonance technique of ultrasonic testing a tunable variable frequency continuous wave oscillator is used to drive a transducer. The oscillator is turned through its tuning range. If the specimen has thickness resonant frequencies within the tuning range of the oscillator, the specimen will vibrate in resonance. When resonance occurs, there is an increase in the energy drawn by the transducer. This increased energy can be indicated by a suitable meter or by an oscilloscope. Thickness resonance occurs whenever the thickness of the specimen is equal to an integral number of half wavelengths of the ultrasonic wave. Figure 8-33 shows some of the different ways in which a specimen can vibrate.

Frequency Modulation. In the frequency modulated type of equipment,[14] only one transducer is used but the ultrasonic energy is being sent continuously and received continuously. The radio frequency signal applied to the transducer is rapidly changing in frequency. An echo which arrives after the transmitted signal leaves the transducer will have an instantaneous frequency differing from that being transmitted. The greater the depth of the flaw, the greater the difference in frequency. This difference in frequency can be measured, giving a measurement of the depth of the flaw. Since energy is being sent continuously, the problem of transmitting a large amount of power in a short interval is eliminated.

Acoustic Image Systems. A number of techniques have been tried for producing a visual or optical image from ultrasonic energy. These techniques include the use of chemical effects, the diffraction of light, the orientation of spherical particles, temperature sensitive compounds, phosphors, and special electronic tubes. Figure 8-34 shows the acoustic image system developed by Pohlman.[15] An ultrasonic wave is passed through the specimen and focused on the image forming cell. The image cell consists of flake-shaped aluminum particles, 5 to 25 μ in size, suspended in a

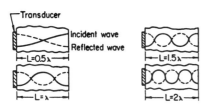

FIG. 8-33. Standing wave patterns in solid material.

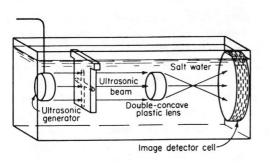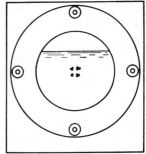

FIG. 8-34. Pohlman acoustic image system.

light viscous liquid. In the absence of any ultrasonic energy, the particles assume random positions as a result of thermal agitation. The particles are oriented by the ultrasonic beam. The particles which are oriented perpendicular to the incident beam reflect considerable light and appear bright. Particles randomly oriented reflect negligible light and appear as a dark background. The degree of orientation depends on the intensity of the ultrasonic beam at a given point in the image. Thus the variation in intensity in the reflected light pattern indicates the amount of ultrasonic energy transmitted through the specimen. Flaws appear as dark spots in the image. Such a cell is limited in its sensitivity and is slow because the alignment of the particles may require several seconds. Also, the particles must be reagitated after each image is formed.

Other techniques for producing acoustical images are briefly mentioned. Rust, Haul, and Studt[16] used a potassium iodide–starch solution to produce an image. The ultrasonic waves liberated iodide, and as a result the potassium iodide–starch solution acquired a blue color. Ernst and Hoffman[17] and Rust[18] have suggested the use of substances whose color changes are sensitive to temperature, which would utilize the thermal effects of an ultrasonic beam. These authors have suggested a number of substances possessing the desired properties. Schuster[19] describes a technique in which the ultrasonic image is projected vertically onto a liquid surface and the resulting surface deformations are observed optically. Marinesco[20] observed that intense ultrasonic waves produced an image on a photographic plate immersed in water. Bennett[21] showed that the ultrasonic waves have a direct effect on photographic emulsions. Schreiber and Degner[22] observed the extinction of phosphorescence by ultrasonic waves. Figures 8-35 and 8-36 shows a system patented by S. Sokoloff[23] for producing an optical image. A piezoelectric plate was used to convert an ultrasonic image into an electronic image. The electronic image was then scanned with an electron beam, which in turn produced an optical image on a cathode ray tube. In 1950 Sokoloff[24] described an ultrasonic microscope which used this electronic method of image conversion. A sound-to-image transducing system was described in a patent[25] issued to G. L. Dimmick. The ultrasonic waves are focused upon a mosaic of transducers, and the individual transducers convert the waves into a multiplicity of discrete electrical signals. A scanning mechanism connects each transducer in turn through an amplifier rectifier to the control grid of a cathode ray Kinescope. The United States Naval Ordnance Laboratory has carried on an extensive project directed toward the development of an electronic method of image conversion. Barium titanate was used as the piezoelectric sensitive element. The tube developed by the Naval Ordnance Laboratory has two distinct

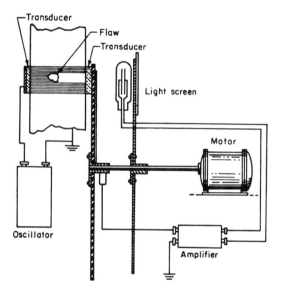

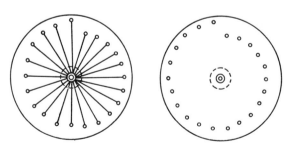

FIG. 8-35. Sokoloff acoustic image system.

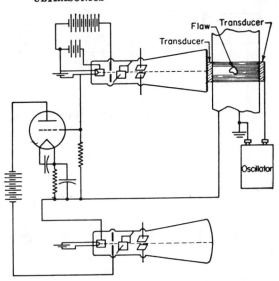

FIG. 8-36. Sokoloff acoustic image system.

sections. In one section the ultrasonic image is converted into an electronic image. A metallic transmission plate is used between the liquid
couplant and the barium titanate transducer. The metal plate serves
as part of the vacuum envelope, helps support the fragile transducer,
and gives the system a better impedance match. In the other section of
the tube is the electron image scanner, which consists of an electron gun.

Firestone Flaw Instrument. In 1942 a U.S. Patent[26] was issued to
Dr. F. A. Firestone for a "flaw detecting and measuring instrument."
Figure 8-37 shows the principle of operation of the instrument: (1)
mechanical vibrations from a transducer are transmitted into the specimen, (2) a portion of the waves have impinged on a defect, (3) an echo is
sent back toward the transducer from the defect, and (4) the remainder of
the waves travel to the opposite side of the specimen and are reflected.
The pattern observed on the cathode ray oscilloscope is shown in Fig.
8-37. An electrical pulse is delivered to the oscilloscope by a timing
device which gives rise to the initial pulse indication. The reflected
energy from both the defect and the opposite side of the specimen is
received by the transducer. The reflected energy is converted into
electrical pulses which appear on the oscilloscope displaced from the
initial pulse. The position of the flaw echo relative to the initial pulse
and back reflection indicates the position of the flaw in the specimen.
The height of the flaw indication gives an approximate measure of the
size of the flaw.

A schematic diagram of the Firestone instrument is shown in Fig.
8-38. The synchronizer is the timing source for the system, sending out
timing signals to the pulse generator, the sweep generator, and the marker

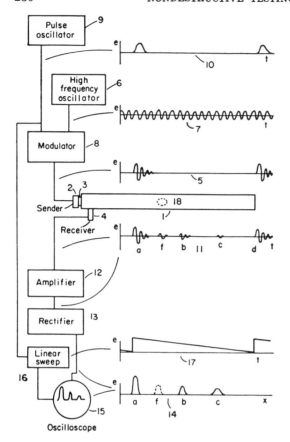

Fig. 8-37. Firestone's flaw detecting device and measuring instrument.

generator. This assures that the sweep of the cathode ray tube starts at the same instant that the transmitted pulse leaves the transducer. The synchronizer is a free running multivibrator. Detailed information concerning the operation of multivibrators can be found in an electronics text.[27] By varying the design parameters, different frequencies of oscillation can be obtained. The pulse generator produces a radio frequency pulse which is applied to the searching unit. The transducer is shock excited, producing a wave train having a duration of a few microseconds.

Figure 8-39 shows a schematic of a shock excited pulser. The fixed condenser C_1 is charged to a given voltage through the high resistance. A variable condenser C_2 and an inductance L are connected in a parallel resonant circuit. The transducer is connected across the parallel circuit. When a positive trigger pulse is applied to the thyraton tube T_1, it begins to conduct and essentially short-circuits C_1. Then C_1 discharges through the parallel resonant circuit, setting up a train of damped free vibrations. The frequency of the waves is determined by L and C_2. The resistance in

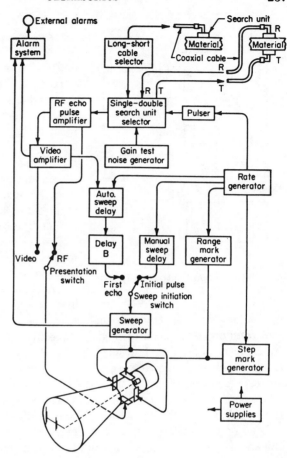

FIG. 8-38. Block diagram of reflectoscope. (*Courtesy Sperry Products, Inc.*)

parallel with L and C provides an adjustable damping resistance to control the pulse length. When the transducer is shock excited, it oscillates at its own natural frequency. The frequency at which such wave trains are produced depends on the frequency of the synchronizer. After the train of waves has been produced, the transducer is inactive for a long period of time and acts as a receiver. The echoes returning to the transducer are converted into electrical signals which are amplified and applied to the deflection plates of the cathode ray tube. In order accurately to locate flaws in the specimen, it is necessary to

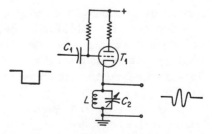

FIG. 8-39. Shock excited pulser.

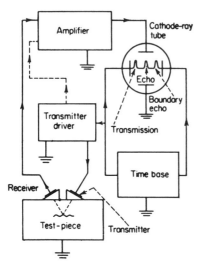

FIG. 8-40. Hughes ultrasonic instrument.

have time marks on the sweep. The marker generator produces square waves which are imposed as vertical deflections of the sweep trace

Henry Hughes and Sons in England has developed an ultrasonic testing instrument. This instrument uses the reflection principle but uses two transducers. The transmitter is mounted on a wedge which directs the ultrasonic beam into the specimen. The receiver is mounted on a similar wedge to make it sensitive to reflected energy coming from the same region in the specimen. Figure 8-40 shows a block diagram of the instrument.

Transmission Technique. The first technique used for ultrasonic inspection was the transmission technique. Continuous waves were sent from one transducer through the specimen and received by a second transducer. Flaws in the specimen tended to reduce the amount of transmitted energy. However, with the continuous waves, standing waves were often set up in the specimen or in the surrounding medium. The energy transmitted was often influenced more by the standing waves than by defects in the specimen. Equation (8-9) shows that the amount of energy transmitted by a thin specimen is dependent on specimen thickness. At the frequencies used for ultrasonic testing, the wavelength in most materials is only a few thousandths of an inch. So if the thickness of the test specimen varied only a few thousandths of an inch, the amount of energy transmitted might vary between wide limits. The effect of standing waves was overcome by modulating or "wobbling" the continuous frequency oscillator. An easier way to overcome the effect of standing waves is to use a pulsed ultrasonic beam.

Erwin[28] describes a transmission system in which provision is made to prevent overloading of the receiving transducer. Schubring[29] describes a system using a wobbled radio frequency oscillator. The specimen to be inspected was placed in a tank of water. The sending and receiving transducers were mounted in the sides of the tank. The electrical signal from the receiving transducer was amplified, rectified, and used to actuate a meter. In addition a relay was put in series with the meter. The relay could be adjusted to trigger at any predetermined amplitude of the transmitted energy, to make a "go-no-go" type of instrument. The General Motors Corporation used this equipment to find pipe and cup

flaws in cylinder head bolts such as shown in Fig. 8-41.   Krautkramer
and Rüdiger[30] describe an apparatus which can be used to test specimens
of thicknesses down to 0.10 in.   The pulse generator was modulated
with a noise potential, producing random variations in the output from
the generator.   These authors report that using their apparatus they
are able to detect fissures, shrinkage, and pores in cemented-carbide tips.

Resonance Technique.   The resonance technique can be used for the
detection of internal flaws.   Detection of flaws is even simpler than
thickness measurement, since only the presence or absence of thickness
indications need be observed.   It may be used on pieces having two
opposite sides which are smooth and parallel, such as plates, sheets,
tubes, blocks, or bars with squared ends.   It may be used on cylinders
$\frac{1}{2}$ to 36 in. in diameter and on spheres.   If, when the transducer is
coupled to one of the surfaces of the specimen, the resonance indications
corresponding to thickness are obtained, then the section under the
crystal is ultrasonically sound.   This means that it has no voids, cavities,
or porosity covering any appreciable portion of the area under the
transducer.   If, however, there are flaws under the transducer and these
flaws are not parallel to the surface, all the thickness indications will be
greatly reduced or will disappear altogether.   This technique for flaw
detection consists of moving the transducer over the specimen and watch-
ing for significant changes in the resonance peaks.   In Fig. 8-42a, the
crystal is over a sound section of a steel block and a large number of
harmonic peaks are visible on the screen.   When the transducer is over a

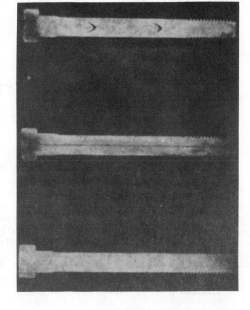

FIG. 8-41. Inspection of
bolts   using   ultrasonic
transmission   technique.
Sound bolt is at the bot-
tom. Pipe flaws were
detected while cup flaws
in the top bolt were very
easily detected. (*Cour-
tesy Research Laborato-
ries, General Motors Cor-
poration.*)

nonparallel saw cut, Fig. 8-42b, the thickness indications on the screen disappear.   Figure 8-43 illustrates a case where the flaw is perpendicular to the contact surface.   If such a flaw is small and perpendicular to the surface, the effect on the thickness indications will be small.   However, if the width of the flaw is large compared to the crystal, an appreciable decrease in resonance indications may be obtained.

A second resonance technique for flaw detection may locate and give information on the depth to the flaw.   It can be applied only to special cases where the flaw is flat and parallel to the crystal contact surface.

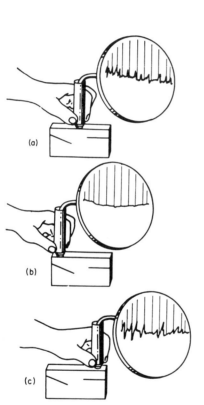

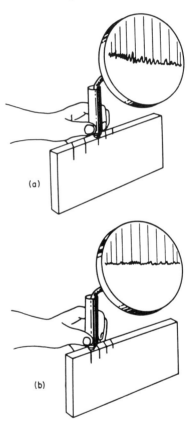

Fig. 8-42. Resonance from solid and laminated material.   If the portion under the crystal is sound, the indication corresponding to total thickness is obtained, as shown at a.   (Black lines in block are saw cuts.)   If the flaw is not parallel to the surface, no thickness indications appear on the screen, as shown at b.   If the flaw is parallel to the contact surface a thickness reading is obtained, as shown at c, characteristic of the depth to the flaw.

Fig. 8-43. Resonance inspection.   When a small flaw lies perpendicular to the surface so that its projected area on the crystal is negligible or small, the effect on the resonance indications is likewise small, as shown at a.   When a perpendicular crack is deeper than about 0.5 in., it may be found by the reduction in amplitude of the resonance indications, as indicated in illustration b.

In this case the metal between the flaw and crystal may vibrate at its own resonant frequencies. The resonance peaks for the total thickness of the part do not appear, and the presence of a flaw is indicated. The depth to the flaw below the crystal contact surface may be determined from the instrument readings. The effect just described is illustrated in Fig. 8-42c. When the transducer is moved over the simulated parallel flaw, the appearance of the harmonic pattern on the screen changes. The resonance peaks are farther apart in Fig. 8-42a because of the greater thickness which is set into resonance. This second technique has some use on parts which do not have an opposite parallel surface, in which case no thickness indications are observed unless a parallel flaw is found. It should be emphasized that in this case a nonparallel flaw will not be detected. Since flaws in general are not parallel to the contact surface, negative results for this inspection do not guarantee freedom from flaws. Typical flaws found using the resonance technique include laminations, shrinkage cavities, porosity, and cracks. Even intergranular cracks and some carbides are detectable.

In addition to testing solid parts, the resonance technique may be applied to testing certain types of bond. In order to detect a faulty bond with a resonance technique there must be an area of actual separation of the bonded material. One typical application of this type is the inspection of a metal sheet copper-brazed to another member, as around the edges and over the rib of a propeller blade. Another typical application is the examination of bond between a silver bearing pad and the backing to which it has been attached by a soldering operation. Butt welds in bars may be inspected from the squared ends. The resonance technique has been used to test the bonds between sheet metal and rubber. Where there is good bond thickness indications are absent or small because of the damping due to the rubber. Where a void exists in the bond the thickness indications appear. The resonance technique has been used in the case where there is a thin layer of low damping cement between two sheets of metal. Greenberg[31] has used the ultrasonic resonance technique to check the quality of brazed circuit-breaker contacts. He found good correlation between the ultrasonic results and distructive shear tests. The resonance technique has been used successfully to test for corrosion. Evans[32] discusses the use of the Audigage in measuring corrosion and evaluating corrosion control systems in ship bulkheads.

Factors which sometimes make it impossible to use the ultrasonic resonance technique are the type of material, excessive thickness variations in the material within the area contacted by the crystal, and closely adherent scale. Excessive thickness variations in the material within the area contacted by the transducer may be due to severe corrosion which creates deep pits.

Resonance Thickness Measurement. Thickness resonance occurs whenever the thickness of the material is equal to an integral number of half wavelengths of the ultrasonic wave. Figure 8-33 shows various standing patterns in material. In a standing wave the points of maximum displacement are referred to as antinodes and the points of minimum displacement as nodes. The distance between adjacent nodes or adjacent antinodes is a half wavelength. In resonance testing there is always an antinode at the transducer and an antinode at the opposite side of the test specimen. In Fig. 8-33 the thickness of the material t is equal to $\lambda/2$, $3\lambda/2$, and 2λ, respectively. Using the expression $V = f\lambda$ one can write for the fundamental frequency at which thickness resonance occurs

$$f_1 = \frac{V}{\lambda} = \frac{V}{2t} \tag{8-20}$$

where f_1 = frequency, cps
$\quad V$ = velocity of sound in material
$\quad t$ = thickness of material

Thickness resonance also occurs at harmonics or multiples of the fundamental frequency such as

$$f_2 = \frac{V}{t} \qquad f_3 = \frac{V}{2t/3} \tag{8-21}$$

Thus $\qquad f_2 = 2f_1 \qquad f_3 = 3f_1 \qquad f_n = nf$

The frequency difference between any two adjacent harmonics is equal to the fundamental frequency. If the fundamental frequency can be determined, this relation can be used to determine the thickness of the material. When two adjacent harmonic frequencies are known, the thickness relationship becomes

$$t = \frac{V}{2(f_n - f_{n-1})} \tag{8-22}$$

Resonance Instruments. One of the instruments using the resonance principle is the Sonigage. The basic circuit of the Sonigage is shown in Fig. 8-44. It consists of a variable frequency self-excited oscillator with a d-c milliameter in the plate circuit. The oscillator generates an alternating voltage which is applied to a transducer. When the oscillator is tuned to a thickness resonant frequency of the material, an increase in the plate current of the oscillator results, which is indicated on the milliameter. At resonance there is

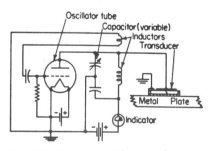

FIG. 8-44. Sonigage. (*Courtesy Branson Instruments, Inc.*)

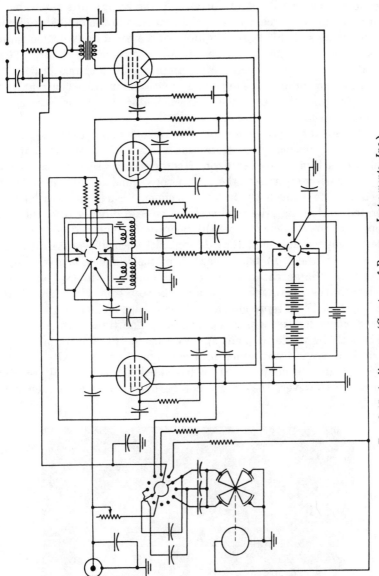

FIG. 8-45. Audigage. *(Courtesy of Branson Instruments, Inc.)*

an increase in the amplitude of vibration in the specimen beneath the transducer. There will be an increase in the energy dissipated, which in turn causes an increase in the plate current in the oscillator circuit. In later versions of this type of equipment a mechanically driven tuning condenser sweeps the frequency of an electronic oscillator. These momentary increases in energy after amplification and rectification are indicated by an audible tone in the earphones and increased current in the output meter in the Audigage. A schematic circuit diagram of the instrument is shown in Fig. 8-45.

Other resonant types of equipment are sold under the trade names Reflectogage, Sonizon, and Metroscope. In each case the amplified resonance pattern is displayed on a cathode ray oscilloscope. Points along the horizontal axis of the screen correspond to frequencies being applied by the transducer to the specimen. Such an instrument is calibrated so that material thickness can be read directly from the oscilloscope. In the Vidigage, Fig. 8-46, instead of a motor-driven capacitor an electrically variable inductor working on the principle of the saturable-core reactor is used. A block diagram of the Vidigage is given in Fig. 8-47. A very large cathode ray tube is employed to give maximum scale length, the advantage of which can be seen from Fig. 8-46, where a scale length of 17 in. represents a thickness range of 0.04 in.

Transducers for Resonance Equipment. The transducer constitutes the heart of the whole resonant system, since it is the medium whereby electrical energy is transformed into mechanical energy. Crystals used in the resonance technique must be driven below their natural resonance frequency. However, they must not be operated at a frequency too

Fig. 8-46. Vidigage. (*Courtesy Branson Instruments, Inc.*)

far from their natural frequency; otherwise a considerable loss of sensitivity will result. In practice a frequency range not greatly exceeding 2 to 1 has proved satisfactory. In most of the commercial instruments each tuning range covers approximately one octave, or a 2 to 1 interval. Additional ranges are incorporated in the instrument by merely switching different inductance coils into the oscillating circuit. For each range a different calibration scale is required in front of the cathode ray screen. To facilitate changing, the

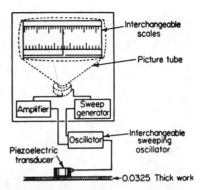

FIG. 8-47. Block diagram of Vidigage. (*Courtesy Branson Instruments, Inc.*)

scales are printed on transparent sheets which can be easily positioned in front of the cathode ray screen. When a flat transducer is used on curved surfaces, only a small area of the face is in contact with the specimen. Consequently, only a small amount of energy is transmitted into the specimen. A larger area of contact, giving a greater transmission of energy, can be achieved by grinding the crystal to fit the surface.

On a thick specimen a series of harmonic peaks will appear. The reading on the scale for one peak will be r_1 and for another peak r_2; then

$$r_1 = \frac{V}{2nf_0} \qquad r_2 = \frac{V}{2mf_0} \qquad (8\text{-}23)$$

where n and m are integers and f_0 is the fundamental frequency. It is not possible to know which harmonic frequencies are present, but each scale reading can be written in terms of the fundamental frequency. If the resonant peaks are adjacent, $n = m - 1$; then, since $f_0 = V/2t$,

$$\frac{V}{2t} = \frac{V}{2(m-1)r_1} = \frac{V}{2mr_2}$$

$$t = (m-1)r_1 \qquad \text{and} \qquad t = mr_2$$

Eliminating m, $\qquad\qquad t = \dfrac{r_1 r_2}{r_1 - r_2}$

If two nonadjacent harmonic peaks are taken, then the above expression becomes

$$t = N \frac{r_1 r_2}{r_1 - r_2} \qquad (8\text{-}24)$$

when N is number of harmonic peaks between r_2 and r_1.

The commercial resonance type instruments give direct readings of thickness from approximately 0.025 to 0.300 in. But by use of har-

Fig. 8-48. Ultrasonic scanner. (*Courtesy Argonne National Laboratory.*)

monics the range can be extended to approximately 4 in. The commercial instruments are provided with removable calibration screens for the different thickness ranges. Calibration of the equipment is checked by the use of a test block of known thickness.

Transmission Systems. McGonnagle and Beck[33] describe an ultrasonic transmission and recording system. No one part of this system is unique, but as a complete system it has interesting features, the most important being that it produces a permanent two-dimensional record of the ultrasonic transmission through a specimen. The system consists of an ultrasonic generator and receiver, a scanning assembly, and a recording system. The ultrasonic generator and receiver used was a commercial Reflectoscope. The instrument was modified so that a negative pulse could be taken from the pulse generator for keying the recording system.

The scanner assembly consists of a rigidly supported track, a tank, transducer carriage, drive motors, and Selsyn motors. Figure 8-48 shows the scanner with the sheet metal side panels removed. The transducer carriage is used to support and carry the two transducers along the specimen. The transducer carriage is moved along the rails by means of a chain drive, the chain being driven by a motor through a variable-speed transmission. This motor is referred to as the translational drive motor. The drive chain also engages a Selsyn master motor.

The transducer assembly is a brass yoke designed to hold the two crystals in a rigid position with the faces of the transducers parallel. In addition to the transducers a fixed speed motor for moving the transducers up and down is mounted on the transducer carriage. A master Selsyn motor is also actuated by the vertical drive motor. This scanning assembly can be used to test either flat or round specimens. Round specimens are mounted in a special rack and rotated in the couplant by means of a chain drive through a variable-speed transmission. In this application, the transducer carriage moves the transducers past the rotating specimen at a fixed elevation.

A recorder, Fig. 8-49, produces a permanent two-dimensional record of the ultrasonic transmission through the specimen. The recorder uses a moist electrosensitive paper which passes between two electrodes. The darkening of the paper is a function of the current flowing between the electrodes and the writing speed. The greater the current flow, the greater the darkening of the paper. The faster the writing speed, the greater the current needed for satisfactory recording. Two electrodes are used in this recorder: a fixed electrode which is the positive electrode and a movable electrode which is the negative electrode. The positive electrode is made from a strip of stainless steel $\frac{3}{64}$ in. thick. The negative electrode is a cylindrical drum on which a spiral turn of nichrome wire is wound. As the moving electrode revolves, the point of intersection between the fixed electrode and the helix travels in a straight line,

Fig. 8-49. Electrosensitive paper recorder. (*Courtesy Argonne National Laboratory.*)

ANL
A N L

FIG. 8-50. Electrosensitive recorder trace. (*Courtesy Argonne National Laboratory.*)

sweeping across the paper. The paper feed motor is actuated by the Selsyn motor connected to the translational drive of the scanner. Thus the length of the recorder trace is equal to the length of the specimen being scanned. A master Selsyn is connected to each of the drive motors which are used either to rotate the specimen in the tank or move the transducers up and down. This master Selsyn motor is connected to a slave Selsyn motor which drives the recorder helix. By proper selection of the diameter of the recording helix the width of the recording is equal to the width (or circumference) of the specimen. Figure 8-50 shows typical recorder traces and the kind of sensitivity obtainable with this recorder using artificial defects.

The signal to be recorded is taken from the vertical amplifier of the Reflectoscope. The amplitude of this signal is a function of the ultrasonic transmission through the specimen. The electronic component of the recording system operates as follows. A negative keying pulse is taken from the synchronizer of the Reflectoscope. The keying pulse activates an electronic gate so that the initial pulse does not reach the recorder. The ultrasonic pulse transmitted through the specimen goes to the recorder.

This equipment has been used to test the integrity of uranium castings,[34] the light areas being areas of poor transmission showing the presence of pipe, blowholes, and porosity and the dark areas being areas of good transmission. Figure 8-51 shows a typical trace of a zirconium clad uranium fuel element. The light areas in the recording indicate regions of poor bond, and the dark areas are regions of good bond in the fuel element. Using a beam having a cross section of $\frac{5}{64}$ in., unbonded areas $\frac{1}{8}$ in. in diameter or larger can readily be found. Using beams of smaller cross section unbounded areas as small as $\frac{1}{32}$ in. have been found.

It is possible to use this technique not only to determine lack of bonding but also to determine the relative degree of bonding of zirconium clad fuel elements. By proper selection of the amplification and recording level, the relative darkening of the paper gives a relative measure of the bond strength. Bond strengths were determined by tensile testing. Figure 8-52 shows test specimens prepared for tensile testing. Comparison of the tensile test with the relative darkening of the recorder trace gives correlation between bond strength and darkening of the paper.

Ross and Leep[35] describe a transmission system in which barium titanate transducers are used. The use of these transducers eliminates the need for the high-voltage power supply required to excite quartz transducers. In the hypersonic analyzer,[36] ammonium dihydrogen phosphate transducers were used. To protect the transducers, the crystals were encased in a rubber jacket filled with oil. The ultrasonic transmission equipment is in general easy to operate and to interpret. One disadvantage of this technique is that it does not indicate the depth of the flaw below the surface unless orthogonal scanning techniques are used. This technique is not so able to differentiate between various types of flaws as is the reflection technique.

The sensitivity and resolution of the ultrasonic transmission method of testing are influenced by such factors as the effective cross section of the ultrasonic beam, the parallelism of the transducers, the presence of air

FIG. 8-51. Recording of nuclear reactor fuel element. (*Courtesy Argonne National Laboratory.*)

FIG. 8-52. Specimens for tensile test of bond. (*Courtesy Argonne National Laboratory.*)

bubbles on the transducers or specimen, the speed of scanning and recording, and the orientation of the specimen with respect to the ultrasonic beam. In order to achieve the maximum sensitivity, the faces of the transmitting and receiving tranducers must be parallel. The effective cross section of the ultrasonic beam can be limited by use of a mask. An air space makes an ideal reflecting surface, and a good mask can be made by incorporating an air space into the mask. Another method of limiting the cross section of the beam while at the same time concentrating the ultrasonic energy is to use lenses. Lenses made of Lucite or Polystyrene or carbon tetrachloride liquid lens have been used. In the contact lens technique using a plano-spherical or plano-cylindrical lens, the plane side of the lens is fastened to the transducer with wax. To minimize acoustic losses the lens material should have an acoustical impedance as nearly equal to that of the transducer as possible. Since aluminum and quartz have nearly the same impedance, lenses made from aluminum have been used successfully. Aluminum also has a high acoustical refractive index when immersed in water (velocity of sound in aluminum $\div$ velocity of sound in water = 3.5). The principles of physical optics carry over into the field of image formation with acoustic lenses. From the usual lens makers formula, the radius of curvature can be calculated for a given focal length. The lenses can be made to give a line focus or a point focus. From optical theory the resolving power of a lens is given by $a = 1.22\lambda/D$, where λ is the wavelength and D the diameter of the aperture. For highest resolving power a high-frequency and large aperture to focal length ratio lens should be used. The aperture ratio is limited by the fact that spherical aberration increases rapidly with the increasing aperture ratio.

Reflection Techniques. In general it is advisable to use the reflection technique wherever it can be applied. However, this technique cannot be used for testing thin specimens. This is due to the fact that in 1 μsec an ultrasonic wave will travel approximately ¼ in. With thin material the reflected and initial pulses will not be spread out sufficiently to see any reflections from defects in the specimen. The reflection or pulse echo technique is the most widely used ultrasonic testing technique. Typical flaws detectable include the following: cracks, folds, inclusions, laminations, partial welds, voids, segregations, shrinks, porosity, and flaking. The ability to interpret the oscilloscope pattern is best developed by experience.

Scanning is the general name given to the process of moving the transducer over or along the test specimen. Scanning can be divided into two categories, contact and immersed. In contact scanning the transducer directly contacts the specimen through a thin film of oil or other suitable couplant. In immersed testing the specimen and the transducer are each immersed in the couplant but are separated from each other.

The contact technique is widely used, because it is well adapted for manual scanning, and the portability of the transducer makes it useful for field and preventative maintenance inspection. Large parts sometimes cannot conveniently be immersed. The advantages of immersed inspection are: it permits inspection of rough surfaces without excessive wear of the transducer; higher frequencies may be used so that smaller defects and defects closer to the surface can be detected; it permits angulation of the transducer, so that the beam can be directed normally into areas often inaccessible by the contact method; and it is easily adapted for automatic or semiautomatic scanning. Some disadvantages of immersed inspection are: the specimen must be immersed in a couplant; sound energy is lost at interfaces, requiring more powerful equipment, precise directioning of the sound beam is required; and diffraction will occur if the sound enters the surface at other than normal incidence and mode conversion may occur.

Types of Scanning. Three types of data presentation are used in reflection testing: the A scan, Fig. 8-53A; the B scan, Fig. 8-53B; and the C scan, Fig. 8-53C. The A scan which is a one-dimensional presentation shows the existence of flaws, shows their position, and gives an estimate of their size. The B scan shows the reflections from the top and bottom of the specimen and from flaws as the transducer moves along a line. The B scan gives a cross-sectional view of the specimen along a given line. The advantage of this type of presentation is that both the length of the flaw and its depth below the surface are revealed. The C scan is analogous to a radiograph, in that it is a projection of the internal details of the specimen into a plane. The C scan can be obtained from an A scan presentation made at a number of points over the surface of the specimen. If sufficient points are plotted, a contour of the flaw area is obtained. In the C scan front and back surface reflections are not used, only the reflections from the flaws.

In an A scan presentation the initial pulse is the indication of the pulse generated for transmission to the transducer. Too long an initial pulse may obscure indications from defects lying close to the test surface. The initial pulse should be adjusted to be as short as possible but still permit sufficient penetration. Selection of the proper frequency may be important in identifying the type of flaw. For example, clusters of porosity, segregates, or

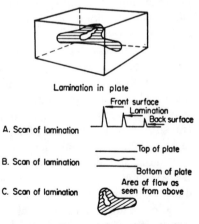

Lamination in plate

A. Scan of lamination

B. Scan of lamination

C. Scan of lamination

Front surface
Lamination
Back surface

Top of plate

Bottom of plate

Area of flaw as seen from above

FIG. 8-53. Types of data presentation.

FIG. 8-54. Defect parallel to transducer face. (*Courtesy Sperry Products, Inc.*)

FIG. 8-55. Defect at angle to transducer face. (*Courtesy Sperry Products, Inc.*)

FIG. 8-56. Indication of flaking. (*Courtesy Sperry Products, Inc.*)

FIG. 8-57. Indication of coarse porosity. (*Courtesy Sperry Products, Inc.*)

FIG. 8-58. Indication of segregation. (*Courtesy Sperry Products, Inc.*)

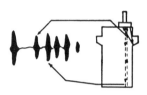

FIG. 8-59. Effect of flange on indications. (*Courtesy Sperry Products, Inc.*)

inclusions may be mistaken for a large defect when using a high test frequency, but inspection using a lower frequency may aid in deciding the type of defect.

Back Reflections. The reflection from the rear surface of the specimen is usually called the back reflection. The appearance of a back reflection on the viewing screen indicates that the vibrations are penetrating the test specimen completely. Reflections from defects will be indicated between the initial pulse and the first back reflection indication. Reflections from shoulders, flanges, or fillets may also be indicated to the left of the back reflection.

Typical Indications. A sharp indication, Fig. 8-54, usually signifies a crack or other defect in a position approximately parallel to the face of the transducer. A bulbous indication, Fig. 8-55, usually signifies a defect whose surface lies at an angle to the face of the transducer. It may be either a crack or a void. A discontinuity at an angle to the transducer will cause a shifting of the indication as the transducer is moved along the specimen. A scattering of sharp indications, which fade in and out as the transducer is moved, usually signifies flaking, Fig. 8-56. A scattering of bulbous indications, which fade in and out as the transducer is moved, usually signifies coarse porosity, Fig. 8-57. Numerous small overlaid indications, which fade in and out and are sharp, usually signify segregation, Fig. 8-58.

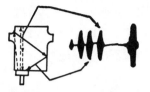

FIG. 8-60. Reflection from fillet. (*Courtesy Sperry Products, Inc.*)

FIG. 8-61. Indication of laminations or lack of bond. (*Courtesy Sperry Products, Inc.*)

FIG. 8-62. Indication of porosity, large grain size, or segregation. (*Courtesy Sperry Products, Inc.*)

FIG. 8-63. Lack of back reflection. (*Courtesy Sperry Products, Inc.*)

In Fig. 8-59, the path of the reflection from the flange is shorter than the path to the back face, and hence will be indicated to the left of the back reflection. This type may be mistaken for a defect indication. Differentiating between the two calls for reasonable skill and judgment on the part of the operator. Multiple reflections may appear on the viewing screen when testing parts having irregular shapes and should not be confused with multiple back reflections. Ultrasonic vibrations tend to spread out, giving reflections from the flange and shoulders. In Fig. 8-60 the path to the fillet is longer than that to the back face of the specimen. Hence the indication from the fillet will fall to the right of the first back reflection indications. Successive back reflection indications are equally spaced. However, a sudden increase in the number of back reflections without a corresponding decrease in the length of the

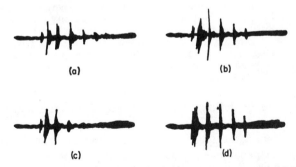

FIG. 8-64. Reflectograms showing effect of microstructure; material, high-speed steel. (*a*) Heated to 2100°F for 15 min, quenched in oil and double tempered at 1075°F; hardness, Rockwell C 60. (*b*) Retained in annealed condition; hardness, Rockwell C 21. (*c*) Heated to 2500°F for 15 min, quenched in oil and double tempered at 1075 F; hardness, Rockwell C 64. (*d*) Heated to 2350°F for 15 min, quenched in oil and double tempered at 1075°F; hardness, Rockwell C 63. (*Courtesy Iron Age.*)

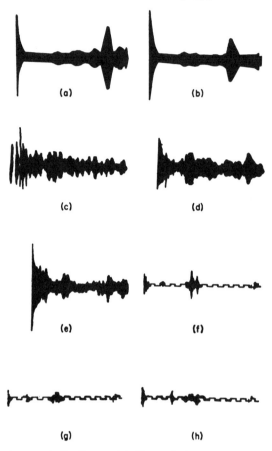

FIG. 8-65. Reflectoscope indications. (a) Shows the back reflection with minor indications. Hot top end. (b) Same as Fig. a but from the bottom of the ingot. (c) Lack of back reflection and sharp indications indicate porosity. (d) The sharp indication between initial pulse and apparent back reflection indicates a secondary pipe. (e) Same as Fig. d, showing continuation of pipe in Fig. d. (f) Indication of small forging bursts at center line of shaft. (g) Indications of porosity at center line of shaft. (h) Indications of porosity and bursts. (*Courtesy American Society for Testing Materials and R. N. Hafemeister.*)

specimen is usually indicative of the presence of laminations, see Fig. 8-61.

Reduction, fading, or loss of back reflection as the transducer is moved along the specimen can be due to a crack at right angles to the face of the transducer. This condition may cause complete lack of back reflection when the transducer is directly above the crack. This is particularly true of such cracks when they extend into, or close to, the surface of the specimen. An angular discontinuity or crack whose upper face is at such an angle as to reflect the vibrations in a direction away from the transducer causes loss of the back reflection. Porous areas in the material may cause the back reflection to fade as the transducer passes directly over the area. The back reflection will reappear when the transducer is moved away from the surface directly above the porosity. Large grain size may affect the back reflection in a manner similar to porosity. Loss of back reflection may occur when the ultrasonic vibrations are traveling completely through the specimen and into the material upon which it is resting. A noticeable decrease in the amplitude or number of successive back reflections may also be indicative of porosity, large grain size, or segregation, Fig. 8-62. If a back reflection fails to appear on the screen,

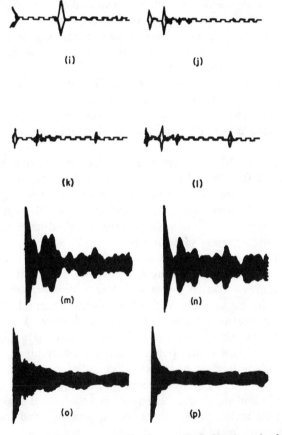

FIG. 8-65 (*Continued*). Reflectoscope indications. (*i*) Indication of a large burst at center line of shaft. (*j*) Evidence of burst and porosity. Back reflection indicates 3-in. bore. Figures *j* to *l* confirm tests made prior to boring. (*k*) General area of Fig. *j*. (*l*) Further exploration of general area of Fig. *j*. (*m*) Shows no evidence of bridging at the hot top. Small pipe indicated. (*n*) Shows evidence of residual pipe. (*o*) No evidence of pipe remains. General porosity. (*p*) Evidence of porosity only. (*Courtesy American Society for Testing Materials and R. N. Hafemeister.*)

Fig. 8-63, regardless of the position of the transducer, it may indicate one of the following conditions. The vibrations are not penetrating to the far side of the material because of poor surface condition. The vibrations are not penetrating to the far side because of the use of too high a frequency, or the opposite side of the specimen is rough or pitted. The transducer is not in good contact, which may be because of poor handling or improper couplant. Laminations in the material close to the testing surface cause loss of back reflection. If the opposite side is angular, it will deflect the reflections away from the transducer.

Care must be taken in attempting to deduce the size of a defect from the amplitude of the reflection on the oscilloscope. The orientation as well as the size of the defect influence the size of the echo. The effect of orientation of the flaw on the A scan presentation is shown in Fig. 8-98. The defect appears far less serious in the side test than it does from the top, because the reflecting surface is greater when the defect is perpendicular to the sound beam.

Pulse Echo Applications. In the following section the application of ultrasonic testing in various industries and in a variety of materials will be briefly discussed. Literature references are included for anyone desiring detailed information on a particular material or in a particular industry.

Hartley and Mull[37] have used the Reflectoscope to test tool steels. They report that this technique can be used to differentiate varying degrees of segregation, voids and segregation, and varying amounts of heat-treating. Figures 8-64a, b, c, and d show reflectograms of four specimens heated to different temperatures. Photomicrographs of the specimens show a variety of structures.

Hafemeister[38] has examined a group of forging ingots using the reflection technique and contact scanning. Tests were made at a frequency of 0.5 Mc. Figures 8-65a to 8-65p are A scan presentations obtained in this study. In Figs. 8-65a and 8-65b the prominent back reflections with minor indications between the initial pulse and back reflection show that this is no evidence of porosity or secondary pipe. Figures 8-65c and 8-65d show center porosity and secondary pipe. Figure 8-65i indicates a large forging burst and Fig. 8-65f a small forging burst. Ward and Borg[39] have used the reflection technique to test ingots of arc cast zirconium and its alloys. Frequencies of 2.25 and 5 Mc were used. They were able to locate inclusions, voids, porosity, and discontinuities in ingots up to 48 in. in length.

Valuev[40] discussed the testing of forging of the type shown in Fig. 8-66, at frequencies of 1.25 and 2.5 Mc. Indications of flaws were obtained when the transducer was moved over the end surfaces. Indications were also obtained by radial scanning of the center section of the forging. This indicated that the flaws were close to the central hole through the forging.   The size of the flaws was estimated from measurements on calibrating specimens cut out of the body of the forgings. Artificial flaws were produced by drilling flat bottom holes of various sizes.   The size of the flaws was estimated to be in the range of 0.023 to 0.031 in.[2]   After complete ultrasonic inspection the forging was sectioned and found to have a large number of metallic inclusions.   These inclusions were 0.004 to 0.100 in. long in a direction at right angles to the axis of the forging and approximately 0.200 in. in the axial direction. Thus the destructive tests showed good correlation with the nondestructive test results.

One of the difficulties in testing castings is due to as-cast surface roughness.   This is especially true in using the contact scanning technique.   The immersed technique does not suffer from this difficulty. However, the immersion of large castings also presents a problem.   The use of a water jet eliminates this difficulty.   This device consists of a tubular nozzle-like extension in front of the transducer.   The extension had a side port through which water is introduced to fill the nozzle and provide a continuous flow from the nozzle.   The stream of water impinges on the specimen, providing a path for the transmission of the ultrasonic energy from the transducer into the test specimen.

Ultrasonic techniques are used for locating upper-fillet and bolthole cracks in rail ends within joint-bar limits.[41]   The reflection technique is used for finding defects in locomotive axles, crankpins, booster axles,

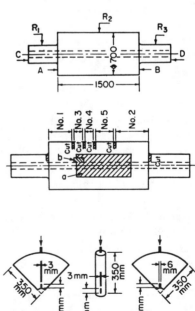

FIG.   8-66.   Testing of forging.   (*Courtesy Henry Brutcher, Altadena, Calif., Translation No. 3685.*)

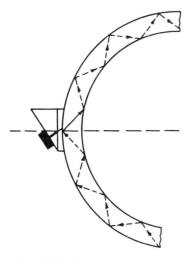

FIG. 8-67. Shear waves in pipe.

diesel driving axles, and diesel traction motor shafts.[42] The results obtained not only have increased with savings in material and labor costs, but have increased the safety of train operation and reduced out-of-service time. In England ultrasonics have been used extensively for finding hairline cracks in steel.[43]

Shear Wave Applications. The next applications to be discussed are those dealing with the application of shear or transverse waves. Ultrasonic shear waves travel a zigzag path, as shown in Fig. 8-31. Shear waves will travel in a pipe by the path shown in Fig. 8-67 and have been used successfully for flaw detection in pipes.[44] The technique used in testing pipe is to hold the transducer firmly against the pipe as the pipe is being rotated. The transducer is slowly moved laterally. This assures complete coverage of the pipe and aids in interpreting the pattern on the oscilloscope screen. Since the pipe is rotating the flaw reflection is readily perceived on the screen by its movement to the right or left, depending upon the direction of the pipe rotation. Any false or spurious indications remain fixed on the screen. Notches on the inside and outside of a pipe are often used as a calibrating standard. Notches 3 per cent of the wall thickness or 0.004 in. deep, whichever is larger, are commonly used.

The technique described above cannot be used on small tubing because of the blanking of the receiver at the instant of pulse transmission. The nondestructive testing group at Knolls Atomic Power Laboratory[45] has developed a technique for testing small-diameter tubing. The shear waves are made to travel parallel to the longitudinal axis of the tubing. A special transducer was made from a commercial shear wave transducer. By rotating the tubing with the transducer located at one end of the tubing and moving a transducer longitudinally approximately ½ in., the length of a tube can be inspected. The results obtained with various artificial defects using this technique are shown in Fig. 8-68.

A technique examining small-diameter tubing circumferentially is to use the shear wave unit shown in Fig. 8-69.[46] The transducer is mounted on a wedge which delays the entrance of each pulse into the specimen. The time delay must be larger than half the duration of the pulse itself. To eliminate the reverberations due to reflections at the interface of the specimen, an absorbing wedge is incorporated in the delayed shear wave

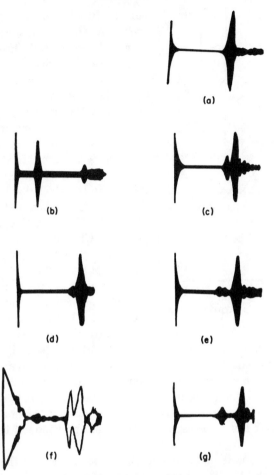

Fig. 8-68. Shear wave testing of small diameter tubing. (a) Reflectogram of tubing shows no defects. Peaks are of initial pulse (*left*) and back reflection (*right*). (b) Solid steel rod of 0.080 in. diameter shows no defects in reflectogram. From left to right are initial pulse, first and second orders of back reflections. Better sound transmission is obtained with solid bars. (c) File marking about 0.005 in. deep, 2 in. from end of tube, and 40° circumferentially is indicated on screen. Middle defect indication is 33 per cent of back reflection. (d) Defect indication is that of notch 0.002 in. deep, 2 in. from end of tube, $\frac{1}{32}$ in. wide, and 180° circumferentially. Second indication from left shows that the defect is 17 per cent of back reflection. (e) Two notches made circumferentially 3 and 4 in. from end of tube, 0.002 and 0.001 in. deep, respectively, are indicated in the reflectogram. Second and third indications show defect side by side. (f) Axial groove in tubing is indicated by second largest peak from left. The groove is about 0.005 in. deep, $\frac{1}{4}$ in. long, and 2 in. from opposite end of tube. (g) Single hole made 4 in. from end of tubing with a No. 80 drill through 0.010-in. wall. Second indication from left shows that defect is 30 per cent of back reflection. (*Courtesy Knolls Atomic Power Laboratory.*)

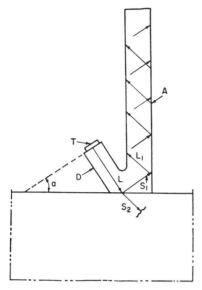

FIG. 8-69. Delayed shear wave unit. (*Courtesy Knolls Atomic Power Laboratory.*)

unit. In the absorbing wedge the reflected beam bounces back and forth along the walls, losing energy by absorption. Using this type of shear wave unit, an artificial defect, a No. 60 hole $\frac{1}{4}$ in. deep, was detected $\frac{1}{32}$ in. below the surface.

A recording system was developed for use with the delayed shear wave transducer. The recording unit comprises two independent channels as shown in the block diagram, Fig. 8-70. The two channels are identical except that they are "active" at different times. Channel 1 is "on" during the period between the end of the initial pulse and the start of the end return; channel 2 is "on" during the period in which the end return is being received. The selected pulses are amplified and applied to a recorder.

The sensitivity of channel 1 is set by making the indication from a standard artificial defect equal to a predetermined deflection of the recorder. The amplitude of the end return (channel 2) gives a constant measure of the test sensitivity. If the coupling between the search unit and the tube decreases, the amount of energy transmitted into the tube is reduced and a correspondingly lower amplitude of the end return is recorded.

A shear wave unit has also been used for inspecting the bore of gun barrels,[47] Fig. 8-71. The shear wave is projected into the wall of the gun barrel and travels circumferentially around the barrel. Any discontinuity in the wall will reflect part of the beam back to the transducer. As the search unit is advanced in the bore, the entire body and the wall are inspected in a spiral manner as the tube rotates. Using this method, it is possible to detect cracks of very small area.

The shear wave technique has been used for detecting welded seams in low-carbon steel pipe.[48] The transducer is shaped to the contour of the pipe and the surface of the pipe scanned by moving the transducers circumferentially around the pipe. In each case a noticeable reflection was present as the transducer approached a position approximately 15° from the weld. This reflection increased in amplitude and moved toward the initial pip as the transducer approached the welded seam.

Usually the shear wave transducer is placed in direct contact with the specimen, but shear waves are also used in immersion testing. The shear

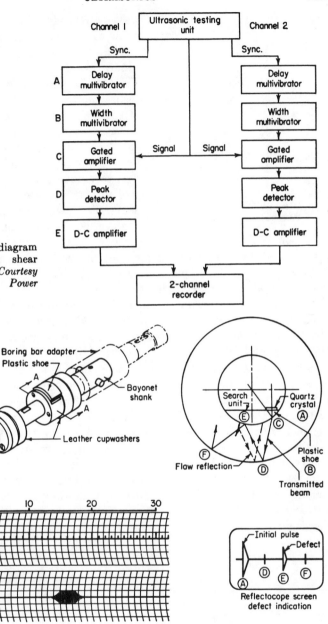

FIG. 8-70. Block diagram for two-channel shear wave unit. (*Courtesy Knolls Atomic Power Laboratory.*)

FIG. 8-71. Shear wave technique for inspecting gun barrels. (*Upper left*) Internal bore angle search unit attached to boring bar. (*Upper right*) Diagram of cross section of gun barrel, showing bath of ultrasonic test beam. (*Lower right*) Diagram of CRT screen indications for visual monitoring. (*Lower left*) Typical diagram with scale or distance markers and defect indication showing length and distance from end of gun tube. (*Courtesy Sperry Products, Inc.*)

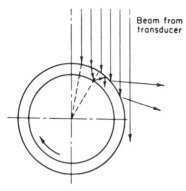

FIG. 8-72. Immersed testing using shear waves.

waves are produced in the specimen by mode conversion, as shown in Fig. 8-72.

Weld Inspection. Ultra sonic techniques are being used to test welds. Figure 8-73 shows two techniques for inspecting welds. The weld bead is often too rough for proper coupling using contact scanning; therefore it can be used only after the bead has been removed by machining. In the first technique defects near the surface will not be resolved. Another reason why the first technique is unsatisfactory is that a crack will give no reflection if the ultrasonic beam is parallel to the plane of the crack. The second technique can sometimes be used and gives satisfactory results. A third technique is shown in Fig. 8-74. Flaws in any plane tend to produce a detectable effect. A fourth technique is shown in Fig. 8-75. It is necessary to move the transducer back and forth while moving the transducer parallel to the weld. Any loose scale, rust, or dirt must be removed for a distance of approximately 6 in. from the weld. It is often advisable to inspect a weld from both sides.

Since an absolute method of judging a defect is not available, standard defects such as saw cuts of various depths are often used. The need to determine the exact size of defects has lead Krächter and Krautkramer[49] to suggest the use of a twin transducer, Fig. 8-76. The two transducers are placed on either side of the weld and are electrically connected in parallel, so that each transmits as well as receives. The pattern observed on the oscilloscope is shown in Fig. 8-76. The defect indication D_1 is the reflection from the defect received by transducer I; D_2 is the reflection from the defect received by transducer II. The transmitted energy received by both transducers is shown midway between the two defect indications. The size of the defect can be estimated by comparing

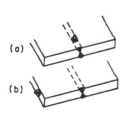

(a)

(b)

FIG. 8-73. Techniques for testing welds. (*Krächter*.)

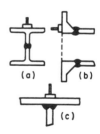

(a)　(b)

(c)

FIG. 8-74. Technique for testing welds.

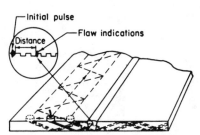

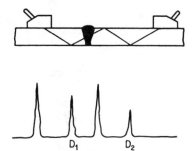

FIG. 8-75. Testing of welds using shear waves.

FIG. 8-76. Twin transducer technique for inspecting welds. (*Courtesy Henry Brutcher, Altadena, Calif., Translation No. 3302.*)

the ratio of the size of the reflected pulse to the size of the transmitted pulse, to the ratio of the size of the defect to the cross section of the beam. Figure 8-77 shows the ratio of defect indication to back reflection.

The shear wave technique can also be used[50,51] for inspecting circumferential pipe welds. In this use, the shear wave is directed axially along the pipe. One major difficulty are the false indications or reflections that come from irregularities at the inner and outer weld surfaces. False indications do not affect the sensitivity of the test but do make interpretation of the results extremely difficult. This difficulty can be reduced by grinding the top and bottom surfaces flush with the bare metal. It is necessary to move the search unit back and forth while it is being moved parallel to the weld. It is also necessary to prepare the surface by removing any loose scale, rust, or dirt on one side of the weld, approximately 6 in. from the weld. Known defects such as slag inclusions, lack of penetration to the root, and cold shuts were deliberately placed into a test weld, and all were located by ultrasonic testing. Minor defects were found in some of the welds, but in all instances the defects were smaller than allowed by the ASME code for piping defects when using radiography. After 1 year of operation, these welds were reinspected and the same defects were located and were of the same magnitude.

Tubing Inspection. Figure 8-78 shows the application of the ultrasonic reflection technique to the testing of welds in tubing. The transducer is set in a plastic wedge at such an angle that surface waves are generated. The transducer is spring-mounted in contact with the tubing at an angle of 120° from the weld

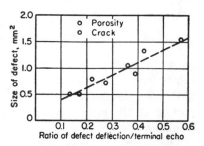

FIG. 8-77. Ratio of defect indication to back reflection. (*Courtesy Henry Brutcher, Altadena, Calif., Translation No. 3302.*)

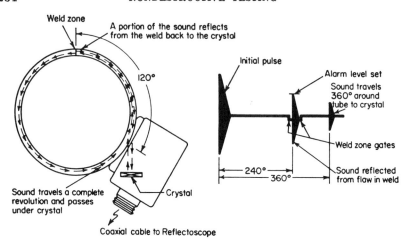

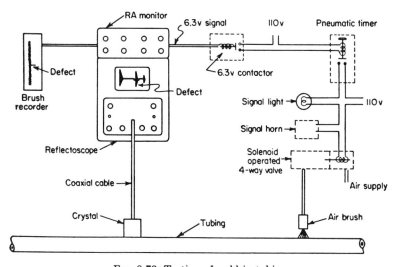

FIG. 8-78. Testing of weld in tubing.

bond. When no defects are present, the wave travels 360° around the tubing back to the transducer, giving an indication on the oscilloscope 360° from the initial pulse. If a defect is encountered at the weld, a portion of the surface wave is reflected and an indication appears 240° from the initial pulse.

The testing of tubing, especially small-diameter tubing, is becoming of increasing importance. The increased interest has come about as a result of increased use of small-diameter tubing in the nuclear energy field and other critical applications. Shear waves can be produced by

means of a transducer producing longitudinal waves
and a plastic wedge. The information given on how
to calculate the wedge angle and the angle which the
sound enters the specimen will now be applied to the
case of tubing.

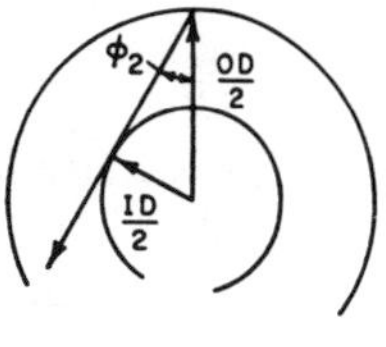

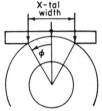

FIG. 8-79. Test angle.

Figure 8-79 shows the optimum entrance angle for
testing small tubing. The numerical value of this
angle can be calculated by using the equation

$$\phi = \frac{\sin^{-1} \mathrm{ID}}{\mathrm{OD}} \qquad (8\text{-}25)$$

After calculating the entrance angle, the proper wedge
angle can be calculated. However, calculations of the
angle for this wedge differ from calculations based
on a flat interface. The curvature of the tube surface
introduces a great variation in the effective wedge
angle. The effect of the tube curvature is similar to
a lens and diverges the sound beam into a wide-angle
cone. Figure 8-80 shows the variation in the wedge
angle caused by the tube curvature. The value of the angle can be
calculated using the equation

FIG. 8-80. Variation of wedge angles due to tube curvature.

$$\phi = \sin^{-1} \frac{W}{2R} \qquad (8\text{-}26)$$

Equations (8-25) and (8-26) show that there is a large variation in the
entrance angle between the front and rear edge of the transducer. The
ultrasonic waves will actually travel in two directions around the tube if
the transducer is located so that the center of the beam impinges on the
tube at the optimum entrance angle. However, by varying the position
of the transducer with respect to the tube, a best position can be found.
One way to vary the position of the transducer with respect to the tube
is to use what is known as a "saddle adapter."
As can be seen in Fig. 8-81, this saddle
adapter consists of a holder for the transducer,
with slotted holes for attaching the contact
shoe. This permits the transducer to be moved
forward or backward on the contact shoe,
varying the position at which the beam strikes
the curvature of the tube surface. The final
adjustment of the transducer is therefore made
by trial and error on the tubing. Figure 8-82
shows typical setups for immersed inspection
of tubing. If it is desired to separate out

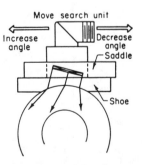

FIG. 8-81. Saddle adapter.

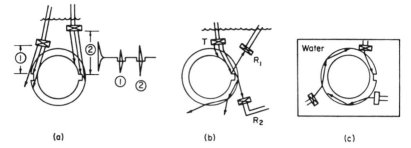

FIG. 8-82. Immersed inspection of tubing.

signals from inner and outer wall defects, a second transducer could be used positioned so that the signals from the two transducers are well separated on the screen or in the alarm or recording equipment.

Oliver and coworkers[52] inspected a large quantity of $\frac{3}{16}$- and $\frac{1}{4}$-in.-diameter tubing having a wall thickness of 0.025 in. Because of the small dimensions, short pulse lengths and wide-band amplification were required. For the high production rate, immersed testing was chosen because of the relative ease of positioning and the advantages obtained with a delay column. Lithium sulfate transducers were chosen because of their greater amplitude in response as compared to conventional quartz crystals. The B scan presentation was used for its fast scanning speed and because of surer and more rapid interpretation. The transducer is aligned relative to the tube so that the sound beam will enter the tube wall and be propagated in a circumferential direction, Fig. 8-83. The waves will strike the tube at all incident angles from 0 to 90°. However, waves with incident angles greater than about 30° will not enter the tube wall.

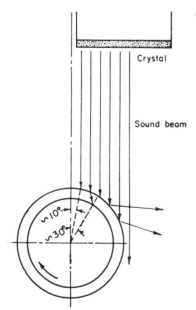

FIG. 8-83. Alignment of specimen and transducer. (*Courtesy Oak Ridge National Laboratory, operated by Union Carbide Corporation for the U.S. Atomic Energy Commission.*)

The uncollimated sound beam covers about $\frac{1}{8}$ in. of the tube length, using only a short chord on the edge of the crystal, Fig. 8-84. Thus this is the limiting factor on linear feed per tube rotation. Collimation, as illustrated in Fig. 8-85, will increase the effective crystal width almost to that of the actual crystal diameter, thus increasing the linear feed potential and subse-

FIG. 8-84. Effective crystal width of un-
collimated beam   (*Courtesy Oak Ridge
National Laboratory, operated by Union
Carbide Corporation for the U.S. Atomic
Energy Commission.*)

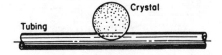

quently the inspection rate. This method propagates sound circumferentially through the walls in one direction only. Thus, to ensure
adequate inspection for less favorably oriented defects, a second
scan must be made sending the sound through the wall in the opposite
direction. The tubing is rotated and the crystal is translated along the
tubing axis.

Surface Wave Applications. Several applications of testing using
surface waves are briefly described. The T-section aluminum stiffeners,
ribs, and spars shown in Fig. 8-86 are 18 to 20 ft long and about ¼ in.

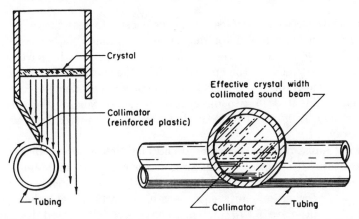

FIG. 8-85. Collimation of sound beam. (*Courtesy Oak Ridge National Laboratory,
operated by Union Carbide Corporation for the U.S. Atomic Energy Commission.*)

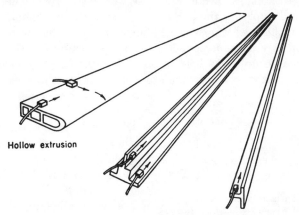

FIG. 8-86. Surface wave testing of aluminum sections.

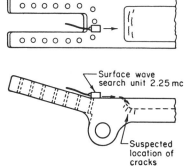

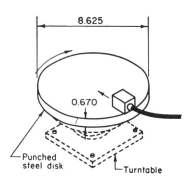

FIG. 8-87. Surface wave testing of winged hinge fitting. (*Courtesy Sperry Products, Inc.*)

FIG. 8-88. Surface wave testing of shell blanks. (*Courtesy Sperry Products, Inc.*)

maximum thickness. Surface waves were used to test these specimens for surface inclusions and for inclusions just under the surface of the material. The surface waves follow the curvature of the part and are able to locate defects on the opposite side of the specimen from the transducer. Orientation of the defects was somewhat random; therefore the test was conducted in two directions on the surface of the specimen. Defects only 0.008 to 0.010 in. in diameter were detected from as far as 12 ft away.

A wing hinge fitting, the exposed end of which is shown in Fig. 8-87, was inspected "in place" on an aircraft for cracks in the depression as indicated. Cracks were easily detected in this area, which is not readily accessible because of adjacent assembled parts. Steel shell blanks punched from plate 0.670 in. thick have been inspected for surface imperfections which show up as folds, seams, and laps in the finished shell case. These blanks, shown in Fig. 8-88, were tested by placing them on a turntable and applying the surface wave transducer manually near one edge. Wrought-steel car wheels, shown in Fig. 8-89, were tested for circumferential cracks using 1-Mc surface waves by rotating the wheel while the search unit was held in contact with the plate near the hub. The waves followed the moderate curvature of the coned plate to the area where cracks had been experienced.

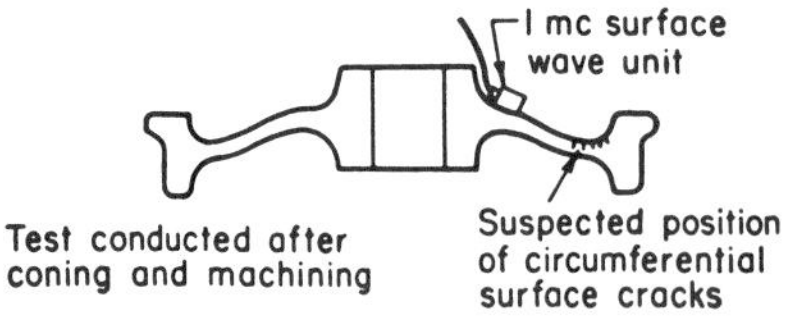

Die block cracks which develop in sharp corners of the impressions frequently are deep enough to remain even after the blocks have been planed flat to receive a new impression. One of the applications for surface wave testing was

FIG. 8-89. Surface wave testing of wrought steel wheel. (*Courtesy Sperry Products, Inc.*)

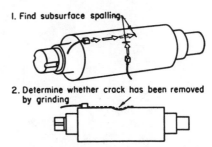

1. Find subsurface spalling

2. Determine whether crack has been removed
   by grinding

FIG. 8-90. Surface wave testing of valve
spring. (*Courtesy Sperry Products, Inc.*)

FIG. 8-91. Surface wave testing of rolls.
(*Courtesy Sperry Products, Inc.*)

to determine whether the blocks had been planed below the depth
of the deepest crack. Prior to sinking a new impression, surface
wave tests in two directions on these blocks were made, assuring
that expensive die work would not be attempted on defective material.
Surface wave testing of aircraft engine valve springs for fatigue cracks is
shown in Fig. 8-90. The slightest surface irregularity, nick, or corrosion
can eventually develop into a serious failure in service. Test with surface
waves allows rejection of springs with incipient failures. The 1-Mc
surface waves are useful in the inspection of millwork rolls for spalling,
as shown in Fig. 8-91. After refacing, testing with surface waves can
determine whether discontinuities still exist just under the surface,
which will cause early spalling when the roll is returned to service.
Prior to this refacing, the depth of cracks can be determined by localized
grinding to a depth where the cracks disappear. Surface waves follow
the contours of the ground-out areas and are used to make sure the cracks

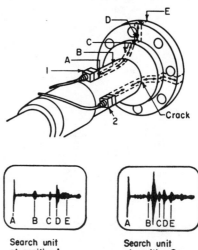

FIG. 8-92. Surface wave
testing of a shaft. (*Courtesy Sperry Products, Inc.*)

Search unit
at position 1

Search unit
at position 2

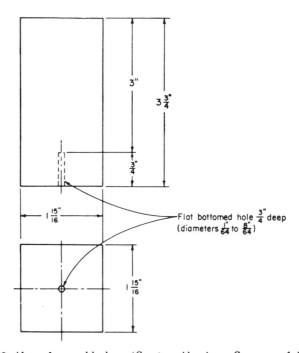

FIG. 8-93. Alco reference blocks.   (*Courtesy Aluminum Company of America.*)

have actually been removed.   The depth of the grind marks then equals
the depth of the cracks.   The fillet areas of many shafts, rolls, and similar
parts may be inspected by projecting the surface waves from an adjacent
flat or cylindrical surface, as shown in Fig. 8-92.   In this composite
illustration, the transducer is shown at two locations: one opposite a
cracked area, the other opposite a sound area.   The accompanying A
scan shows the indications received and identifies their origin.

This brief review of the applications of surface wave testing indicates
that the development and uses of this new technique have only begun.
For example, surface waves can easily be produced in wire as small as
0.010 in. in diameter, which opens up interesting possibilities for continuously monitoring small-diameter materials as they are produced.
Many other possibilities and applications will suggest themselves to the
nondestructive testing engineer as more experience with these intriguing
waves is accumulated.

Bradfield[53] describes four improvements in the pulse echo techniques:
a mode changer made in the form of a laminated assembly which gives
better efficiency and discrimination than the usual Lucite wedge, circuitry
for piezoelectric transducers to give improved discrimination, a monitoring device to reduce the effect of rough surfaces, and a device by which
an ultrasonic beam can be steered to desired locations.

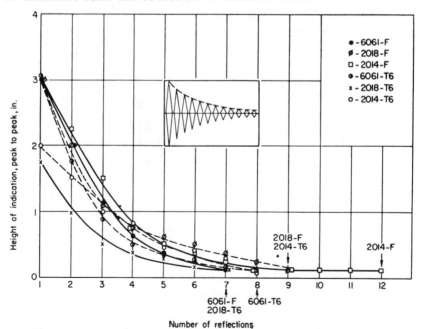

FIG. 8-94. Height of successive back reflections received from 5-in. diameter by
4-in.-long blocks of several aluminum alloys. (*Courtesy Aluminum Company of
America.*)

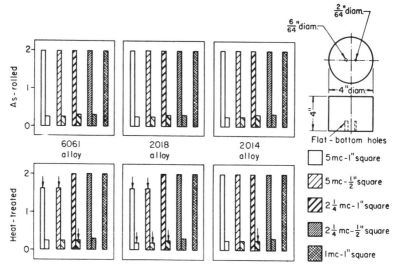

FIG. 8-95. Height of reflections from various aluminum alloys.   (*Courtesy Aluminum Company of America.*)

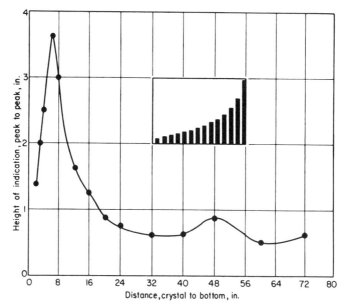

FIG. 8-96. Effect of distance on height of indications received from 4/64-in.-diameter flat-bottomed hole in 2014-F rolled bar, using a 5-Mc 1/8-in.-diameter crystal.   (*Courtesy Aluminum Company of America.*)

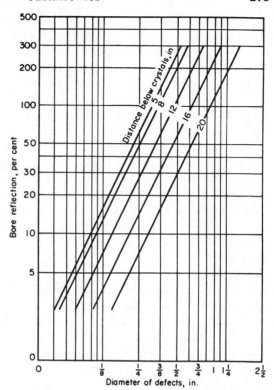

Fig. 8-97. Relationship between defect reflection and minimum size of defect. (*Courtesy General Electric Company.*)

Test Standards. In order to obtain consistent test results between instruments, manufacturers, fabricators, and users as well as to provide a common basis for expressing results of ultrasonic testing, some kind of standardization procedure is often required.   Reference blocks are sometimes used as a method of standardization.   The reference blocks most widely used are the Alcoa Ultrasonic Standard Reference Blocks.[54] Figure 8-93 shows a sketch of one of these blocks.   A complete set consists of eight blocks, each containing a single flat-bottomed hole.   The holes range in diameter from $\frac{1}{64}$ to $\frac{8}{64}$ in., in $\frac{1}{64}$-in. steps.   Figure 8-94 shows the amplitude of successive back-reflection indications in a single reference block.   Figure 8-95 shows the variation in the height of the back reflections obtained from blocks of various aluminum alloys. Figure 8-96 shows the effect of distance between the crystal and the hole in a reference block.   It will be observed that as the distance becomes smaller the height of indication increases to a point and then suddenly decreases.   This point of inflection corresponds to the end of a zone of interference frequently called the "close field zone."   The extent of this close field zone from the surface of the material varies with crystal size and frequency.

Flaw Size. The shape of the specimen, defect orientation, and surface condition are some of the factors that affect the indication of the size of a flaw found by ultrasonics. Serabian[55] has made a detailed study of some of these effects in large rotor forgings. He used amplitude of the back reflection from the bore in a defect free region as the standard. He produced artificial defects by drilling flat-bottomed holes of various diameters into a sectioned rotor such that the smooth plane surface of the bottom of the hole simulated a defect oriented normal to the central portion of the testing beam. The resulting indications are shown in Fig. 8-97 as a function of defect size. The effect of defect orientation on the reflected energy is shown in Fig. 8-98. The fact that the ultrasonic beam spreads out from the transducer also has an effect on the amount of energy reflected by a defect. By moving the transducer so that a point of maximum and a point of minimum indication were found, Serabian determined a defect orientation correction factor for defects found in large rotor forgings.

Tests on specimens of various shapes with artificial flaws and surfaces show that it is possible to estimate the size of a flaw from the ratio of flaw echo to back wall echo, provided that the following conditions are observed. A flaw must be located from several directions and the receiver sensitivity must be kept constant. The ratio of flaw echoes varies for quartz crystals of different diameters. The surface of the specimen must be flat. The nature of the surface finish affects only the intensity of the reflected energy. The back surface must be approximately parallel to the testing surface. In testing round bodies with a line contact only,

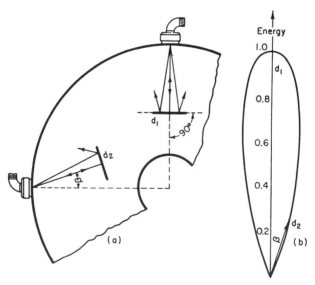

Fig. 8-98. Effect of defect orientation on amplitude of the reflected energy.

flaw size estimation at present depends entirely on values obtained from experience. In this, care must be taken in determining the position of the flaw It is important to remember that in spite of the elimination of many influencing factors it remains a difficult task, demanding great care, to make a definite estimate of flaw size. A large number of examples obtained by an experienced operator from cut-up specimens makes it possible to estimate the most probable size of flaws.

In a further attempt to standardize the ultrasonic contact scanning, the subcommittee VI of ASTM Committee E-7 has drafted a tentative Recommended Practice for Ultrasonic Testing by the Reflection Method Using Pulsed Longitudinal Waves Induced by Direct Contact and a tentative Recommended Practice for the Resonance Method of Ultrasonic Testing.

GENERAL REFERENCES

American Society for Testing Materials: Symposium on Nondestructive Tests in the Field of Nuclear Energy, *ASTM, Spec. Tech. Pub.* 223, 1958.

American Society for Testing Materials: Symposium on Ultrasonic Testing, *ASTM, Spec. Tech. Pub.* 101, June 28, 1949.

Bergmann, L.: "Der Ultraschall," 5th ed., p. 405, S. Hirzel Verlag, Leipzig, 1949.

Cady, W. G.: "Piezoelectricity," McGraw-Hill Book Company, Inc., New York, 1946.

Carlin, Benson: "Ultrasonics," 2d ed., McGraw-Hill Book Company, Inc., New York, 1960.

Ewing, W. Maurice, Wenceslas S. Jardetzky, and Frank Press: "Elastic Waves in Layered Media," McGraw-Hill Book Company, Inc., New York, 1957.

Hanstock, R. F.: "The Nondestructive Testing of Metals," The Institute of Metals, London, 1951.

Hueter, T. F., and R. H. Bolt: "Sonics," John Wiley & Sons, Inc., New York, 1955.

Kinsler, Lawrence E., and Austin R. Frey: "Fundamentals of Acoustics," John Wiley & Sons, Inc., New York, 1950.

Kolsky, H.: "Stress Waves in Solids," Oxford University Press, New York, 1953.

Mason, W. P.: "Piezoelectric Crystals and Their Applications to Ultrasonics," D. Van Nostrand Company, Inc., Princeton, N.J., 1950.

Morse, Philip M.: "Vibration and Sound," 2d ed., McGraw-Hill Book Company, Inc., New York, 1948.

Rayleigh, John William Shutt (Lord): "The Theory of Sound," 2d ed., vols. I and II, Dover Publications, New York, 1945.

Richardson, E. G.: "Technical Aspects of Sound," vol. II, Elsevier Publishing Company, Amsterdam, 1957.

Richardson, E. G.: "Ultrasonic Physics," Elsevier Publishing Company, Amsterdam, 1952.

Stephens, R. W. B., and A. E. Bate: "Wave Motion and Sound," Edward Arnold & Co., London, 1950.

Vigoureux, P.: "Ultrasonics," John Wiley & Sons, Inc., New York, 1951.

SPECIFIC REFERENCES

1. American Society for Welding: "Welding Handbook," 3d ed., American Welding Society, New York, 1950.

2. Stonely, R.: Elastic Waves at the Surface of Separation of Two Solids, *Proc. Roy. Soc. (London)*, A106, p. 416, 1924.

3. Ernst, P. J.: Ultrasonic Lenses and Transmission Plates, *J. Sci. Instr.*, vol. 22, pp. 238–243, 1945.

4. Knott, C. G.: Reflexion and Refraction of Elastic Waves with Seismological Applications, *Phil. Mag. S.* 5, vol. 48, no. 29, July, 1899.

5. Firestone, F. A.: Tricks with the Supersonic Reflectoscope, *J. Soc. Nondestructive Testing*, vol. 7, no. 2, Fall, 1948.

6. Worlton, D. L.: Nondestructive Grain Size Measurements with Ultrasonics, *J. Soc. Nondestructive Testing*, vol. 13, no. 6, pp. 24–26, November–December, 1955.

7. Grossman, Nicholas: New Methods Determine Grain Size Ultrasonically, *Iron Age*, Dec. 31, 1953.

8. Mason, W. P.: "Piezoelectric Crystals and Their Applications to Ultrasonics," D. Van Nostrand Company, Inc., Princeton, N.J., 1950.

9. Kolsky, H.: "Stress Waves in Solids," Oxford University Press, New York, 1953.

10. Bergmann, L.: "Der Ultraschall," 5th ed., p. 405, S. Hirzel Verlag, Leipzig, 1949.

11. Cook, E. G., and H. E. Van Valkenburg,: Surface Waves at Ultrasonic Frequencies, *ASTM Bull.* 198, pp. 81–84, May, 1954.

12. Firestone, F. A.: Patent 2,536,128, Method and Means for Generating and Utilizing Vibrational Waves in Plates, Jan. 2, 1951.

13. Lamb, Horace: On Waves in an Elastic Plate, *Proc. Roy. Soc. (London)*, ser. A, vol. 93, p. 114, 1917.

14. Erdman, D. C.: Ultrasonic Flaw Detector, U.S. Patent 2,593,865, Apr. 22, 1952.

15. Pohlman, R.: Internal Examination of Materials by Ultrasonic-Optical Pictures, *Z. angew. Phys.*, vol. 4, pp. 181–187, 1948.

16. Rust, H., R. Haul, and H. J. Studt: Employment of Chemical Reactions to Obtain Pictures of Acoustic Phenomena, *Naturwissenschaften*, vol. 36, pp. 374–375, 1949.

17. Ernst, P. J., and C. W. Hoffman: New Methods of Ultrasonoscopy and Ultrasonography, *J. Acoust. Soc. Am.*, vol. 24, pp. 207–211, 1952.

18. Rust, H. H.: Ultrasonic Image Conversion by Means of a Thermally Induced Colour Change, *Angew. Chem.*, vol. 64, pp. 308–311, 1952.

19. Schuster, K.: Ultrasonic Image Converter, Unclassified Intelligence Report, Germany/GDR Serial 507–55, 1955.

20. Marinesco, N., and M. Rezzini: Impression of Photographic Plates by Ultrasonics, *Compt. rend.*, vol. 200, pp. 548–550, 1935.

21. Bennett, G. S.: New Method for Visualization and Measurement of Ultrasonic Fields, *J. Acoust. Soc. Am.*, vol. 24, pp. 470–474, 1952.

22. Schreiber, H., and W. Degner: A New Technique for Producing Acoustic Optical Pictures, *Ann. Phys.*, vol. 7, pp. 275–278, 1950.

23. Sokoloff, S.: Means for Indicating Flaws in Materials, U.S. Patent 2,164,125, June 27, 1939.

24. Sokolov, S. Ya.: Ultrasonics and Its Applications, *Puroda*, vol. 43, no. 3, pp. 21–34, 1954. Brutcher Translation 3532.

25. Dimmick, G. L.: Sound-to-Image Transducing System, U.S. Patent 2,453,502, Nov. 9, 1948.

26. Firestone, F. A.: Flaw Detecting Device and Measuring Instrument, U.S. Patent 2,280,226, Apr. 21, 1942.

27. Moskowitz, Sidney, and Joseph Racker: "Pulse Techniques," Prentice-Hall, Inc., Englewood Cliffs, N.J., 1951.

28. Erwin, W. S.: Ultrasonic Transmission Testing, U.S. Patent 2,655,035, Oct. 13, 1953.

29. Schubring, N. W.: Ultrasonic Transmission Tester Speeds, Simplifies Production Inspection, *Iron Age*, vol. 176, pp. 87–90, Aug. 4, 1955.
30. Krautkramer, J. H., and O. Rüdiger: An Ultrasonic Apparatus for Nondestructive Testing of Materials, *Arch. Eisenhüttenw.*, vol. 20, no. 11-12, pp. 355–358, 1949. Brutcher Translation 2879.
31. Greenberg, H.: Ultrasonic Inspection Checks Quality of Brazed Joints, *Materials & Methods*, p. 102, June, 1952.
32. Evans, D. J.: Corrosion Evaluation of Ship Bulkhead and Hull Plating by Audigage Thickness Measurements, *Natl. Assoc. Corrosion Engrs.*, vol. 11, p. 23, 1955.
33. McGonnagle, W. J., and W. N. Beck: An Ultrasonic Scanning and Recording System, *Proc. ASTM*, vol. 56, 1956.
34. McGonnagle, W. J.: Nondestructive Testing of Reactor Fuel Elements, *Nuclear Sci. and Eng.*, vol. 2, pp. 602–616, 1957.
35. Ross, J. D., and R. W. Leep: Ultrasonic Transmission Tester, *J. Soc. Nondestructive Testing*, vol. 15, no. 3, pp. 152–154, May–June, 1957.
36. Shaper, H. B.: Hypersonic Nondestructive Material Testing, *Instruments*, vol. 19, no. 6, pp. 327–330, July, 1946.
37. Hartley, J. C., and E. K. Mull: Ultrasonic Testing of Tool Steels, *Iron Age*, May 19, 1949.
38. Hafemeister, R. N.: The Ultrasonic Testing of Forging Ingots, *ASTM Bull.* 197, p. 52, April, 1954.
39. Ward, F. W., and J. O. Borg: Ultrasonic Inspection of Arc-Cast Zirconium and Its Alloys, *U.S. Bur. Mines Rept.* 5126, March, 1955.
40. Valuev, D. P., and others: Ultrasonic Detection of Flaws in Forgings, *Vestnik Mashinostroeniya*, vol. 34, no. 11, pp. 63–64, 1954. Brutcher Translation 3685.
41. Hall, E. D.: Ultrasonic Testing in Railroad Work, Symposium on Ultrasonic Testing, *ASTM, Spec. Tech. Pub.* 101, June 28, 1949.
42. Hall, E. D.: Eliminating Axle Breakage, *Ry. Mech. Eng.*, December, 1947.
43. Desch, C. H., D. O. Sproule, and W. J. Dawson: The Detection of Cracks in Steel by Means of Supersonic Waves, Paper No. 17/1946 of the Alloy Steels Research Committee, February, 1946.
44. Moriarty, C. D.: Ultrasonic Flaw Detection in Pipes by Means of Shear Waves, *ASME Trans.*, vol. 73(3), pp. 225–235, April, 1951.
45. Pardus, A. J.: Ultrasonic Testing Used for Small Diameter Testing, *Iron Age*, Jan. 29, 1953.
46. Fleischmann, W. L., and H. A. F. Rocha: Testing of Small-Diameter Tubing with Automatic Recording Ultrasonic Equipment, *ASME Paper* 55-S-23, 1955.
47. Smack, J. C.: Immersed Ultrasonic Inspection with Automatic Scanning and Recording or Warning Signal, *J. Soc. Nondestructive Testing*, vol. 12, pp. 29–33, May–June, 1954.
48. Wilkinson, Walter D., and William F. Murphy: "Nuclear Reactor Technology," chap. 20, D. Van Nostrand Company, Inc., Princeton, N.J., 1958.
49. Krächter, H., and J. H. Krautkramer: *Schweissen und Schneiden*, vol. 5, no. 8, pp. 305–314, 1953. Brutcher Translation 3302.
50. McGonnagle, W. J.: Ultrasonic Shear Wave Testing, *Metal Prog.*, October, 1956.
51. Pollock, W. A.: Backing Ring Elimination Permits Ultrasonic Testing and Avoids Cracking at Piping Welds, *Welding J.*, vol. 34, pp. 954–960, October, 1955.
52. Oliver, R. B., R. W. McClung, and J. K. White: Immersed Ultrasonic Inspection of Pipe and Tubing, *J. Soc. Nondestructive Testing*, vol. 15, no. 3, pp. 141–144, 1957.
53. Bradfield, G.: Improvements in Ultrasonic Flaw Detection, *J. Brit. Inst. Radio Engrs.*, vol. 14, no. 7, pp. 303–308, July, 1954.

54. Cline, C. W., and J. B. Morgan: Standardization in Ultrasonic Testing, *J. Soc. Nondestructive Testing*, vol. 13, no. 4, pp. 23–27, July–August, 1955.
55. Serabian, S.: Influence of Geometry upon Ultrasonic Defect Size Determination in Large Rotor Forgings, *J. Soc. Nondestructive Testing*, vol. 14, no. 4, pp. 18–21, July–August, 1956.

ADDITIONAL REFERENCES

Betz, C. R.: Curved Crystal Developments in Ultrasonic Resonance Testing, *J. Soc. Nondestructive Testing*, vol. 10, no. 4, pp. 28–31, Spring, 1952.

Bratt, M. J., and V. I. E. Wiegand: Detection of Flaws in Jet Engine Parts by Ultrasonics, *J. Soc. Nondestructive Testing*, vol. 13, no. 5, pp. 45–47, September–October, 1955.

Erdman, D. C.: Ultrasonic Inspection Using Automatic Recording and Frequency Modulated Flaw Detector, *J. Soc. Nondestructive Testing*, vol. 11, no. 8, pp. 27–31, November–December, 1953.

Erwin, W. S.: Supersonic Measuring Means, U.S. Patent 2,431,233, Nov. 18, 1947.

Erwin, W. S., and G. M. Rossweiler: The Automatic Sonigage, *Iron Age*, p. 48, July 24, 1947.

Erwin, W. S., and G. M. Rossweiler: "Ultrasonic Resonance Applied to Nondestructive Testing," *Rev. Sci. Instr.*, vol. 18, no. 10, pp. 750–753, October, 1947.

Firestone, F. A.: The Supersonic Reflectoscope for Internal Inspection, *Metal Prog.*, vol. 48, p. 505, September, 1945.

Firestone, F. A., and J. R. Frederick: Refinements in Supersonic Reflectoscopy, Polarized Sound, *J. Acoust. Soc. Am.*, vol. 18, no. 1, pp. 200–211, July, 1946.

Firestone, F. A.: Surface and Shear Wave Method and Apparatus, U.S. Patent 2,439,130, Apr. 6, 1948.

Frank, R. H.: Ultrasonics, *Steel*, June 16, 1952.

Lutach, A.: Nondestructive Ultrasonic Testing by the Pulse-Echo Technique, *Arch. Eisenhüttenw.*, vol. 23, nē. 1–2, pp. 57–65, 1952. Brutcher Translation 3070.

Marinesco, N., and J. J. Trillot: Action of Ultrasonics on the Photographic Plates, *Compt. rend.*, pp. 196, 858, 1933.

Marinesco, N.: The Law of Darkening of Photographic Plates by Ultrasonics, *Compt. rend.*, p. 201, 1935; pp. 202, 757, 1936.

Martin, Erich: The State of Ultrasonic Testing in Germany and the Application of This Method by the German Federal Railways, *J. Soc. Nondestructive Testing*, vol. 14, no. 4, pp. 26–31, July–August, 1956.

Pringle, F. E.: Simac-Sonic Inspection, Measurement and Control, *J. Soc. Nondestructive Testing*, vol. 14, no. 3, pp. 22–25, May–June, 1956.

Smack, John C.: Ultrasonic Weld Inspection, *Welding Engr.*, May, 1949.

Smack, J. C.: Immersed Ultrasonic Inspection with Automatic Scanning and Recording or Warning Signal, *J. Soc. Nondestructive Testing*, vol. 12, no. 3, pp. 29–33, May–June, 1954.

Snowdon, R. W.: Sonic Tests Spot Flaws in Heavy Forgings, *Iron Age*, Apr. 13 and 27, 1950.

Tarr, Levi: Experiences with Ultrasonic Reflectoscope Inspection of Main Seam Welds of Seven Large Spheres, *ASTM Bull.* 196, pp. 54–60, February, 1954.

CHAPTER 9

DYNAMIC TESTING

As was mentioned in the last chapter, one of the oldest nondestructive tests involves striking the specimen with a hammer and listening to the sound produced. The existence of a flaw is indicated by an "off tone" ring and a rapid diminishing of sound intensity. This method is not too reliable because the frequency of the sound emitted is somewhat dependent on the way the specimen is supported. The way in which the specimen is struck determines whether the fundamental, a harmonic, an overtone, or a combination of frequencies is produced. Likewise, the force used to strike the specimen may influence the frequency or frequencies produced. The natural frequencies of some specimens may be beyond the audible range, but this difficulty can be overcome by use of suitable electronic instruments. Sound waves in the audible range spread around the small flaws, so that only gross flaws are found by this technique. The rate at which the sound intensity of a vibrating specimen decreases is a characteristic of the condition of the material. The unaided human is not a good judge of decay time; differences as large as 20 per cent can be determined, but better resolution is difficult. In recent years, the rate of attenuation of vibrations in solids has been the subject of many investigations. The use and significance of attenuation measurements in nondestructive testing will be considered later in the chapter. Tests involving the measurement of natural frequencies of specimens will be considered first. Both kinds of tests are nondestructive, because the vibration amplitudes necessary for frequency or attenuation measurements are usually so small that the stresses produced in the specimen have no or a negligible effect on the material.

Natural Vibrations. Longitudinal, transverse, or torsional vibrations can be excited in rods or bars. Every specimen has certain characteristic frequencies at which it can be made to vibrate. These characteristic frequencies are functions of the size, shape, mass, elastic properties, and mode of vibration produced in the specimen. For specimens of simple shape, it is possible to derive relations between the various parameters and the frequencies for simple modes of vibration. When the specimen is of so complex a shape that such relations cannot be derived mathematically, an empirical relationship can be found.

279

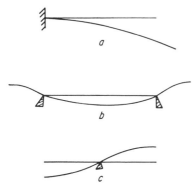

FIG. 9-1. Three ways to suspend specimens.

Derivations of these relationships can be found in any standard textbook on sound or vibrations.[1,*] Only the simple case of longitudinal vibrations in a rod or bar will be derived here. As the longitudinal wave moves along such a specimen, the displacement of the particles is parallel to the length of the specimen. When a specimen is vibrating at any of its characteristic frequencies, standing waves are set up in the specimen. If one end is clamped, the clamped end becomes a node in the standing wave, no matter in what mode the specimen is vibrating. In a standing wave, nodes must always be separated by antinodes, and antinodes must always be separated by nodes. In a standing wave the distance between adjacent nodes or antinodes is always one-half a wavelength. Figure 9-1 shows three ways of suspending the specimen as well as various possible modes of longitudinal vibration.

For case c, $$\lambda = 2L \tag{9-1}$$

and $$f = \frac{V}{2L} \tag{9-2}$$

The velocity of propagation of longitudinal waves in a bar or rod is given by

$$V = \left(\frac{Y}{\rho}\right)^{1/2} \tag{9-3}$$

where L = length of specimen
Y = Young's modulus
ρ = density of material
Equation (9-2) becomes

$$f = \frac{1}{2L}\left(\frac{Y}{\rho}\right)^{1/2} \tag{9-4}$$

For case b

$$\lambda = L \tag{9-5}$$

For the general case of a specimen clamped in the middle or $L/4$ from each end,

$$f = \frac{N}{2L}\left(\frac{Y}{\rho}\right)^{1/2} \tag{9-6}$$

* Superscript numbers indicate Specific References listed at the end of the chapter.

where N is an integer, its value depending on the node of vibration. For case a,

$$\lambda = 4L \qquad (9\text{-}7)$$

giving the general equation for a specimen clamped at one end,

$$f = \frac{M}{4L}\left(\frac{Y}{\rho}\right)^{\frac{1}{2}} \qquad (9\text{-}8)$$

where M is an integer, but can have only odd number values, depending on the mode of vibration.

The following list gives the characteristic natural frequencies for various types of vibrations, shapes of specimens, and method of support. It should be noted that in some cases all the frequencies higher than the fundamental frequency, called "overtones," are harmonics of the fundamental. In some cases, all the harmonics are present, in other cases only the odd number harmonics are present. In the case of transverse vibrations the overtones are not harmonics.

Longitudinal Vibration

General formula:

$$w_n = \left(n + \frac{1}{2}\right)\pi\sqrt{\frac{AE}{\mu_1 l^2}}$$

Longitudinal vibration of cantilever:

$$w_n = 2\pi f_n$$

where A = cross section
E = modulus of elasticity
μ_1 = mass per unit length
$n = 0, 1, 2, 3$ = number of nodes

Transverse Vibration

General formula:

$$w_n = a_n\sqrt{\frac{EI}{\mu_1 l^4}}$$

where EI = bending stiffness of the section
l = length of the beam
μ_1 = mass per unit length = W/gl
a_n = numerical constant, different for each case in the list that follows

Cantilever or "clamped-free" beam	$a_1 =$	3.52	
	$a_2 =$	22.4	
	$a_3 =$	61.7	
	$a_4 =$	121.0	
	$a_5 =$	200.0	
Simply supported or "clamped-clamped" beam	$a_1 =$	$\pi^2 =$	9.87
	$a_2 =$	$4\pi^2 =$	39.5
	$a_3 =$	$9\pi^2 =$	88.9
	$a_4 =$	$16\pi^2 =$	158.
	$a_5 =$	$25\pi^2 =$	247.
"Free-free" beam or floating ship	$a_1 =$	22.4	
	$a_2 =$	61.7	
	$a_3 =$	121.0	
	$a_4 =$	200.0	
	$a_5 =$	298.2	
"Clamped-clamped" beam has same frequencies as "free-free" beam	$a_1 =$	22.4	
	$a_2 =$	61.7	
	$a_3 =$	121.0	
	$a_4 =$	200.0	
	$a_5 =$	298.2	
"Clamped-hinged" beam may be considered as half a "clamped-clamped" beam for even a numbers	$a_1 =$	15.4	
	$a_2 =$	50.0	
	$a_3 =$	104	
	$a_4 =$	178	
	$a_5 =$	272	
"Clamped-free" beam or wing of autogyro may be considered as half a "free-free" beam for even a numbers	$a_1 =$	0	
	$a_2 =$	15.4	
	$a_3 =$	50.0	
	$a_4 =$	104	
	$a_5 =$	178	

Moduli. When a stress is applied to any isotropic solid, two kinds of deformations are possible: an elongation or contraction combined with a change in cross-sectional area and a relative movement of planes parallel to the applied stress. The elastic constants or moduli express the resistance of a material to these deformations: Young's modulus Y expresses the resistance to elongation, and the modulus of rigidity G expresses the resistance to shear. Poisson's rate σ expresses the ratio of the cross-sectional contraction to the elongation. Young's modulus, the modulus of rigidity, and Poisson's ratio are related as shown in the following equation:

$$Y = 2(1 + \sigma)G \qquad (9\text{-}9)$$

This equation is good only for isotropic materials and is not applicable to such materials as wood and certain types of stone. Additional elastic moduli can be computed from the equations

$$E_f = f^2 C W \qquad (9\text{-}10)$$
$$E_s = f'^2 D W \qquad (9\text{-}11)$$

where E_f = dynamic modulus of elasticity in flexure

E_s = dynamic modulus of elasticity in shear

f = resonant frequency in flexure

f' = resonant frequency in torsion

C = a factor depending upon shape and size of specimen, its mode of vibration, and Poisson's ratio

D = a factor depending on dimensions of specimens and mode of vibration

W = weight of specimen

Pickett[2] has presented a discussion of the two factors C and D and equations and graphs from which these factors can be determined.

Methods of Excitation. The method of starting vibrations by impact generally excites more than one natural frequency, so that a pure tone is not obtained. Several other methods of excitation will be discussed. Mechanical coupling of the specimen to a subsidiary vibrator, such as a piezoelectric transducer, the nickel tube of a magnetostriction oscillator,[3] or an electromagnetic vibrator,[4] can be used as a means of excitation. The attachment of an auxiliary vibrator may alter the natural frequency of the specimen under examination. A convenient method of exciting small vibrations are the electromagnetic methods for magnetic materials and the induced-current methods for nonmagnetic metals of good conductivity. Since no physical contact is required between the specimen and exciting system, the specimen is free to vibrate at its own natural frequency. For magnetic material, the region of maximum vibration amplitude is positioned to bridge the poles of an electromagnetic coil carrying alternating current. The resonant frequency may be found by varying the frequency of the alternating current to obtain a maximum vibration amplitude.

Nonmagnetic metals, having good conductivity, can be set into vibration by passing an alternating current through a coil placed near the specimen. Induced currents are set up in the specimen and produce a reaction between the magnetic field of the coil. This reaction produces an alternating force on the specimen and causes it to vibrate. Individual arrangements of the coil and magnetic field depend on the form of the specimen and the mode of vibration required. The maximum alternating force is obtained when the magnetic field is perpendicular to the path of the induced current. The path of the induced current is parallel to the coil through which the primary current flows. The relative direction of induced current, magnetic field, and force are shown in Fig. 9-2.

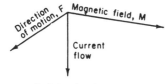

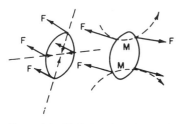

FIG. 9-2. Left-hand rule and excitation of specimen.

Elastic Constants. Schneider and Burton[5] and Reynolds[6] have used ultrasonic methods to determine the elastic constants of solids. If the longitudinal and shear wave velocities can be measured for a given specimen, the elastic constants can be determined. The expressions for the longitudinal V_L and shear wave or transverse V_t velocities are

$$V_L = \left[\frac{(1 - \sigma)Y}{(1 + \sigma)(1 - 2\sigma)\rho} \right]^{\frac{1}{2}} \quad (9\text{-}12)$$

$$V_t = \left(\frac{G}{\rho} \right)^{\frac{1}{2}} = \left[\frac{Y}{2(1 + \sigma)\rho} \right]^{\frac{1}{2}} \quad (9\text{-}13)$$

Schneider and Burton have used the rotating plate technique to determine elastic constants. In this technique the variation in the intensity of the transmitted energy is measured as the angle between the incident beam and the specimen is varied. Figure 9-3 shows what happens when a longitudinal wave strikes a specimen at an angle Θ. Mode conversion occurs and a longitudinal and shear wave are each transmitted into the specimen.

$$\frac{\sin \Theta}{\sin \Theta_L} = \frac{V}{V_L} \quad (9\text{-}14)$$

$$\frac{\sin \Theta}{\sin \Theta_s} = \frac{V}{V_s} \quad (9\text{-}15)$$

where V is the velocity of longitudinal waves in the first medium, and V_L and V_s are the velocity of longitudinal and shear waves in the specimen. As Θ increases, Θ_L increases until at some angle of incidence Θ', $\Theta_L = 90°$, and the longitudinal waves are totally reflected. As Θ is further increased, an angle of incidence Θ'' is found at which the shear waves are totally reflected. Equations (9-14) and (9-15) become under these conditions

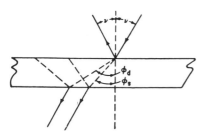

$$V_L = \frac{V}{\sin \Theta'} \quad (9\text{-}16)$$

$$V_s = \frac{V}{\sin \Theta''} \quad (9\text{-}17)$$

FIG. 9-3. Mode conversion.

By determining experimentally the values of Θ' and Θ'' and using Eqs. (9-16) and (9-17), the value of V_L and V_s can be computed.

When the incident beam strikes the front surface of a specimen, part of the energy is reflected and part transmitted. Likewise, at the rear surface of the specimen, part of the energy is reflected. Consequently, there are waves existing in the specimen which can interfere both constructively and destructively. The transmitted energy goes through definite maxima and minima as a result of interference between incident waves and the wave trains within the specimen. Unfortunately, these interference patterns can lead to considerable difficulty in locating the angle of total reflection for the longitudinal waves. In general, however, the intensity at the angle of total reflection is practically zero, and is much less than the interference minima. In addition to these types of interference, the longitudinal and shear waves can interact with each other to give another series of maxima and minima.

It can be shown that the position of interference maxima for longitudinal waves is given by

$$M\lambda_L = 2d \cos \Theta_L \qquad (9\text{-}18)$$

Using Eq. (9-14) the above equation can be written as

$$\sin^2 \phi = \left(\frac{V}{V_L}\right)^2 - \left(\frac{\lambda}{2d}\right)^2 M^2 \qquad (9\text{-}19)$$

Likewise relations for the shear waves can be written

$$M\lambda_s = 2d \cos \Theta_s \qquad (9\text{-}20)$$

or

$$\sin^2 \phi = \left(\frac{V}{V_s}\right)^2 - \left(\frac{\lambda}{2d}\right)^2 M^2 \qquad (9\text{-}21)$$

For the interaction of the longitudinal and shear waves

$$\frac{M}{d} = \cos \frac{\theta_L}{\lambda_L} - \cos \frac{\phi_s}{\lambda_s} \qquad (9\text{-}22)$$

where λ = wavelength in first medium

d = thickness of specimen

M = order of interference

λ_s = wavelength of shear waves in specimen

λ_L = wavelength of longitudinal waves in specimen

The number and position of the interference maxima vary with sample thickness. Figure 9-4 shows the transmission curve for an aluminum

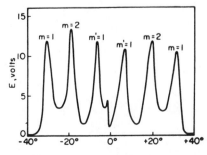

FIG. 9-4. Transmission curve for aluminum, 0.813 mm thick, in water at 5 Mc. (*Courtesy Journal of Applied Physics.*)

TABLE 9-1. VALUES OF MODULI DETERMINED BY ROTATING PLATE METHOD

Material	Tank liquid	V, cm/sec	V_L, cm/sec	V_S, cm/sec	$Y \times 10^{11}$ dynes/cm²		$G \times 10^{11}$ dynes/cm²		σ	d, in.
					Obs.	Handbook	Obs.	Handbook		
Aluminum	H_2O	1.5×10^5	7.05×10^5	2.82×10^5	6.45	6.96	2.14	2.37	0.40	1/16
	H_2O	1.5×10^5	7.05×10^5	2.99×10^5	6.73	6.96	2.42	2.37	0.38	1/8
Copper	H_2O	1.5×10^5	4.82×10^5	2.28×10^5	12.2	12.1–12.8	4.52	4.24	0.346	1/8
	CCl_4	9.22×10^4	5.96×10^5	2.26×10^5	12.3	12.1–12.8	4.44	4.24	0.376	1/8
Steel	H_2O	1.5×10^5	6.15×10^5	2.84×10^5	18.4	19.2	6.73	7.79	0.364	1/16
	$C_6H_5CF_3$	9.85×10^4	6.31×10^5	2.72×10^5	17.1	19.2	6.16	7.79	0.386	1/16

TABLE 9-2. ELASTIC CONSTANTS FOR VARIOUS THERMOSETTING AND THERMOPLASTIC RESINS

Material	Tank liquid	$V \times 10^5$, cm/sec	$V_L \times 10^5$, cm/sec	$V_S \times 10^5$, cm/sec	Y		G		σ	d, in.
					dynes/cm²	Psi	dynes/cm²	Psi		
Melmac resin 1079*	$C_6H_5CF_3$	0.985	3.69	1.72	1.21×10^{11}	1.75×10^6	4.44×10^{10}	6.43×10^5	0.360	1/16
Melmac resin 26-8B*	$C_6G_5CF_3$	0.985	3.69	1.72	1.21	1.75	4.44	6.43	0.360	1/8
Melmac resin S-6003*	$C_6H_5CF_3$	0.985	4.58	2.09	2.40	3.48	8.75	1.27×10^6	0.369	1/8
Glass (lantern slide)	$C_6H_5CF_3$	0.985	5.93	2.85	5.49	7.95	2.03×10^{11}	2.94	0.350	1/16
Glass (window)	$C_6H_5CF_3$	0.985	6.79	3.26	7.16	1.04×10^7	2.66	3.86	0.344	1/8
Beetle resin 20RC-4*	$C_6H_5CF_3$	0.985	3.96	1.74	1.26	1.83×10^6	4.55×10^{10}	6.60×10^5	0.381	1/8
Laminac resin 4116*	$C_6H_5CF_3$	0.985	2.52	1.29	4.83×10^{10}	7.00×10^6	1.83	2.67	0.322	1/8
Lucite†	CH_3OH	1.13	2.63	1.28	5.64	8.16	2.15	3.12	0.310	1/4
Polystyrene†	CH_3OH	1.13	2.30	1.18	3.91	5.66	1.48	2.14	0.322	1/4

* Thermosetting resin.
† Thermoplastic resin.

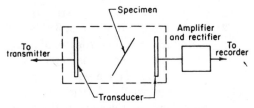

FIG. 9-5. Block diagram of equipment for measuring elastic properties by the rotating technique.

FIG. 9-6. Deviation of ultrasonic beam by specimen.

specimen. The small numbers adjacent to the interference maxima refer to the order of interference. In practice, the assignment of order numbers is a difficult problem since there is no arbitrary method for making this determination. Table 9-1 lists moduli determined by this technique.

Figure 9-5 shows a block diagram of equipment used for this type of measurement. Burton and Schneider used a frequency of 5 Mc. There are two effects which must be considered in work of this type. The first is the deviation of the beam as it passes through the sample as shown in Fig. 9-6. This difficulty can be overcome by moving the receiving transducer laterally. A second difficulty arises because the beam from the transducer is not uniform across the beam but is rather a diffraction pattern. Bez-Bordili[7] and Sanders[8] have shown empirically that if the product of specimen thickness in centimeters and frequency in megacycles is greater than 4 Mc-cm the derived expressions are valid. Schneider and Burton got consistent results from aluminum when the frequency thickness product was 1.5 Mc-cm or larger. Schneider and Burton have used this technique to measure the elastic constants for several thermo-

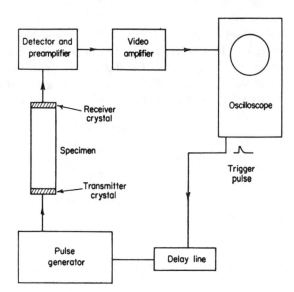

FIG. 9-7. Apparatus for velocity measurements. (*Courtesy Knolls Atomic Power Laboratory.*)

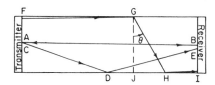

FIG. 9-8. Wave propagation in a cylindrical specimen.

setting and thermoplastic resins. Their results are given in Table 9-2.

Reynolds[6] in his work used cylindrical rods whose diameters were large in comparison with the wavelength and with ends accurately perpendicular to the axis of the specimen. Suitable transducers were attached to the ends, as shown in Fig. 9-7. The cross-sectional area of the transducers was approximately equal to the cross-sectional area of the rod. A short longitudinal wave pulse sent out from the transmitter can reach the receiver by a number of paths, as shown in Fig. 9-8. The first pulse received is the pulse which traveled along the path AB. Reflection may take place at the cylinder wall as shown in path CDE, which is due to the fact that the transmitter sends out a divergent beam. The effect of this is to delay the arrival of the pulse at the receiver by a time which is proportional to the increase in path length over the direct path length AB. Reference to Fig. 9-8 shows that the increase in path length will be small and has no fixed value. The only observable effect is lengthening and distorting of the received pulse. As a result of reflections at the sides of the cylinder, shear waves are generated by mode conversion. In addition, reconversion to the longitudinal mode occurs as shown in path $FGHI$. The reflected shear waves originate at various points along the cylinder wall. These waves are propagated across the rod at the angle given by Eq. (9-23) for the case of grazing incidence.

$$\frac{\sin \Theta_L}{\sin \Theta_t} = \frac{V_L}{V_t} \tag{9-23}$$

This effect delays the arrival of the pulse at the receiver by a fixed length of time. This delay is equal to the difference between the transit time of the pulse along paths GH and JH. The pulse along GH travels with the shear wave velocity, while the pulse along JH travels at the longitudinal velocity. Referring to Figure 9-3, it can be seen that

$$\sin \Theta = \frac{V_t}{V_L} \tag{9-24}$$

If the rod diameter is D

$$GH = \frac{D}{\cos \Theta} \tag{9-25}$$

and

$$JH = D \tan \Theta \tag{9-26}$$

the time delay is

$$\Delta t = D \left(\frac{1}{V_t \cos \Theta} - \frac{\tan \Theta}{V_L} \right) \tag{9-27}$$

$$\Delta t = \frac{D}{V_t V_L} (V_L{}^2 - V_t{}^2)^{\frac{1}{2}} \tag{9-28}$$

For a given rod the time delay Δt is a constant since the exact point at which reflection and conversion to the shear mode take place has no appreciable effect on the total path length. Multiple simple reflections between the cylinder ends give paths such as $ABAB$ and $ABABAB$. There is also the possibility of having any combination of the paths mentioned above.

The pulses arriving at the receiver end of the rod may be amplified electrically and presented on an oscilloscope as functions of time. The relative time of arrival of various received pulses may be measured and wavelength velocities computed from the time data and the rod dimensions if the path of the various pulses can be properly identified. For details of the calculations the reader is referred to the original paper.[6]

Given the density of a solid and using the experimentally determined velocities of waves in the solid, the elastic constants and Poisson's ratio for the solid can be computed from equations (9-9), (9-12), and (9-13).

Damping. When a solid specimen vibrates, its free oscillations decay even when isolated from their environment. Some of the energy is always converted into heat. The various mechanisms by which this transfer of energy occurs are collectively termed "internal friction." The damping factor Q is given by the following equation, provided that the damping is small:

$$Q = \frac{2\pi W}{\Delta W} \tag{9-29}$$

where W = total energy of vibration per unit volume per cycle

ΔW = damping capacity—the part of the energy per unit volume per cycle used to overcome the internal friction

A number of techniques have been developed for investigating internal friction, but most of them fall into one of the following six groups:

1. Determination of amplitude decay in free vibrations: This method consists of determining the time for the amplitude to decrease by a specific fraction. The ratio between the amplitudes of successive free oscillations is constant. The natural logarithm of this ratio δ, called the logarithmic decrement, is taken as a measure of the internal friction. The damping factor Q and the logarithmic decrement δ are related by the equation

$$Q = \pi/\delta \tag{9-30}$$

It is the most suitable technique for very low stresses. The apparatus is simple and easy to operate. External losses are difficult to overcome, the stress distribution is not uniform, and it is inconvenient to change frequencies.

2. Determination of the hysteresis loop in the stress-strain curve during forced vibrations: In this technique stress and strain are observed

during cyclic loading. The area enclosed on the resulting plot is used to compute the internal energy loss. The technique is best suited to high stress studies and has the advantage of more uniform stress distribution. The apparatus is complicated and difficult to use, precision is low, and the frequency range is limited to low values.

3. Determination of the resonance curve during forced vibration: In this technique the specimen is subjected to an alternating force of variable frequency and the amplitude observed. The two frequencies for which the amplitude is one-half that at resonance are found. The lower the internal friction of the specimen, the sharper the resonance peak. If ΔF is the change in the impressed frequency necessary to change the amplitude from half its maximum value at one side of the resonant frequency to half its maximum value on the other side, $\Delta F/F$ is a measure of the internal friction. The damping factor can be determined from the shape of the resonance curve and is given by

$$Q = \frac{F}{F_2 - F_1} \tag{9-31}$$

where F = resonant frequency
F_2 = frequency higher than resonant frequency which has an amplitude 0.707 that of resonant frequency
F_1 = frequency lower than resonant frequency which has an amplitude 0.707 that of resonant frequency.

The apparatus is simple, easy to use, and can be used over a broad frequency range. Any shape cross section may be used. It may not be used if the logarithmic decrement is large or frequency or amplitude dependent. Parasitic losses are difficult to remove.

4. Determination of energy absorption during forced vibration: In this, the specimen is oscillated and the rate of energy absorption determined from the power input from the oscillator or the temperature rise of the specimen. From this and the frequency, the damping capacity may be computed. The technique is easy to use, high stresses can be achieved but the precision is low, the stress is not usually uniform, and the method is not suitable for materials with small internal friction.

5. Determination of mechanical impedance during forced vibrations: In this technique the specimen is driven electromechanically. The impedance it presents to the driver is measured as a function of frequency. The impedance is also determined as a function of frequency when the specimen is clamped so that it is unable to vibrate. The difference of these two impedances is a true electrical impedance called the motional impedance. This difference may be used to compute the mechanical impedance of the specimen if the electromechanical coupling constant is known. The technique is not difficult to use, but precision is low and direct calibration is difficult. The technique is not applicable unless the log decrement is small and is independent of frequency and amplitude.

6. Determination of sound wave propagation constants: In this group of techniques the propagation of sound waves is observed directly or by interferometric means. Pulses are used for the direct observations; otherwise standing wave systems are employed. These methods are the most suitable for the very high frequencies, but they are limited to low stress levels. It is very difficult to prepare specimens for the higher frequencies, and the interpretation of the data is frequently difficult.

Internal Friction. Internal friction in solids may be produced by several different mechanisms. In all cases, however, mechanical energy is transformed into heat. One such process depends on the inelastic behavior of the material. With metals, however, it is the thermal losses which are generally more important. Zener[9] discusses several different thermal mechanisms which will result in the dissipation of the mechanical energy into heat. Changes in the volume of a solid will be accompanied by changes in temperature. When a solid is compressed, its temperature will rise, and when it is extended its temperature will fall. Also there is thermal loss due to thermal conduction to the surrounding air. There is another type of thermal loss which will not be discussed here, but it has been discussed by Landau and Rumer[10] and by Gurevich.[11]

Kê[12] has investigated internal friction produced by "viscous slip" at the crystal boundaries of polycrystalline metals. Experiments indicate that the metal at the boundaries of the crystals behaves in a viscous manner. There are two other processes which occur in crystalline solids when they are deformed which result in internal friction. One process is the movement in the crystals of regions of disarray which are known as "dislocations." The other process is the ordering of solute atoms on the application of stress, which occurs when there is an impurity dissolved in the crystal lattice.

The most direct method of defining internal friction is as the ratio $\Delta W/W$, where ΔW is the energy dissipated in taking a specimen through a stress cycle and W is the elastic energy stored in the specimen where the strain is a maximum. This ratio is sometimes called the "specific damping capacity" or the "specific loss" and can be measured for a stress cycle without any assumptions being made about the nature of internal friction. The value obtained is, however, generally found to depend on the amplitude and the speed of the cycle, and often also on the past history of the specimen.

It can be shown that for a specimen set into vibration and then allowed to vibrate freely that the amplitude of vibration is reduced in one period of the oscillation by a factor $\exp(N/2Mf)$, where N is a damping term, M depends on the shape and mass of the specimen, and f is the frequency of oscillation. The ratio between successive free vibrations is constant, and the natural logarithm of this ratio is the logarithmic decrement δ.

$$\delta = \frac{N}{2Mf} \qquad (9\text{-}32)$$

Instead of vibrating freely, the specimen can be driven. By supplying energy at a rate that is just sufficient to balance the rate of energy loss due to damping, the specimen can be made to vibrate at a constant amplitude. If E is the maintained level of vibrational energy (proportional to the square of the vibration amplitude) and $f\Delta E$ is the rate at which energy is supplied, f being a natural frequency of the specimen, the damping capacity is

$$D = \frac{\Delta E}{E} \tag{9-33}$$

If the supply of energy is cut off and the amplitude falls from a value A_0 to A_n in t sec, the damping capacity is given also within certain limitations by the relation

$$D = \frac{2 \log_e (A_0/A_n)}{ft} \tag{9-34}$$

Damping capacity is calculated from Eq. (9-34) after measuring a decay time.

One of the difficulties in the measurement of internal friction by the resonance method is loss of energy at the supports, and in most investigations of this type, specimens are suspended by fine wires or threads. Even so, some energy will travel along the suspension. If suitable precautions are taken to eliminate all extraneous damping, both the internal friction and the elastic constants of a specimen may be determined by the resonance method.

Experimental Techniques for Internal Friction. Quimby[13] was among the first to employ the resonance method for measuring internal friction in solids. He used a quartz crystal to produce longitudinal oscillations in rods. The crystal was cemented to one end of the specimen, and the amplitude of the vibration was measured with a Rayleigh disk[14] suspended near the other end. He worked with specimens of copper, aluminum, and glass at frequencies around 40 kc/sec. Wegel and Walther[15] employed both longitudinal and torsional oscillations with cylindrical rods of metals at frequencies between 100 and 10,000 cps. The oscillations were generated electromagnetically by the eddy currents induced at one end of the specimen, and the amplitude was measured by the current induced in a coil which vibrated in a stationary magnetic field at the other end of the specimen. Randall, Rose, and Zener[16] used a similar method in their investigation on the relation between internal friction and grain size.

Bancroft and Jacobs[17] and Parfitt[18] used an electrostatic method of generating longitudinal oscillations in metal bars, the amplitude being detected with a condenser microphone. Bordoni[19] has described an electrostatic generator with a condenser microphone to detect the vibrations. The condenser microphone operates in a radio-frequency oscillator, and by using a frequency modulation detector, displacements in which the average movement of the surface is only a fraction of 1 A can be measured.

At frequencies in the range of 12 to 120 kc/sec Nolle[20] used a resonance technique in which the sample of rubber was held against the nickel rod of a magnetostriction oscillator. The presence of the sample shifted the resonant frequency of the nickel rod slightly and broadened the resonance peak.

Förster[21] has described an apparatus with which the values of both modulus of elasticity and damping can be measured rapidly and accurately. The test specimen is suspended from two thin wires, as shown in Fig. 9-9. The wires are each connected with a leaf spring; the latter, in turn, is connected to an electrodynamic system. The specimen is excited in its resonant frequency through the suspension wire by the driving system. Vibrations are transmitted to the receiver system, where they are transformed into electrical vibrations. The deflection given by the measuring instrument after suitable amplifications is proportional to the amplitude of the vibration.

The following relationship exists between the modulus of elasticity and the resonant frequency of a cylindrical bar:

$$E \text{ (kg/mm}^2) = 1.6388 \times 10^{-8} \left(\frac{L}{d}\right)^4 \left(\frac{g}{L}\right) F^2 \qquad (9\text{-}35)$$

where L = length of test bar, cm
d = diameter cm
g = weight, g
F = resonant frequency

The modulus of elasticity, therefore, is obtained by determining the dimensions of the test bar, its weight, and its resonant frequency. By measuring the resonant frequency and the breadth of the resonance

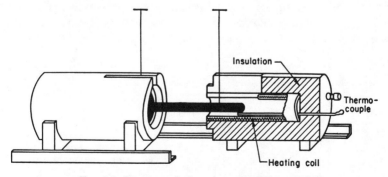

FIG. 9-9. Förster technique for suspending specimen.

curve, the damping can be calculated from

$$\text{Damping} = 1.814\,\frac{\Delta F}{F} \tag{9-36}$$

where ΔF designates the breadth of the resonance curve at the point at which the amplitude has decreased to half the maximum value (breadth at half amplitude). To determine ΔF, it is necessary to measure two frequencies at which the amplitude has dropped to half its value at resonance.

Another way of determining the damping is to measure the time within which the amplitude has dropped from a given value to half this value; the so-called half-amplitude time (t_H) is related to damping by the following expression:

$$\theta = \frac{0.6931}{Ft_H} \tag{9-37}$$

Consequently, the damping value can be obtained, first, by measuring the resonant frequency and the time of decay.

Fusfeld[22] describes an apparatus for measuring internal friction which satisfies the conditions of low induced stress, ease of mounting specimens, rapidity of measurements, and has an accuracy of at least 1 per cent. The mechanical arrangement of specimen, drive, and detector units is shown in Fig. 9-10. The specimen, a bar of uniform cross section, is supported by two vertical wires. An alternating voltage is applied to the driving coil, which induces eddy currents in the specimen. The field of the magnet interacts with the induced eddy currents to produce an axial force on the end of the specimen. The resulting motion is transmitted through the specimen with the frequency of the driving voltage and at the velocity of sound in the specimen. At the opposite end of the specimen, the vibrations cause eddy currents to set up which, in turn, induce a voltage in the coil surrounding the specimen. The induced voltage is amplified and rectified. The rectified output is fed to two trigger circuits, each of which is designed to produce a voltage pulse when the rectified signal drops below some particular value. Both circuits activate a counter circuit.

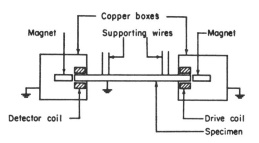

FIG. 9-10. Apparatus for internal friction measurements. (*Courtesy Herbert Fusfeld.*)

To measure the damping, the driving amplifier is shut off electronically so that the specimen is left in free vibration. As the amplitude of vibration decays, the amplifier output decreases to a given value and fires one of the trigger circuits to start the counter. When the vibration decays further, the amplifier output decreases further, and at a given level fires the other trigger circuit which stops the counter. The counter thus registers the time interval for the amplifier output to decay between the given voltages. The damping capacity of the system is related to the measured quantity by the expression

$$\delta = \log \frac{V_1/V_2}{ft} \qquad (9\text{-}38)$$

where δ = damping capacity
V_1 = voltage to start counter
V_2 = voltage to stop counter
f = frequency of vibration
t = interval of time registered on counter

Another technique used for investigating internal friction is the measurement of the attenuation of stress waves as they travel through a solid. The advantages that the wave propagation technique has over the other methods which have been described are: a range of frequencies can be covered with a single specimen, it is easier to reduce extraneous losses at supports, and in nondispersive media the technique is capable of an extremely high degree of accuracy. Bradfield[23] states that the elastic constants of metals can be measured to 1 part in 4,000 with ultrasonic pulses. The disadvantages of wave propagation techniques are: the apparatus needed is generally complicated, it is not always easy to be sure that one particular type of wave is being generated, and in dispersive media the interpretation of the results is often difficult.

Structure Dependence. The damping capacity is dependent on various conditions as well as the metallurgical condition of the specimen. A number of these will be listed, but for detailed explanations, literature references will be given. The damping capacity is affected by the magnitude of the strain, frequency of vibration, temperature, composition, grain size, heat-treatment, aging, cold work, and state of magnetization for ferromagnetic material. The magnitude of these effects varies greatly, depending on the specific material. The damping capacity of an alloy is generally less than that of its constituent elements. In the solid-solution range of composition, damping capacity decreases as the concentration of the solute increases. The first additions of solute are the most effective in causing this decrease. The decrease in damping capacity of steel with increasing carbon content has been investigated by Hatfield and his coworkers,[24] by Förster and Köster,[25] and by Brick and Phillips.[26]

The effect of additions of aluminum, zinc, or tin to copper within the solid solution range has been investigated by Guillet.[27]

The detection of flaws offers a special field of application for damping measurements. Any inhomogeneity may be expected to increase the damping capacity, since energy should be dissipated at the site of the flaw either as a result of stress concentration or solid friction. The size of the defect that can be detected depends on whether the energy dissipated at the flaw is a significant proportion of the total normal energy dissipation in the specimen. With cylindrical pieces, it is possible to determine the position of a crack in the piece. A specimen free of stresses has the same value of damping in all directions. A specimen with stresses or discontinuities has an anisotropy of damping.

Measurements of damping can give information on the origin of defects, such as the formation of quenching cracks. For example, an increase in damping was found in a steel specimen which had been quenched. The increase in damping was attributable to the formation of microquenching cracks owing to excessive quenching temperature. Such cracks are not easily found in the structure, and the time at which they are formed is difficult to determine. Damping measurements may be used for determining hardening ranges and the effect of multiple hardening.

Intergranular corrosion increases damping. Damping measurements, therefore, constitute an excellent means of following intergranular corrosion. In many cases, the beginning of intergranular corrosion can be established after a few minutes of attack by a corrosive solution. Measurements of damping have one great advantage over other methods in that the corrosion process can be observed on a single test specimen as it is not destroyed in the test.

The most extensive investigation of the use of damping measurements for flaw detection is that reported by Frommer and Murray.[28] Transverse and longitudinal modes of vibration were found to be less suitable for this purpose than the torsional mode. A special system was devised for suspending bars several feet in length and several inches in diameter. Production material in the form of cast ingots, forged billets, and extrusions was examined. Their technique was used to test magnesium alloy bars, 80 in. long and $7\frac{1}{2}$ in. in diameter at one end, tapering to a diameter of 3 in. at the other end. One bar contained a longitudinal crack in the middle region of the specimen. During the test the damping capacity was measured at the fundamental frequency and at several harmonic frequencies. Figure 9-11 shows the test results for a sound specimen and

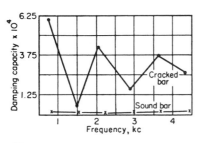

FIG. 9-11. Damping of cracked bar.

for a cracked specimen. Using torsional vibration, the sensitivity with which flaws can be detected is least when they are situated in the center of the bar. The effect of small cracks on the characteristic damping curve is very marked and in certain cases affords a convenient indication of failure by fatigue. High-strain damping is very sensitive to the presence of cracks, and this fact has been applied successfully in the routine detection of season cracks in brass cartridge cases. The technique consists in exciting longitudinal resonance vibrations by producing an eddy current loop around the mouth of the case, which is situated in the radial field of a magnet. In operation, the power required to produce a stress amplitude of 10,000 psi is measured and is found to rise appreciably when the case is cracked.

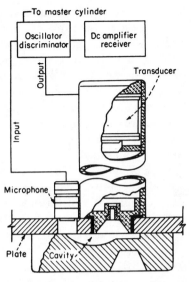

Fig. 9-12. Cavitometer. (*Courtesy Electronics.*)

Other Dynamic Techniques. Three other interesting dynamic and sonic testing techniques are included in this chapter. Industrial testing of cylinder head tolerances in automobile manufacturing has been speeded up considerably through the application of an electronic volume comparator known as the Cavitometer.[29] The system is shown in Fig. 9-12. All cavities within a certain type of throat opening have a natural resonant frequency. A master cavity of known volume is used for comparison with an unknown cavity, and the two cavities are made to resonate at their natural sonic frequency. The difference in frequency can be interpreted in terms of difference in volume as compared with the master cavity. The volume varies as the reciprocal of the square of the frequency. The difference in sonic frequency may be translated into a difference in electrical frequency and used to deflect a pointer on a dial which is calibrated in terms of volume. Any change in ambient temperature or humidity affects equally the natural period of both the cylinder head being tested and the master cylinder head cavity.

Sonic testing of concrete has been recognized for several years as a means for studying the quality of concrete. Leslie and Cheesman[30] have developed the Soniscope, shown in Fig. 9-13. The Soniscope is designed to measure group velocities through as much as 50 ft of concrete. Tests on bridges, dams, highways, and other concrete structures indicate that the test will identify concrete of very good or very poor quality. Changes in group velocity with time appear to have particular significance in studying the performance of concrete, increasing group velocity indicating improvement in quality, decreasing velocities indicating deterioration.[31-34]

Kesler and Higuchi[35] have determined the compressive strength of concrete by measuring sonic properties. Both the modulus of elasticity and the shape of the resonant curve were measured. Figure 9-14 shows a schematic diagram of their apparatus. The frequency counter was used to measure the frequencies because of the accuracy required in the quantity f. Specimens were 6- by 12-in. cylinders. From their test data they obtained a set of curves from which the strength of concrete could be predicted. The accuracy of the prediction is generally within an error of 5 per cent. This accuracy can be obtained without knowledge of the age, mix, or moisture content of the concrete. These results also indicate why use of the modulus of elasticity alone is not sufficient to predict accurately the strength of concrete.

The Stanford Research Institute has developed an instrument known as the STUB meter for measuring the strength of adhesion bonds. Commercial instruments of this type are sold under the trade name of Ultra-Sonic Bond Analyzer or Coindascope. The transducer is tightly coupled to the test specimen and thus imparts some vibrational energy to the specimen. In the STUB-meter technique, the system is continuously

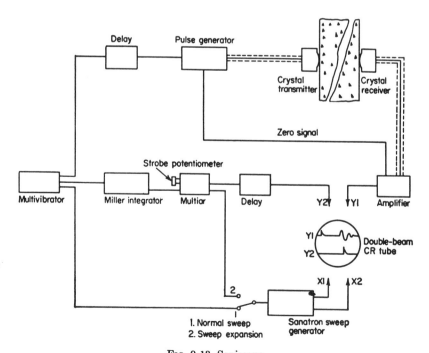

FIG. 9-13. Soniscope.

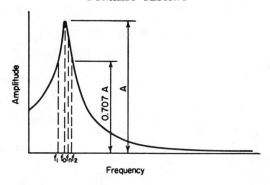

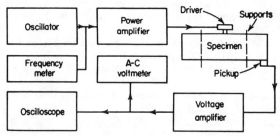

FIG. 9-14. Block diagram of apparatus for measuring compressive strength of concrete. (*Courtesy Kesler and Higuchi.*)

supplied with vibrational energy. The system behavior is examined as a function of the driving frequency, the mechanical resonance of the system being of particular interest. The range of the driving frequency is so chosen that the resonances that exist are those of the transducer itself. The amplitude and frequency of these resonances change when the transducer is coupled to the specimen. The effects of these contributions are shown by the idealized resonance curves of Fig. 9-15. Changes in system mass and stiffness affect the frequency at which resonance occurs, while changes in loss affect the amplitude of the response at resonance. The change in system losses, however, may be indicative of bond quality. The experimental work to date indicates that there is a generally recognizable relationship between bond quality and the STUB-meter display.

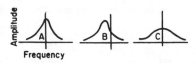

FIG. 9-15. Idealized resonance curves. (*A*) Initial system: transducers. (*B*) Composite system: measurement and stiffness changes affect position on frequency axis. (*C*) Composite system: changes in loss affect amplitude.

SPECIFIC REFERENCES

1. Morse, Philip M.: "Vibration and Sound," 2d ed., McGraw-Hill Book Company, Inc., New York, 1948.
2. Pickett, Gerald: Equation for Computing Elastic Constants from Flexural and Torsional Resonant Frequencies of Vibration of Prisms and Cylinders, *Proc. ASTM*, vol. 45, p. 846, 1945.
3. Quimby, S. L.: New Experimental Methods In Ferromagnetism, *Phys. Rev.*, vol. 39, pp. 345–353, 1932.
4. Zener, C., and R. H. Randall: *Trans. AIME*, vol. 137, p. 41, 1940.
5. Schneider, W. C., and C. J. Burton: Determination of the Elastic Constants by Ultrasonic Methods, *J. Appl. Phys.*, vol. 20, pp. 48–58, 1949.
6. Reynolds, M. B.: The Determination of the Elastic Constants of Metals by the Ultrasonic Pulse Technique, *Trans. Am. Soc. Metals*, vol. 45, pp. 839–861, 1953.
7. Bez-Bordili, W.: Uber ein Ultraschall-totalreflektrometer Zur Messung von Schallgeschwindigkeiten Sowie der Elastic Konstantes Fester Korpes, *Z. Physik*, vol. 96, pp. 761–786, 1935.
8. Sanders, F. H.: Transmission of Sound through Thin Plates, *Can. J. Research*, vol. 17A, no. 9, pp. 179–193, September, 1939.
9. Zener, C.: "Elasticity and Anelasticity of Metals," University of Chicago Press, Chicago, 1948.
10. Landau, L., and C. Rumer: Schallabsorption en sesten Über Külorperm, *Physik. Z. Sowjetunion*, vol. 11, pp. 18–25, 1937.
11. Gurevich, L.: On the Absorption of High Frequency Sound in Metals, *J. Phys. (U.S.S.R.)*, vol. 9, pp. 383–384, 1945.
12. Kê, T. S.: Experimental Evidence of the Viscous Behavior of Grain Boundaries in Metals, *Phys. Rev.*, vol. 71, no. 8, pp. 533–546, Apr. 15, 1947.
13. Quimby, S. L.: On The Experimental Determination of the Viscosity of Vibrating Solids, *Phys. Rev.*, vol. 25, pp. 558–573, April, 1925.
14. Stephens, R. W. B., and A. E. Bate: "Wave Motion and Sound," p. 247, Edward Arnold & Co., London, 1950.
15. Wegel, R. L., and H. Walther: Internal Dissipation in Solids for Small Cyclic Strains, *Physics*, vol. 6, pp. 141–157, 1935.
16. Randall, R. H., F. C. Rose, and C. Zener: Intercrystalline Thermal Currents as a Source of Internal Friction, *Phys. Rev.*, pp. 56, 343, 1939.
17. Bancroft, D., and R. B. Jacobs: An Electrostatic Method of Measuring Elastic Constants, *Rev. Sci. Instr.*, vol. 9, pp. 279–281, September, 1938.
18. Parfitt, G. G.: Energy Dissipation in Solids at Sonic and Ultrasonic Frequencies, *Nature*, vol. 164, no. 4168, pp. 489–490, Sept. 17, 1949.
19. Bordoni, P. G.: Metodo elettroacustico per ricerche sperimentali sulla elasticila, *Nuovo cimento*, vol. 4, no. 3-4, pp. 177–200, 1947.
20. Nolle, A. W.: Methods for Measuring Dynamic Mechanical Properties of Rubber-like Materials, *J. Appl. Phys.*, vol. 19, no. 8, pp. 753–774, August, 1948.
21. Förster, F., and W. Köster: A New Method of Measurement for Determining Modulus of Elasticity and Damping, *Z. Metallk.*, vol. 29, pp. 109–115, 116–123, 1937.
22. Fusfeld, H. I.: Apparatus for Rapid Measurement of Internal Friction, *Rev. Sci., Instr.*, vol. 21, no. 7, pp. 612–616, July, 1950.
23. Bradfield, G.: Precise Measurement of Velocity and Attenuation Using Ultrasonic Waves, *Nuovo cimento Suppl.* 2, vol. 7, pp. 162–181, 1950.
24. Hatfield, W. H., G. Standfield, and L. Rotherham: *Trans. North East Coast Inst. Engrs. & Shipbuilders*, vol. 58, p. 273, 1942.

25. Förster, F., and W. Köster: Modulus of Elasticity and Damping in Relation to State of the Material, *J. Inst. Elect. Engrs.*, vol. 84, pp. 558–564, 1939.
26. Brick, R. M., and A. Phillips: Fatigue and Damping Studies of Aircraft Sheet Materials, *Trans. Am. Soc. Metals*, vol. 29, pp. 435–469, June, 1941.
27. Guillet, L., Jr.: Influence de la composition chimique et de la structure de certain allaiges metalliques sur leur capacité d'amortissement, *Rev. mét.*, vol. 43, pp. 265–267, 1946.
28. Frommer, L., and A. Murray: The Influence of the Heat Treatment of Steel on the Damping Capacity at Low Stresses, *J. Iron Steel Inst.*, vol. 151, pp. 45–53, 1945.
29. Sound Waves Test Cylinder Heads, *Electronics*, pp. 99–100, October, 1951.
30 Leslie, J. R., and W. H. Cheesman: An Ultrasonic Method of Studying Deterioration and Cracking in Concrete Structures, *Proc. Am. Concrete Inst.*, vol. 46, 1949.
31. Whitehurst, E. A.: Soniscope Tests Concrete Structures, *J. Am. Concrete Inst.*, vol. 47, p. 433, February, 1951.
32. Whitehurst, E. A.: Use of the Soniscope for Measuring Setting Time of Concrete, *Purdue Univ., Eng. Expt. Sta., Bull.* 74, September, 1952.
33. Whitehurst, E. A.: The Soniscope—A Device for Field Testing of Concrete, *Purdue Univ., Eng. Expt. Sta., Bull.* 71, 1951.
34. Ratchelder, G. M., and D. W. Lewis: Comparison of Dynamic Methods of Testing Concretes Subjected to Freezing and Thawing, *Purdue Univ., Eng. Expt Sta., Bull.* 99, January, 1954.
35. Kesler, Clyde E., and Yoshiro Higuchi: Determination of Compressive Strength of Concrete by Using Its Sonic Properties, presented at 56th Annual Meeting of the ASTM, June, 1953.

ADDITIONAL REFERENCES

Bond, W. L.: The Mathematics of the Physical Properties of Crystals, *Bell System Tech. J.*, vol. 22, pp. 1–72, 1943.
Gold, L.: Evaluation of the Stiffness Coefficients for Beryllium from Ultrasonic Measurements in Polycrystalline and Single Crystal Specimens, AECD-2644.
Hughes, D. S., W. E. Pondrom, and R. L. Minas: Transmission of Elastic Pulses in Metal Rods, *Phys. Rev.*, vol. 75, pp. 1552–1556, 1949.
Huntington, H. B.: Ultrasonic Measurement on Single Crystals, *Phys. Rev.*, vol. 72, pp. 321–331, 1947.
Koster, W.: Modulus of Elasticity and Damping of Iron and Iron Alloys, *Arch. Eisenhüttenw.*, vol. 14, pp. 271–278, 1940. Brutcher Translation 1310.
Laquer, H. L.: Elastic Constants and Sound Velocities III, The Elastic Constants of Uranium, AECD-2606.
Prescott, J.: Elastic Waves and Vibrations of Thin Rods, *Phil. Mag.*, vol. 33, pp. 703–754, 1942.
Brennan, J. N.: Large Amplitude Vibrations of Rods and Tubes at Audio-Frequencies, *J. Acoust. Soc. Am.*, vol. 25, pp. 610–616, 1953.

CHAPTER 10

MAGNETIC METHODS

Inhomogeneities such as blowholes, cracks, and inclusions in a magnetic material produce a distortion in an induced magnetic field. The path of the magnetic flux is distorted because the inhomogeneities have different magnetic properties than the surrounding material. All magnetic methods of nondestructive testing employ some means by which this distortion, often called leakage flux, can be measured or detected.

One simple way of detecting distortion in a magnetic field is to move a compass over the magnetized specimen. The needle will align itself with the magnetic field and thus indicate any distortion in the field. This technique suffers both from low sensitivity and difficulty in applying it to a large-scale rapid testing. Search coils can be used to scan the specimen. When magnetic flux is "linked" to a search coil, an induced voltage wi'l be generated, provided that there is relative motion between the specimen and the search coil. This can be done either by moving the magnetized specimen through the search coil or moving the search coil over the specimen. Flux distortion due to a flaw will vary the induced voltage generated in the coil. One company uses this technique to inspect pipe. The pipe is magnetized in a longitudinal direction, using an electrical current. The search unit is propelled along the outside of the pipe. Flaws in the pipe cause variation in flux, which in turn causes a change in the induced voltage. Using this technique, it is reported that it is possible to differentiate between inside and outside fatigue cracks in pipe. Another way in which distortion of a magnetic field may be detected is by using a fine magnetic powder. Since this technique is widely used it will be described in detail.

Magnetic Particle Technique. Magnetic particle inspection is a relatively easy and simple technique. It is almost completely free from any restriction as to size, shape, composition, and heat-treatment of a ferromagnetic specimen. There are two essentially basic steps in the magnetic particle inspection technique: magnetization of the material and the application of magnetic particles. The finely divided magnetic particles or powder can be either dry or suspended in liquid. If the flaw is a surface flaw or lies sufficiently close to the surface, there will be set up a pair of magnetic poles which act like small magnets. The magnetic

302

powder is attracted and held by the leakage flux, thus forming a visible indication of the location and extent of the defect. The surface condition of the object being inspected may affect the sensitivity of the method especially for locating subsurface flaws. The surface must be clean, dry, and free of slag or rust. Wire brushing or sandblasting will usually clean the surface sufficiently. In the case of an excessively rough surface, grinding or machining is advisable. Surface defects usually produce powder patterns which are sharp and tightly held with a heavy build-up of powder. Subsurface defects usually give less sharply defined powder patterns since the powder is less tightly held. However, experienced inspectors can evaluate the severity of a flaw by the nature of the powder pattern. Generally speaking, all surface and near-surface defects commonly encountered in ferromagnetic materials are detectable by magnetic particle inspection. The types of defects include quenching cracks, thermal cracks, seams, laps, laminations, grinding cracks, overlaps, nonmetallic inclusions, fatigue cracks, hot tears, and plating cracks.

Types of Defects. Since some of the defects mentioned above have not been previously discussed, a brief discussion of these follows. Grinding cracks, Fig. 10-1, are cracks of varying sizes in hard surfaces produced during grinding operations frequently by a glazed or improper wheel. Grazing results in increased pressure, which in turn causes overheating. Grinding cracks are similar in appearance to thermal cracks and at times are probably identical. They usually occur in groups and may be roughly parallel to each other. Grinding cracks are very sharp at the bottom and are extremely damaging to surfaces on which they occur. This is especially true if the surfaces are highly stressed, since the presence of any stress raiser is undesirable from a fatigue standpoint. The narrower the cracklike discontinuity, the sharper the notch and the greater the stress concentration.

Forging laps or folds, Fig. 10-2, as their name implies are usually a thin fold of metal squeezed together or lapped over during the forging operation, from a fissure which may be either open and shallow or tight

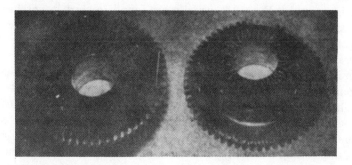

FIG. 10-1. Grinding cracks. (*Courtesy Magnaflux Corporation.*)

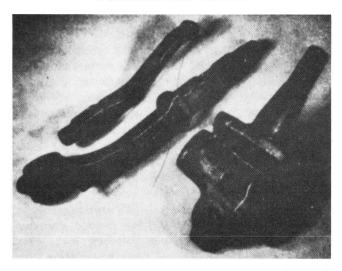

Fig. 10-2. Cracks in forged parts.   (*Courtesy Magnaflux Corporation.*)

lipped and deep.   Laps run with the grain flow which may not bear any relation to the actual direction of the part.   Forging laps are usually not sharp, and therefore care must be taken in attempting to locate them. Tears, overfill, and scratches frequently owe their origin to the billet mill and occur in common with certain kinds of folds and are usually evident on visual inspection.   Machining cracks are usually found in billets or locations where sudden changes of section occur, but they are to be distinguished from both quenching cracks and toolmarks which are ordinarily not seen by this technique.   Certain types of seams are really longitudinal streaks of inclusions in the direction of rolling.   Inclusions are not always rolled out, however, and may take many and varied unrelated shapes.   They are relatively small and shallow.   In welds there may be surface cracking in the cast metal, in the metal adjacent to the weld, in the junctions of the two, or in the boundaries of the decarburized area.

Fatigue cracks, Fig. 10-3, usually are found to start in notches.   The formation of such notches may be due to 'a number of causes.   One class of notches has its origin in accidental blows received in handling or in scratches, abrasions, and nicks.   Another class of notches has its origin in design which may be responsible for sharp changes of section or contour such as fillets, thread routes, oil holes, keyway, and splines. Metallurgical defects such as surface inclusions, decarburized surface, or hydrogen embrittlement are occasionally responsible for subsequent fatigue cracks.   Still another class of notches results from processing and includes grinding marks, grooves from machining operations, file marks, stencils, seams, quenching cracks, grinding cracks, forging laps, and

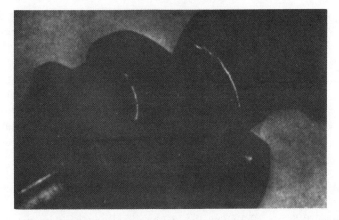

FIG. 10-3. Fatigue crack in crankshaft.   (*Courtesy Magnaflux Corporation.*)

bursts.   The parts in which stress raisers occur do not fail under a single
load or a few loads of the kind which occur in normal service.   They
occur after these loads have been repeated many hundreds or thousands
of times.   A fatigue crack following its formation does not always progress immediately to complete failure particularly if the stresses are
reduced.

Subsurface defects form a large and important class of flaws including
hairline and other seams, porosity, cold shuts, shrinkage cavities, and
inclusions.   The desirability of locating them increases as they approach
the surface or other locations where stress may be increased.   Shatter
cracks or forging flakes, Fig. 10-4, have their origin in the processing of
the metal after the forging.   Forging cracks may be caused by forging
either too hot or too cold, too rapidly or too slowly and are simply the
results of overloading the metal when it is in a condition when it cannot

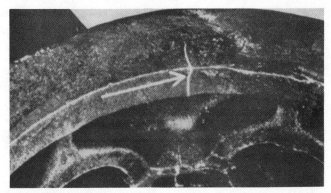

FIG. 10-4. Malleable casting cracks.   (*Courtesy Magnaflux Corporation.*)

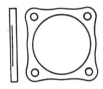

FIG. 10-5. Laminations.

take the treatment. One cause for not being able to take the forging treatment may be the presence of a nonmetallic inclusion. The forging cracks may have any direction and are not necessarily related to the grain structure of the finished part. Forging defects such as bursts, ruptures, cold shuts, and some subsurface laps usually lay some distance below the surface and may not be susceptible to satisfactory location by the magnetic particle inspection method unless they appear at the end of the forging or at locations where deep cuts are made and hence brought to the surface. Laminations, Fig. 10-5, in sheet are caused by pipe, by large areas of rolled-down nonmetallics, or by unwelded blowholes from the original ingot.

Figures 10-6 and 10-7 show the leakage field produced by a surface and a subsurface flaw. The magnitude of the leakage field is affected by a number of factors. These factors include the intensity of the magnetizing force, the permeability of the material, and the shape of the specimen. The shape, size, location, and orientation of the inhomogeneity also affect the magnitude of the leakage field. The size of the flaw that can be detected depends on the strength of the leakage field produced by the defect. The strength of the leakage field decreases rapidly with increasing depth of the defect below the surface. Depth is a relative factor from the standpoint of magnetic particle inspection. A flaw $\frac{1}{8}$ in. below the surface is relatively nearer the surface in a 2-in. plate than the same flaw in a $\frac{1}{4}$-in. plate. Roughness of the surface of a specimen is usually no problem in locating surface defects, but for locating subsurface defects it is helpful to have a smooth surface. The dry method is more sensitive than the wet method for locating subsurface defects.

Magnetization. A magnetic field can be set up in a magnetic material in either of three ways: by passing electric current directly through all or a portion of the specimen, by passing electric current through a conductor surrounding or in contact with the specimen, and by magnets. A conductor carrying an electric current is surrounded by a magnetic field

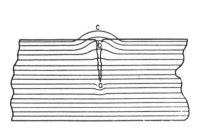

FIG. 10-6. Leakage field due to surface flaw.

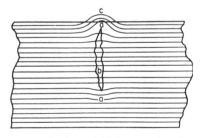

FIG. 10-7. Leakage field due to subsurface flaw.

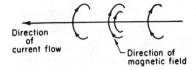

FIG. 10-8. Right-hand rule.

FIG. 10-9. Circular field between electrical contacts.

which forms closed circles in a plane at right angles to the direction of current flow. A useful rule for determining the relative direction of current and magnetic field is the so-called "right-hand rule." If one pretends to grasp a current carrying conductor with the right hand, so that the outstretched thumb points in the direct of current flow, the fingers point in the direction of the magnetic field.

Figure 10-8 illustrates the right-hand rule and shows the magnetic field surrounding a conductor carrying an electric current. This method produces what is known as "circular magnetization." In some cases it may be impossible or impractical to magnetize the specimen as a whole. In this case the specimen may be magnetized by passing current through areas or sections of the specimen by means of contacts or prods. This produces a circular field between the contact points, as shown in Fig. 10-9. Care must always be taken that the contact areas are sufficiently clean for passage of the high-amperage current without arcing or burning. It is advantageous to use low-voltage equipment for this application in order to prevent burning of the part under investigation. The magnetizing current should not be turned on until after the prods have been properly positioned on the surface, and the magnetizing current should be turned off before the prods are removed.

Figure 10-10 shows the magnetic lines of force in a solenoid. When ferromagnetic material is placed in a solenoid, a magnetic field is set up in the specimen. This is often referred to as "longitudinal magnetization." A convenient way of producing a longitudinal field under shop or field conditions is to wrap a number of turns of flexible cable about the specimen. The fewer the turns used, the higher amperage current required. Flexible cable is also widely used for magnetizing large specimens. If the specimen is hollow, such as a cylinder, circular magnetism can be produced by passing current through a central conductor, Fig. 10-11. The part to

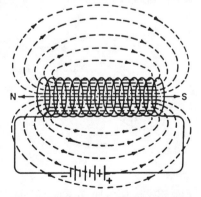

FIG. 10-10. Magnetic field in solenoid.

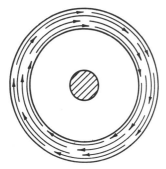

Fig. 10-11. Circular magnetization by use of a central conductor.

be magnetized should generally be placed as near the conductor as is convenient or practical. Material to be magnetized can be placed beside a single conductor carrying a heavy low-voltage current. This type of magnetic field has some of the characteristics of both circular and radial fields. This is referred to as "parallel field magnetization." Two magnetizing forces may be imposed simultaneously on the same area of a magnetizable specimen. When this is done a magnetic field results whose magnitude and direction are determined by the two imposed fields. The resultant field will vary both in direction and magnitude if one or both of the magnetizing components in the fields are alternating. Under the proper phase relationships a field may be formed of a kind which is known as a "rotating vector field." The theoretical advantage of applying such fields is that they cross the major axis of nearly all defects at an angle favorable for location.

For maximum sensitivity, the specimen should be magnetized with the direction of the magnetic field perpendicular to the defect. Figure 10-12 shows how a circular field will locate a defect lying parallel to the long axis of the specimen but will not detect a transverse defect. Figure 10-13 shows how a longitudinal field will locate a transverse flaw. Direct or alternating current can be used for magnetizing the specimen. The intensity of the magnetic field depends on the magnitude of the current. The source of current should be of relatively low voltage in order to minimize danger to the operator or damage to the specimen. Direct current produces fields penetrating deeply into metal. Alternating currents, because of the skin effect, produce fields which are confined

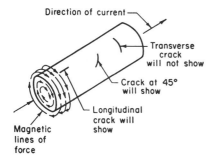

Fig. 10-12. Circular magnetization for longitudinal flaw.

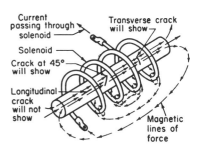

Fig. 10-13. Longitudinal magnetization for transverse defect.

to the surface of the metal. Consequently, alternating currents are best suited for locating surface flaws. Alternating-current sources may be used to advantage when only surface flaws are to be found and subsurface defects are to be ignored.

When a specimen of magnetizable material is placed across the poles of a permanent or electromagnet having a U-shape core, the specimen completes the magnetic path. This permits the flux to travel by a return path through the specimen instead of through the air.

In general the current values used for magnetizing currents are not critical. If too weak a current is used, the resultant field will not be strong enough to form a pattern. If too strong a current is used, dense accumulations of particles may result, which makes the pattern difficult to interpret or may even obscure it. The magnetizing current used for inspecting welds varies from 600 to 2,000 amp, depending on plate thickness and prod spacing. In many cases the magnetizing current requirements are determined by specifications, standards, or purchase orders. When such instructions are not available, current requirements should be determined from experience or by experimentation.

The direct current used for magnetization may be steady direct current obtained from batteries, generators, or full-wave rectified alternating current. Pulsating direct current may be obtained by half-wave rectification of single-phase alternating current. Motor generator sets used for welding can be used as sources of high-amperage low-voltage current. Welding sets have a maximum output of approximately 300 amp. However, with a few turns of a large size flexible cable 1,200 amp-turns or more can be obtained. Generators can be connected in series to obtain larger currents.

Continuous or Residual Magnetization. The magnetic particles can be applied either while the magnetizing current is flowing or after the current has ceased to flow. The first technique is referred to as the "continuous technique" and the latter as the "residual technique." In the continuous technique the magnetizing operation is conducted simultaneously with the application of the inspection medium. In the residual technique only the magnetization retained in the specimen following the application of the magnetizing current is used to attract, orient, and hold the magnetic particles. The continuous technique is the more sensitive. When the magnetizing current is flowing there are often leakage fields present because of conditions other than a defect. This is especially true when a current carrying conductor is wrapped around the specimen to be magnetized. In the continuous technique these leakage fields may be indicated as well as the leakage fields due to flaws. The residual magnetic field is relatively weak compared with the field present when current is flowing. Consequently, the residual technique has a lower sensitivity than the continuous technique. However, the possibility of the false indications is eliminated with the residual technique.

Magnetic Particles. The magnetic powder can be used either in the form of a powder (dry technique) or in a liquid suspension (wet technique). The material used for the dry technique consists of finely divided ferromagnetic particles having high permeability and low retentivity. The particles may be coated to have greater mobility and colored to give maximum contrast on the specimen. The dry powder can be applied above the specimen. This gives the particles an opportunity to line up in indicating patterns as they approach the surface of the specimen. Excess powder can be removed by the use of a low-velocity air stream. In the wet technique finely divided particles of red or black iron oxide are suspended in a light petroleum distillate or in water. The suspension is applied either by spraying or by immersion of the test specimen. Sometimes a fluorescent magnetic powder is used instead of the regular magnetic powder. In this case provision must be made for observing the specimen under ultraviolet light. The fluorescent powder indications are easily seen on rough surfaces. Fluorescent powders give increased sensitivity especially for the detection of subsurface defects.

The dry particle technique gives the greater sensitivity for deep subsurface defects. The dry magnetic powders should be used on rough surfaces as they have less tendency to be held by surface toughness. One method of obtaining mobility with dry powder is to disperse the powder uniformly in the form of a cloud. The second method of increasing mobility is to use mechanical vibrations. This is usually done by tapping the specimen being inspected while the cloud of powder is falling upon it. Particles often can be moved to facilitate pattern formation by the use of variable fields such as rectified or pulsed direct current. A permanent record of magnetic particle indications may be made by photographs or by transfers. Transfers may be made by carefully pressing transparent pressure sensitive tape down over the defect indication. The tape is then removed with the magnetic particles adhering to it. The tape then may be placed on a piece of white paper or in a report to form a permanent record of the flaw.

In some applications it is necessary to remove the residual magnetic field after the inspection has been carried out. This is particularly true in such structures as aircraft where the magnetic field might cause compass error or affect sensitive electrical instruments. The most practical and convenient method of demagnetizing a specimen is to subject the specimen to the action of a magnetic field which is continually reversing in direction and at the same time gradually decreasing in strength. Alternating 60-cycle currents may be employed for demagnetizing in a great number of cases. The usual procedure is to draw the specimen through a solenoid while the current is flowing. Special semiautomatic

FIG. 10-14. Semiautomatic equipment for production testing. (*Courtesy Magnaflux Corporation.*)

or automatic equipment may be used for rapid and economic inspection of parts on a production basis. Figures 10-14 and 10-15 show equipment for production testing using the magnetic particle technique.

The advantages of magnetic particle inspection are: (1) It can be used on any specimen that is made of magnetic material. (2) It is a positive method of finding all cracks and cracklike defects which are on or near the surface. (3) The technique is flexible and with the use of portable equipment may be applied almost anywhere. (4) Both the equipment and manpower requirements are low compared to almost any other non-destructive test and the cost per specimen is usually low compared to almost any other nondestructive test.

Guides in Application. Magnetic particle inspection should always be performed before an acid etch. The etch tends to open up narrow defects, making them wider and at the same time rounding the sharp corners. This shape of defect is not favorable for the best results with magnetic particle technique. Seams may be easy or difficult to detect, depending on the material and sharpness of the edges. Forging laps are not usually sharp and therefore may be difficult to locate. Tears, over-fills, and scratches are always sharp and the direction of magnetization is not critical. Laminations are detectable only in locations where the lamination can be made to produce a leakage field which can penetrate

FIG. 10-15. Automatic equipment for production testing. (*Courtesy Magnaflux Corporation.*)

to the surface. This means practically that only the edges of a plate can be examined. Subsurface defects in welds such as lack of fusion, lack of penetration, porosity, slag inclusions, cold shuts, and shrinkage cracks can be satisfactorily located in many cases when not too deep below the surface. The possibility of detection is not very good beyond $\frac{1}{2}$ in. below the surface unless the defects are extensive. For best results the defects should not be deeper than $\frac{1}{4}$ in. below the surface. Magnetic particle inspection may produce indications of discontinuities in the magnetic field which may or may not be defects. Consequently, it becomes necessary to have an experienced operator to determine what types of discontinuities are being indicated.

Limitations. Some limitations of the magnetic particle technique are listed below. Magnetic particle inspection is applicable only to magnetic material. There are a number of factors such as sharpness, depth, direction, and orientation which determine whether a subsurface defect can be located. Pinpoint inclusions are hard to locate. Hairline cracks are hard to locate unless they are only slightly below the surface. False indications may also occur when there is a local leakage field across a sharp contour. A false defect indication is obtained at the edge of a line of tightly adherent scale. Sudden or abrupt changes in permeability can give false indications. Occasionally cold-working causes an abrupt enough change in permeability to produce a powder pattern. In weld inspection an indication is frequently obtained at the boundary of the cast metal and the base metal. Other indications in the form of lines may appear at the edges of the decarburized zones. These appearances indicate an abrupt change in permeability. At the junction of two different metals of different permeability a magnetic particle indication will invariably be obtained. Another frequent false indication is the so-called magnetic fragment; this is usually caused by careless handling of the component after magnetization. A metallic object being moved across the magnetized specimen normally appears as a blurred line. Such a false indication is often called magnetic writing.

It should be reemphasized that there are a number of variables always present when the magnetic particle technique is used. Presence of a particle pattern indication is not always indicative of metallic discontinuity or objectional defect. The operator should have considerable experience and training so that defects as well as false indications can be recognized. Only quantitative results can be obtained with the magnetic particle technique. Only an estimation of the size and severity of a defect can be made. For this reason the skill and experience of the operator are of greatest importance in the testing operation. As in all methods of nondestructive testing, known or artificially produced defects are needed in order to determine the best testing technique.

Test Block. Harrer[1,]* suggests the following test block for determining the various parameters in the magnetic particle technique. A series of cuts $\frac{1}{32}$ in. wide and at varying depths are made on one side in a 1-in. plate approximately 24 in. square. To the bottom of the cut side a butt strap is welded to cover the slots. Inspection is then performed on the uncut side of the plate. This is not the optimum in arriving at procedural standards. However, it has proved to be a good test block for determining sensitivity.

ASTM Standards. In an effort to standardize the testing procedure the ASTM[2] has issued a standard for dry particle magnetic particle inspection (E109-57T). A committee under the sponsorship of the American Society for Testing Materials has prepared Tentative Reference Photographs for Magnetic Particle Indications on Ferrous Castings (E125-56T). This collection covers types and degrees of defects occurring in ferrous castings detectable by the dry powder technique. These photographs are intended to assist in the classification of these defects and to be compared with the indications observed on actual castings. Forty-seven reference photographs are included covering linear discontinuities (hot tears and cracks), shrinkage, inclusions, internal chills and unfused chaplets, porosity, weld discontinuities, false indications, and magnetic anomalies. Figures 10-16 and 10-17 show typical reference photographs.

* Superscript numbers indicate Specific References listed at the end of the chapter.

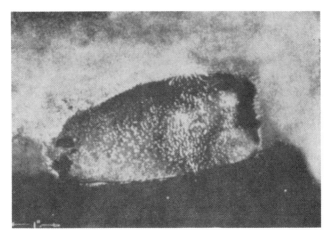

FIG. 10-16. Reference photograph, porosity.   *(Courtesy American Society for Testing Materials.)*

FIG. 10-17. Reference photograph, inclusions in welds.   *(Courtesy American Society for Testing Materials.)*

Suggested Standards.   Magnetic particle inspection standards have been discussed by Caine.[3]   Tiny cracklike discontinuities in an area of high stress and oriented in the proper direction can cause service failures. The narrower the cracklike discontinuity, the sharper the notch and the greater the stress concentration.   A casting which has an indication of a crack through a hole, as in Fig. 10-18, will fail in service.   Indications found on the inner surface, as in Fig. 10-18b, have been proved by service to have no detrimental effect on serviceability.   Acceptance or rejection in Fig. 10-19 is based solely on the location and orientation of the defect.

Several magnetic particle inspection standards for crankshafts have been issued. These standards are summarized in Fig. 10-20. Indications of quite large cracklike defects are allowable. The results are summarized in Fig. 10-21. Quite large indications are permissible in the web and the fillets of the I-beam junctions, unless they are close to the section change from the I beam to bearing sections. Any type of transverse indication in the raised members of the I-beam section are potentially dangerous.

Caine makes tentative recommendations for evaluation of magnetic particle testing. He divides the magnetic particle indications into two categories: dangerous and benign indications. Relatively dangerous magnetic particle indications are:

1. Indications in a uniform section, perpendicular to tensile stress or parallel to bending and torsional stresses.

2. Indications extending around a corner, as in Fig. 10-18. Such indications are particularly dangerous if the indication is in a raised member and transverse to the stress. Internal corners formed by holes, like those in Figs. 10-18 and 10-20, fall in this category.

3. Indications in fillets and changes of sections which impose stress concentration, like those in Fig. 10-20. Peripheral indications are particularly dangerous; longitudinal indications are less so.

4. Indications, especially transverse, at the ends of stiffening members, supporting brackets, or webs.

Relatively benign magnetic particle indications (in areas of low stress) are:

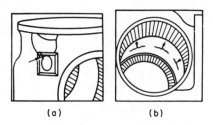

(a)   (b)

FIG. 10-18. Magnetic particle acceptance standards for castings. Crack through hole will grow and cause a fatigue failure. Defects on inner surface of the same casting do not cause service failure. (*Courtesy John B. Caine and Foundry.*)

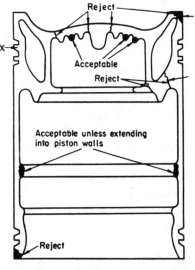

FIG. 10-19. Magnetic particle acceptance standards for pistons. In piston, acceptance or rejection is based solely on location and orientation of defects. The X at left shows area where cracks are unacceptable. (*Courtesy John B. Caine and Foundry.*)

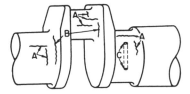

FIG. 10-20. Magnetic particle acceptance standards for crankshafts. (A) Longitudinal or transverse indications 1 to .1½ in. long and ⅟₁₆ in. deep are acceptable away from fillets or oil holes. Up to four shorter indications per bearing are acceptable. (B) Unless indications in fillets are under ¼ in. long, crankshaft must be rejected. (C) Crankshaft must be rejected when indications at the oil holes are not less than ½ in. long. (*Courtesy John B. Caine and Foundry.*)

1. Indications in a uniform section parallel to a tensile stress or perpendicular to bending and torsional stresses.

2. Indications, especially longitudinal, in platelike members stiffened by ribs. Indications in depressed fillets of shapes like those in Fig. 10-21.

3. Indications away from fillets or any stress raisers, as in Fig. 10-20, particularly if longitudinal. The acceptability of peripheral indications of this nature depends not only on the length and depth of the discontinuity, but also on the design itself.

Other Magnetic Techniques. Coils and probes have been used to detect the leakage field. The Förster probe[4] is a special test coil system for the measurement of magnetic leakage flux. Fal' Kevich and associates[5] describe a technique for the continuous inspection of butt welds. In this technique the leakage flux is recorded on a magnetic tape. The tape is passed through a recording head and produces an induced electromotive force which after amplification produces a pattern on an oscilloscope screen. The magnitude of the defect can be determined from the shape of the oscilloscope pattern. The author claims that this technique reveals all possible flaws such as cracks, lack of fusion, porosity, and inclusions.

Förster uses a similar technique which he calls the "magnetographic method."[6] In this technique the magnetized specimen is brought into close contact with a rubberlike band which has been impregnated with iron oxide powder. The recording tape is then scanned with a Förster probe.[4] The induced leakage flux is reproduced quantitatively on a cathode ray screen. Finally the recording tape is passed through an erasing system, allowing the tape to be reused as often as desired.

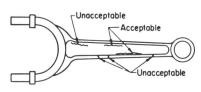

FIG. 10-21. Magnetic particle acceptance standards for connecting rods. Acceptability depends on location and orientation. (*Courtesy John B. Caine Foundry.*)

Förster Probe. The Förster probe has been used for sorting different types of steel. The procedure is as follows: a permanent bar magnet is pressed down onto the surface of the specimen, then released. The spring-mounted magnet returns to a position far enough back so that its field no longer influences the two probes which measure the residual field of the point pole. The strength of this field is read from a meter. Figure 10-22 shows the equipment.

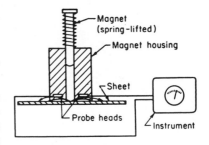

Kodis and Shaw[7] describe an instrument for detecting gun barrel cracks. The gun barrel is magnetized in a circular manner by passing current through the tube bore. A search coil is moved close to the bore surface at a uniform speed. Signals from the search coil are amplified and recorded.

Fig. 10-22. Förster probe.

Farrow[8-11] describes a technique and apparatus for testing magnetic materials. Förster has recently worked with indium arsenide, using the Hall effect to detect or measure magnetic fields. Cook[12] in England has used indium antimonide for the same purpose.

Plated Specimens. Chrome plating of high-strength steels can cause cracking of the steel after plating. In some instances cracking has occurred during the plating process itself. Since the cracks are hidden by the plating, it is essential to inspect the plated specimen. The United States Air Force[13] has reported on a study to determine the effect of plating thickness on the sensitivity of the magnetic particle technique. The experimental procedure used in this investigation is briefly described. Several ⅜-in. diameter fatigue specimens with transverse cracks were butt-welded together. Quenching after welding introduced more cracks and enlarged some of the initial cracks. The composite rod was inspected before plating. Then chromium was plated on the specimen to a total thickness of 0.0066 in. in three separate but identical operations. After each operation the rod was inspected. Full-wave rectified direct current and a solenoid were used in the test. After testing, metallographic specimens were made, and at a magnification of 100 × the width of the cracks was determined. Table 10-1 shows the results of this investigation. From this investigation it may be concluded that (1) plating affects the sensitivity for locating defects in the base metal, (2) the width of the crack has little effect on the ease of detection under the plating, the depth of the crack being the most important factor, and (3) a practical limit of plating thickness is 0.0045 in. if it is desired to find defects effectively in the base metal.

Magnetic Analysis. A magnetic technique known as "magnetic analysis testing" is now widely used in the steel industry to inspect bar stock and tubing for uniformity of quality and freedom from flaws. Magnetic analysis involves the examination of steel for variations in nonmagnetic properties by means of magnetic determinations. Changes in physical properties and chemical constituents vary the magnetic properties of ferrous material. Residual surface stresses may also change

TABLE 10-1. EFFECT OF CHROMIUM PLATING THICKNESS ON MAGNETIC PARTICLE INSPECTION

| Crack measurement, mils | | 350 amp | 520 amp | | 800 amp | | | 1,100 amp | | 2,600 amp | |
| Width | Depth | Chromium thickness, 2.2 mils | Chromium thickness | | Chromium thickness | | | Chromium thickness | | Chromium thickness | |
		2.2 mils	2.2 mils	4.4 mils	2.2 mils	4.4 mils	6.6 mils	4.4 mils	6.6 mils	4.4 mils	6.6 mils
5.6	54.6	Visible	Visible	Visible	Visible	Visible	......	Visible	Visible	Visible	Visible
0.1	3.8	*			Visible						
Narrow	0.4	Visible	Visible		Visible	Faint		Visible	Faint	Visible	Faint
0.2	1.5										
3.4	19.0	Faint	Faint		Visible	Visible		Visible	Visible	Visible	
1.4	3.4	Faint	Faint		Visible					Visible	
0.4	1.9										
0.4	1.4										
0.4	1.3										
0.5	1.9										
2.5	13.4	Faint	Visible	Visible	Visible	Visible		Visible	Faint	Visible	Visible
0.4	2.3		Visible		Visible					Visible	
0.4	1.9		Visible		Visible					Visible	
0.7	3.0		Visible		Visible					Visible	
0.5	38.0	Visible	Visible	Visible	Visible	Visible		Visible	Visible	Visible	Visible
0.6	4.3	Visible	Visible		Visible	Visible				Visible	Faint
0.3	7.5	Visible	Visible		Visible					Visible	
0.1	5.0	Visible	Visible		Visible			Visible		Visible	Visible
0.4	12.4	Visible	Visible	Visible	Visible	Visible		Visible		Visible	Visible
0.6	17.0	Visible	Visible	Visible	Visible	Visible		Visible	Visible	Visible	Visible
0.1	4.5	Visible	Visible		Visible	Visible		Visible		Visible	Visible
0.1	3.6				Visible						
Narrow	4.3				Visible			Faint		Faint	
Narrow	3.4				Visible	Faint		Visible	Visible	Visible	Faint
4.3	112.0	Visible	Visible	Visible	Visible	Visible	Visible	Visible	Visible	Visible	Visible

* Defect not observed.

318

the character and magnitude of the magnetic properties. In the magnetic analysis technique, the specimen is placed in an alternating magnetic field and the effect of the specimen on the voltage induced by the field is observed. The voltage changes are related to the properties and conditions of the specimen. It is necessary in this type of testing to make sure that the test specimens are uniform in all properties which might affect the magnetic properties except those which are to be investigated. For example, in bar stock and tubing the specimens should have uniform cross section, diameter, shape, and wall thickness and have been processed in the same manner. The correlation between residual stress and variations in the magnetic properties is utilized both in flaw detection and identification of material. For example, in flaw detection a defect when subjected to cold-working stresses causes a local relief in the overall stress pattern. This relief causes a change in the magnetic properties in the defective area, which makes it possible to locate the defect. Identification of material may be based on over-all or local variations in the residual stress pattern.

The degree of induced magnetization depends upon the electrical and magnetic properties of the specimen, the frequency waveform, and the strength of the alternating field. The specimen in turn reacts on the magnetizing field. This fact can sometimes be used to emphasize or minimize conditions and properties important to the test. Different frequencies of the magnetizing field can be used; however, 60-cycle fields are often used because this frequency is low enough to give sufficient penetration for surface inspection and at the same time high enough to permit efficient instrumentation. The magnitude of the magnetizing field is one of the factors in this technique which may have to be controlled within wide limits. For example, good results in flaw detection are often obtained when the specimen is magnetized just below saturation. Then the presence of certain types of flaws in the magnetized specimen may cause it to become saturated, which can be detected by distortion in the waveform. When different materials are exposed to the same magnetic field, the resultant waveforms usually differ in amplitude and phase of the fundamental frequency components, as well as in specific waveform distortions. Amplitude and phase determinations can be made using conventional electronic circuits. Investigation of waveform distortions usually requires special electronic circuits. Variations in the nonmagnetic properties or conditions may give waveform distortions which are greater for some portions of the induced voltage wave than for others. Although it is generally not possible to correlate a particular waveform with a specific property or condition, it has been found that most changes in these properties can be checked by measurements which are restricted to relatively short portions of successive wave cycles.

Figure 10-23 shows a cathode ray presentation of a single cycle of a

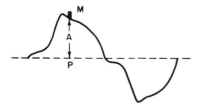

FIG. 10-23. Waveform analysis.

typically distorted 60-cycle wave. The amplitude and phase of any point on the waveform can be measured using the circuit shown in Fig. 10-24. By means of gating circuits the signal induced in the secondary coil is interrogated for very short periods of time during each cycle. The position in the cycle at which this interrogation takes place can be adjusted by the operator. The output of the gate amplifier is a short pulse, which is proportional to the amplitude of the induced signal. This pulse is amplified and measured by a vacuum tube voltmeter. This technique of waveform analysis is used mainly for separating specimens of different grades and for identication of material.

For flaw detection a null technique is usually used. In this technique one section of a specimen is compared with another section of the same specimen. The differential wave resulting from comparing the two sections is used as the indicating wave. For example, identical sections each free of flaws give approximately a straight line. The manner of investigating the differential wave is similar to the technique discussed above. Only that portion of the differential wave which contains significant test information is examined. The circuit used is similar to that discussed previously, except that it is designed for faster response for detection of short length defects. It is possible by combining a single primary coil and a number of secondary coils to inspect simultaneously for flaws and variations in physical, chemical, or metallurgical properties.

Magnetic analysis is used in the steel industry for production testing of bars, tubes, and wires of almost all shapes and sizes. Cracks, seams, and laps can be detected by magnetic analysis techniques. Surface defects which have a depth of not less than 0.001 in. per $\frac{1}{16}$ in. diameter of the specimen may be found. Shallower flaws may be indicated depending on the type of material. Grades and heats of steel differing by a few points in carbon or alloy components can be separated. Specimens varying in hardness by only 2 to 3 points Rockwell C can be separated. Specimens that have been processed slightly differently or just enough to produce minor stress variations are easily identified. Magnetic analysis is used to examine seamless tubing for outside and inside surface flaws. Butt-welded tubing is tested for open and weak welds as well as for inclusions and burned sections.

Physical Property Determinations. Magnetic testing can also be used to determine physical properties of magnetic material. Good correlation has been found between magnetic properties of a material and the physical properties such as hardness, strength, and composition. The most

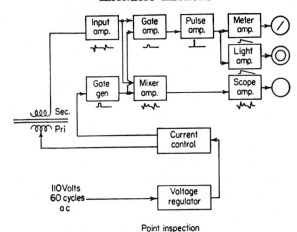

Point inspection

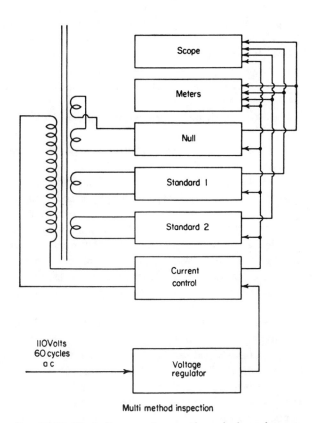

Multi method inspection

FIG. 10-24. Block diagram of magnetic analysis equipment.

commonly used type of apparatus utilizes the transformer principle. An energizing coil is used to establish an alternating magnetic field in the specimen. The voltage induced in the secondary coil placed around the specimen is analyzed. A comparative method is often used in which the induced voltage from a standard specimen is compared with the induced voltage from the test specimen. Only a difference in voltage is measured which is a measure of the difference in the magnetic properties of the two specimens. A meter may be used to show the difference in voltage, or the information may be presented on a cathode ray oscilloscope. The latter type of presentation has the advantage that both the harmonic content and amplitude of the harmonics may be observed. It is possible to have a case in which the specimens show no difference in induced voltage, although there is a difference in magnetic properties. This may happen, for example, when a difference in permeability may be counter-balanced by a difference in hysteresis. It can be shown theoretically that any condition which affects the magnetic properties of the material will vary the induced voltage. The resistive component, amplitude, and phase angles of the various harmonic components will vary with variations in the magnetic properties. For practical purposes only the amplitude and phases of the fundamental and third harmonic need be used. Thus by making a harmonic analysis of the induced secondary voltage, a correlation with metallurgical and physical properties can be made.

Magnetic Comparators. Several types of comparators are in use for comparing magnetic material. One type known as the G.E. Metals Comparator consists of an oscillator, a balancing network, and an indicating instrument. It is basically an impedance comparator. It requires one test coil in which the specimen is to be tested. The impedance of the coil varies with the electrical and magnetic properties of the specimen. The frequency range is 50 to 10,000 cps. Another type of magnetic comparator consists of a bridge circuit having two inductive test coils, two equal capacitors, a copper oxide rectifier, and an indicating instrument. The specimen is placed in one coil, and its variation from a standard specimen is indicated.

A dynamic stress applied to a ferromagnetic test specimen will cause a change in magnetic flux in the specimen. The Cyclograph, to be described and discussed in detail in Chap. 12, has been used in stress analysis of this nature. Magnetic and eddy current losses in the specimen placed in the test coil decrease the oscillator output.

One foundry uses a comparative magnetic technique for detecting mottle in malleable iron castings. Mottle is free graphite precipitated by the iron in cooling. The technique used is to compare the casting against a casting free of mottle selected as a standard. The standard casting and the casting being tested are each used as the core of a trans-

former.   Each of the primary windings has the same number of turns
and is connected in series aiding so that both the standard and test
casting will be affected equally by any voltage fluctuations which may
occur.   The number of ampere-turns (the product number of turns in
the primary wing and the amperes of current flowing) is selected so that
the castings are completely saturated.   The secondary windings are
connected in series opposing.   If there is no difference in the magnetic
properties of the two castings, there will be no current flow.   Figure
10-25 shows a schematic diagram of the instrument.   In setting up
limits for such a test, a number of defective castings were broken up.
The current reading was checked against the degree of mottle.   From
such information it was possible to select which castings are considered
unsatisfactory.   The sensitivity of such a technique can be increased
by using a large number of fine wires in the secondary winding and ampli-
fying the difference in current by means of commercially available
instruments.

Magnetic permeability and magnetic retentivity are used as a basis for
magnetic testing.   The difference in retentivity has been used as a
technique for separating improperly annealed Arma steel castings from
those having the desired hardness.   The magnetized castings are dropped
through a coil, and the induced current is measured.   The magnitude of
the induced current depends on the magnetism retained by the casting.

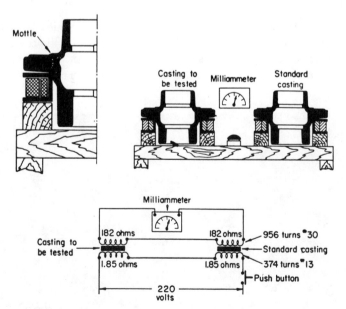

FIG. 10-25. Block diagram of instrument for detecting mottle.   (*Courtesy Research
Laboratories, General Motors Corporation.*)

FIG. 10-26.
Transverse fissure in steel
rail.

Railroads have continual testing problems since they need to locate rails which have become potentially unsafe through the development of internal fissures. These fissures are attributed to fatigue failure of the steel forming the head of the rail. The transverse fissures, Fig. 10-26, are potentially the most dangerous of the various types of defects. Generally they grow slowly in a plane normal to the length of the rail and spread outward from a nucleus until a complete rupture results. Rate of growth is unpredictable and always constitutes a potential danger.

Figure 10-27 shows a Sperry rail car for locating defective railroad rails. In the Sperry method of testing rails a heavy low-voltage current is sent through rails, setting up a strong magnetic field around the head of the rail. A defect in the rail deflects the current in the railhead at the location of the defect, and thus causes a corresponding distortion of the magnetic field at the defect. A series of searching coils suspended at a constant distance above the head of the rail detect any variation in the magnetic field caused by the presence of fissures within the railhead. A variation in the magnetic field induces a voltage in the search coil. The main rail current system consists of a homopolar generator with a rated output of 8,000 amp at 1.8 volts. Current is introduced into rails through a series of brushes. Current flows forward to another set of brushes.

The searching unit consists of a number of search coils, their axis at right angles to the rail. Each of the various sets of coils consists of a pair of coils wound with opposing outputs. Such an arrangement prevents generation of signal due to fluctuations in the rail current. Spacing

FIG. 10-27. Sperry rail car. (*Courtesy Sperry Products, Inc.*)

of each pair is such that local variations in field strength due to defects will affect each coil or pair unequally and generate a signal. The signal from each set of coils is amplified. A permanent record of the signal is made by a suitable recording system. The level of detectability can be set for the amplifiers. The size of the flaw which can be detected varies with different types and weights of rails.

As pointed out in the introduction, if more than one type of test can be performed on a specimen it is advisable to compare results. Such an example is the Sperry rail car service. When the indication of a flaw is recorded on the tape, a visual inspection is also made. If the visual test does not indicate a flaw, an electrical resistance test is made. This test will be discussed in the chapter on electrical testing methods.

GENERAL REFERENCES

American Society for Testing Materials: *ASTM Spec. Tech. Pub.* 85, 1949.

Cavanagh, P. E., E. R. Mann, and R. T. Cavanagh: Magnetic Testing of Metals, *Electronics*, vol. 19, pp. 114–121, August, 1946.

Clarke, J. E., R. A. Peterson, and T. J. Dunsheath: Magnetic Particle Inspection of Welded Pipe and Tubing, *J. Soc. Nondestructive Testing*, vol. 9, no. 4, pp. 7–13, Spring, 1951.

Doane, F. B., and C. Betz: "Principles of Magnaflux Inspection," 3d ed., Photopress, Chicago, 1948.

Edsall, H. L.: Magnetic Analysis Inspection of Metals, *Materials & Methods*, vol. 22, pp. 1731–1735, December, 1945.

McBrian, R., Testing with Magnaflux on the D. & R.G.W., *Railway Eng. and Maintenance*, vol. 45, pp. 137–139, February, 1947.

Spoley, V. L.: Current Applications of Magnetic Analysis Inspection, *J. Soc. Nondestructive Testing*, vol. 8, no. 4, pp. 20–23, Spring, 1950.

Thomas, W. E.: Castings Industry Applications of Magnetic Particle Inspection, *Trans. Am. Foundrymen's Soc.*, vol. 55, pp. 482–488, 1947.

Zuschlag, T.: Magnetic Analysis Inspection in the Steel Industry, Symposium on Magnetic Testing, *ASTM, Spec. Tech. Pub.* 85, pp. 113–122, 1949.

SPECIFIC REFERENCES

1. Harrer, John R.: Greater Acceptance of Welding through the Use of Inspection Methods, *Welding J.*, vol. 36, no. 3, pp. 252–256, March, 1957.
2. American Society for Testing Materials: Tentative Method for Dry Powder Magnetic Particle Inspection, E109-55T.
3. Caine, John B.: Magnetic Particle Inspection Standards, *Foundry*, pp. 84–89, December, 1955.
4. Förster, F.: A Method for the Measurement of DC Magnetic Fields and DC Field Differences and Its Application to Nondestructive Testing, *J. Soc. Nondestructive Testing*, vol. 13, no. 5, pp. 31–41, September–October, 1955.
5. Fal'Kevich, A. S.: Magnetographic Method of Inspection, *Svarochnoe Proizvodstvo*, no. 7, pp. 10–12, 1955; Brutcher Translation 3729.
6. Förster, F.: Electromagnetic Methods of Nondestructive Testing, *Tech. Mitt.*, vol. 50, no. 4, pp. 162–174, 1957; Brutcher Translation 3956.

7. Kodis, R. D., and R. Shaw: Crawler Detects Gun Barrel Cracks, *Electronics*, vol. 24, pp. 93–95, July–September, 1951.

8. Farrow, C.: Nondestructive Electrical Testing of Metals, U.S. Patent 2,415,789, Feb. 11, 1947.

9. Farrow, C.: Method and Apparatus for Determining Phase Shift, U.S. Patent 2,416,517, Feb. 25, 1947.

10. Farrow, C.: Method and Apparatus for Magnetic Testing, U.S. Patent 2,434,203, Jan. 6, 1948.

11. Farrow, Cecil, and Horace C. Knerr: U.S. Patent Re. 21,003, Feb. 14, 1939.

12. Cook, W. G.: Aero Engine Division, Rolls Royce Limited, private communication.

13. Steindorf, W., and B. Cohen: Effective Thickness of Chromium Plate on the Sensitivity of Magnetic Particle Inspection, *WADC Tech. Rept.* 57–342, October, 1957.

CHAPTER 11

ELECTRICAL METHODS OF
NONDESTRUCTIVE TESTING

Electrical methods of nondestructive testing have been used to find laminations in rolled sheet, defects in castings, faulty metal-bearing lining adhesions, poorly brazed joints, defective spot welds, splits in bimetallic strip and cracks in metallic articles, soundness of welded joints, cracks in enamel coatings, cracks in electrical bushings, and insufficient or defective electrical insulation. In addition, these techniques have been used to sort or identify metal stock, measure film thickness, check chemical composition, and check or identify heat-treated metal. The various techniques can be divided into the following: electrical resistance, triboelectric effect, thermoelectric effect, and static field.

Electrical Resistance. Electrical resistance techniques are based on the principle that the electrical resistivity in the immediate neighborhood of a defect differs from that in solid or "good" metal. Electric current flowing in a metal specimen produces a potential difference between any two points on the specimen. The presence of a defect will change the potential difference between the two points. Clean scale-free surfaces and pressure contacts are required to prevent contact overheating or burning of the specimen. Surface contact resistance is an important factor in such tests. Such resistance may cause a greater variation in resistance than defects. Special electrodes have been designed to reduce the disturbing effect of surface contact resistance to a negligible factor.

In a metal specimen of uniform dimensions the resistance between any two points a fixed distance apart will be a constant. The fundamental equation for determining electrical resistance of a material is

$$R = \frac{\rho L}{A} \qquad (11\text{-}1)$$

where R = resistance of specimen
ρ = resistivity of specimen
L = length of specimen
A = cross-sectional area of specimen

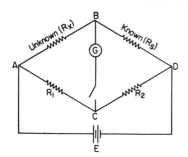

FIG. 11-1. Schematic of Wheatstone bridge.

The unit of resistance is the ohm. Resistivity is given in many handbooks and reference books, and the unit used is ohm-centimeters. In using the above equation the length L must be expressed in centimeters and the cross-sectional area in square centimeters. The value of the resistivity depends on the material and its physical condition. The magnitude of the resistance can be determined by means of a Wheatstone bridge or a Kelvin double bridge. In both cases the unknown resistance is compared with a known standard resistance.

Wheatstone Bridge. A schematic diagram of a Wheatstone bridge is given in Fig. 11-1. The battery E sends a current through the unknown R_X and then through the known resistance R_S and through resistance R_1 and R_2. A current sensitive device such as a galvanometer is connected between points B and C. In using the bridge, resistances R_2 and R_1 are set at suitable values. Resistance R_S is manipulated until there is no current flowing through the current sensitive device. The bridge is then said to be balanced. When the bridge is balanced, the following relationship exists between the resistances:

$$R_X = R_S \frac{R_1}{R_2} \qquad (11\text{-}2)$$

A commercial Wheatstone bridge is shown in Fig. 11-2. The resistances

FIG. 11-2. Commercial Wheatstone bridge. (*Courtesy Leeds and Northrup*)

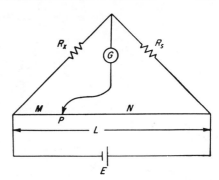

Fig. 11-3. Slide wire
Wheatstone bridge.

R_S, R_1, and R_2, the battery and galvanometer are built into the instrument. Binding posts are provided for external corrections to the resistance being measured. A simple type of Wheatstone bridge is the slide wire bridge shown in Fig. 11-3. In this bridge, the resistances R_1 and R_2 are replaced by a length L of German silver wire or other high resistance alloy. A sliding contact connects the galvanometer to the movable point P. If the wire is of uniform cross section, the resistances of the segments M and N are proportional to their lengths. The bridge equation for a balance now becomes

$$R_X = R_S \frac{M}{N} \tag{11-3}$$

The limit of the sensitivity of measurement with a Wheatstone bridge is determined by the sensitivity and resistance of the galvanometer, the ratio of the resistance in various arms of the bridge, and the currents which may safely flow through the resistances in the bridge. It can be shown that for maximum sensitivity the resistance of the galvanometer should be equal to the resistance which is being measured. It can also be shown that the sensitivity of the bridge is a maximum when the resistance in the four arms is equal. The current through the unknown resistance should be as large as the resistance will permit without undue heating.

The General Motors Corporation[1,*] uses a Kelvin bridge for checking cracks in camshafts. The cracks occur at the junction of the bearing and the shaft. The specially designed electrode to bridge the fillet between the bearing and shaft is shown in Fig. 11-4. A camshaft which was known to be good was used as a standard to set up the equipment. Figure 11-4 also shows the apparatus. The resistance of the fillet is balanced against the resistance of a known length of heavy copper wire. There are fillets having different radii on the same shaft, and conse-

* Superscript numbers refer to Specific References listed at the end of the chapter.

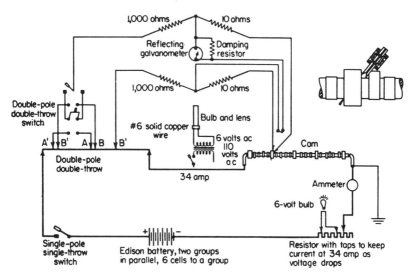

Leads and Northrup variable resistors

FIG. 11-4. Apparatus for checking camshafts. (*Courtesy Research Laboratories, General Motors Corporation.*)

quently the resistance is different at different fillets. This variation in resistance is compensated for by selecting the proper length of wire for each fillet.

The Wheatstone bridge can be used to make reliable measurements of resistance of more than 1 ohm. It is unsuitable for comparing resistances of the order of 0.01 ohm or less for two reasons. The bridge cannot be made as sensitive as is required for measuring small resistances. In addition it is difficult to avoid contact and junction resistance of the same order of magnitude as the resistance to be measured. Since in inspecting metal specimens the resistance to be measured is small, greater precision can be obtained by using a Kelvin double bridge.

Kelvin Double Bridge. The Kelvin bridge is so designed as to avoid the difficulties due to contact resistance and still have high sensitivity. A Kelvin double bridge is shown in Fig. 11-5. The low resistance to be measured R_X is connected in series with a low resistance standard R_S. The ends of the two low resistances are connected by resistances r_1, r_2, r_3, and r_4 to form a double pair or ratio arms. Commercial instruments are provided with a variable standard resistance so that the bridge can be balanced. When the bridge is balanced, no current flows through the galvanometer and the following relationship holds:

$$\frac{r_1}{r_2} = \frac{r_3}{r_4} = \frac{R_X}{R_S} \tag{11-4}$$

FIG. 11-5. Kelvin double bridge.   (*Courtesy Leeds and Northrup.*)

if the contact resistance is neglected.   If the contact resistance r' is taken into account, the relationship becomes

$$X = \frac{r_1}{r_2} s - \left(\frac{r_1}{r_2} - \frac{r_3}{r_4}\right)\left(\frac{r'r_4}{r_3 + r_4 + r'}\right) \tag{11-5}$$

The sensitivity of such a bridge will be maximum when the value of the standard resistance is several times that of the unknown resistance and a low-resistance galvanometer with a high figure of merit is used.   The value of r should be smaller than the galvanometer resistance, and the current should be as high as possible without heating the resistance.   A special form of the Kelvin double bridge known as the Electell Crack Detector[2] has been used to measure the resistance of heavy parts for detecting the propagation of fatigue fractures.

Potentiometer.   A potentiometer is a device for comparing voltages. A simple potentiometer is shown in Fig. 11-6.   The operation of a potentiometer is explained as follows.   The current is first standardized by throwing switch S to position 1 and adjusting the battery current by means of rheostat R until galvanometer G shows that balance has been

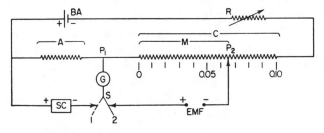

FIG. 11-6. Schematic of potentiometer.

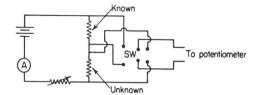

FIG. 11-7. Measuring low resistance with potentiometer.

attained. The voltage drop across resistor A therefore matches the open circuit voltage of the standard cell (SC) of approximately 1.0193 volts. To measure an unknown voltage the switch is set at position 2. The movable contact P_2 is moved along the measuring slide wire to a point where the galvanometer again gives a null indication. The voltage drop from P_1 to P_2 on the slide wire then matches the open circuit value of the measured voltage. The scale position of P_2 on the slide wire is then directly proportional to the measured voltage. Since the same current flows through both A and M, the measured voltage is compared to the standard cell voltage by the resistance ratio M/A. Although the potentiometer is primarily used for comparing voltage, it can easily be adapted to measure current, resistance, and power.

Figure 11-7 shows a potentiometer arrangement for measuring low resistance. Two batteries E_1 and E_2 are required. One battery is to maintain a steady current through the unknown and the standard resistance in series. The other battery is used to maintain a current through a comparison circuit. The same current flows through the two resistors; therefore the difference in potential across the resistors is directly proportional to their resistances. The difference in potential across the unknown and the standard resistance is measured using a potentiometer. The following relation exists between the resistances and measured voltages:

$$R_X = R_S \frac{V_X}{V_S} \tag{11-6}$$

Electrical Resistance Techniques. Electrical resistance techniques have been used for testing the homogeneity and soundness of large turbo-alternator rotors. Electrodes were firmly attached to the ends of the rotor shaft and a current of 10,000 amp direct current was passed through the metal. Potential contacts were placed equidistant along the surface of the rotator. A voltmeter could be switched in turn between each pair of electrodes. With a rotor free of defects, the voltmeter gives equal readings between all pairs of electrodes. Defects in the metal give nonuniform values of potential difference between pairs of electrodes. A similar technique has been used for testing wire, bars, and tubes.

During inspection of railway rails by the Sperry rail testing cars, when defective rails are found, it is the usual practice to carry out a further test. One of these is the electrical resistance technique. Current electrodes are attached at the ends of a measured length of rail, and the potential drop is determined. By this technique it is possible to locate the position of a rail fissure accurately and obtain a rough estimate of its size.

Welds have been tested using the electrical resistance technique. Cracks invisible to the naked eye and about $\frac{1}{16}$ in. deep have been detected as well as undercuts and lack of adhesion of the weld metal. Sciaky[3] describes an apparatus for checking the quality of spot welds by measuring the potential drop across the weld. In this technique an electric current is made to flow transversely through the pieces assembled by welding. The difference of potential produced between the two surfaces is measured. One pair of electrodes is used to carry the current, and a separate pair of electrodes is used to measure the potential drop. The distance between the current electrodes and the potential electrodes is fixed. The difference of potential produced by defective welds is much higher than that produced by good welds.

With castings made to very close dimensional and composition tolerances, it is possible to adopt a simple testing technique to give an indication of the casting quality. A constant voltage is applied across the casting and a reading of the current value made. The presence of a nonconducting defect in the casting reduces the conductive cross section proportionally, increasing the resistivity and lowering the current. A number of sound specimens are essential in order to standardize the apparatus.

Potential Drop Technique. It is frequently important to measure the depth of cracks which have been detected by some other method. Hirst[4] has described an instrument for measuring the depth of transverse fatigue cracks in 5-in.-diameter railway axles. A probe consisting of three contact points arranged in a line and mounted $\frac{1}{4}$ in. apart is used. The probe is placed on the defective area, with the crack between two of the contact points. An alternating voltage is applied to the ends of the specimen. The d-c voltage between each pair of contacts is compared after being amplified and rectified. The instrument can be calibrated against saw cuts of known depth in a specimen of the material. In common with most electrical resistance tests, this instrument does not take into account variations in the electrical and magnetic properties of different specimens of the same design. Sensitivity for flaw detection is influenced by the orientation of the plane of the crack with respect to the direction of current flow.

The following technique is a variation of the technique discussed above. The current is introduced by two electrodes A and B, Fig. 11-8, from a low-voltage d-c source. Two potential electrodes are connected to a sensitive galvanometer or similar detector. The approximate lines of

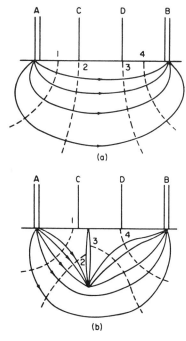

FIG. 11-8. Principle of electrical resistance measurements.

current flow and lines of equal potential are shown in Fig. 11-8. When this fixed electrode system is moved over the surface of the metal, the deflection of the galvanometer remains constant provided that the current is constant. It is essential that all the electrodes make good contact with the surface. If a crack lies between the electrodes as in Fig. 11-8b, the current and equipotential lines are redistributed. Thus the potential drop between C and D will be increased. Sensitivity of detection depends on the depth of the crack compared with the depth of penetration of the current flow.

Electrode Assemblies. Combined current and potential electrodes have been developed. One such electronic assembly is shown in Fig. 11-9. Each electrode is insulated from the others and can be moved independently into and out of contact with the surface being examined, but the relative spacing is fixed. Another type of electrode assembly is shown in Fig. 11-10. Each electrode assembly in Fig. 11-10 is a dual unit consisting of a current electrode completely surrounding a potential electrode. These two electrodes in each unit are electrically insulated. The current electrode terminates in a replaceable tip, which is in contact with the surface of the test specimen. The potential electrode projects through the bore in this sleeve, in which it can move freely. Tips have been made of chrome-plated brass, molybdenum, and a silver-molybdenum alloy, Fansteel-E. The diameter of the face of a tip can be made as small as 1/8 in. Both the current electrode and the potential electrode are spring-loaded to maintain adequate contact with the test specimen. The contact force exerted by these springs can be adjusted independently.

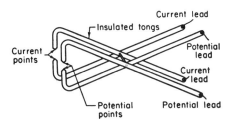

FIG. 11-9. Electrode assembly.

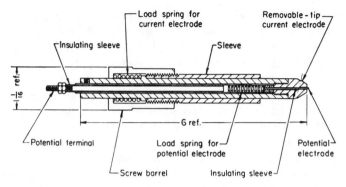

Fig. 11-10. Section through Marburger electrode. (*Courtesy Argonne National Laboratory.*)

Electrode Potential Technique. The special electrodes described above were used in a test which could differentiate between areas of intimate metal-to-metal contact with no bond and areas where the components were fused together. This technique has been called the "electrode potential method" by Marburger.[5] The method makes use of the fact that the electric resistivity in the immediate neighborhood of a poor bond differs from that in solid metal. However, it uses potential difference directly as an indicator of poor bond. The special electrodes reduced the disturbing effect of surface contact resistance to a negligible factor and made continuous scanning of a test specimen possible. The fundamental principle on which the electrode potential method is based may readily be explained by reference to Fig. 11-11a. As indicated in this figure, C and C' represent current electrodes, P and P' represent potential electrodes, and D represents a thin metal specimen. The

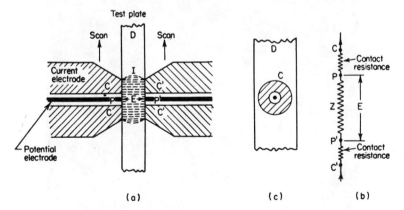

Fig. 11-11. Principle of electrode potential method. (*Courtesy Argonne National Laboratory.*)

potential electrodes are electrically insulated from the current electrodes. In operation, a 60-cycle sinusoidal current denoted by I is sent through the sample between electrodes C and C'. A difference in potential denoted by E is set up between the opposite faces of the specimen. The two pairs of electrodes and the sample constitute a four-terminal resistor, as may be seen more easily from the equivalent circuit diagram of Fig. 11-11b in which the parts are rearranged. From this figure it is apparent that the surface contact resistances between the current electrodes and the sample are included in the current circuit between C and C' but are not included in the potential circuit between the electrodes P and P' and the sample, and that they are in series with the potential measuring circuit whose internal resistance is very high. Variations in these contact resistances are therefore of negligible importance.

Voltages developed across the specimen are amplified for satisfactory operation of an oscillograph or recorder. A rectifier filter was used to rectify the 60-cycle output voltage. The output of the filter circuit was an undirectional voltage whose magnitude varied with the a-c voltage developed in the sample between the potential electrodes. Figure 11-12 is a photograph of the scanner used in connection with these electrodes.

An estimate of the approximate size of the smallest defect detectable by the electrode potential method can be made. The impedance between electrodes pressing on opposite sides of a flat specimen is practically a pure resistance. The value of the equivalent resistance R can readily be found by Eq. (11-1). The actual current flow lines through the specimen form a barrel-shaped pattern, as indicated by Fig. 11-11. This

FIG. 11-12. Electrode potential scanner. (*Courtesy Argonne National Laboratory.*)

TABLE 11-1. TEST PLATE MEASUREMENTS

Section	Thickness of specimen, in.		
	0.249	0.128	0.0625
A_e, cross-sectional area of equivalent conducting cylinder, in.^{2}..............................	0.0419	0.0255	0.0222
$0.2A_e$, equivalent projected area of smallest detectable void, in.^{2}..............................	0.00838	0.00510	0.00444
D, diameter of smallest detectable void, if circular, in...	0.033	0.026	0.024

actual pattern may be considered as replaced by an equivalent cylinder. The length of the cylinder would be the same as the thickness of the specimen. The effective cross-sectional area A_e and the thickness would be of such size as to make the electric resistance from end to end of the equivalent cylinder the same as the actual resistance. The cross-sectional area of this hypothetical cylinder would be

$$A_e = \frac{\rho t}{R} \qquad (11\text{-}7)$$

where ρ = resistivity of specimen
t = thickness of specimen
R = equivalent resistance of specimen

If it is assumed that a 20 per cent change in the potential difference can be measured, then it follows that, on the average, the smallest defect that can be detected must have a projected area normal to the flow lines equivalent to that of a complete void whose area is $0.2A_e$. Measurements on a zirconium test plate, using ⅛-in. electrodes, yielded the values ($\rho = 47 \times 10^{-6}$ ohm-cm) given in Table 11-1. The data show

TABLE 11-2. ELECTRODE POTENTIAL METHOD RESULTS ON KNOWN DEFECT BLOCKS

Thickness of block, in.	Specimen A		Specimen B
	Large defect	Small defect	
0.250	NR*	NR	NR
0.160	NR	NR	NR
0.120	PR	NR	NR
0.090	R	NR	NR
0.060	R	R	NR
0.030	R	R	NR

* NR, not revealed; PR, partially revealed, and R, revealed.

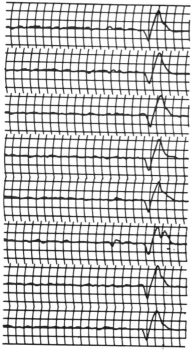

Fig. 11-13. Electrode potential results. (*Courtesy Argonne National Laboratory.*)

that the sensitivity of this technique is greater for thinner specimens. In practice, the smallest detectable defect may be even smaller than indicated by the preceding analysis. A defect midway between the electrodes may escape detection, while the same size defect nearer either electrode may be detected. It is also probable that changes in potential difference smaller than 20 per cent are detectable.

Figure 11-13 shows typical results obtained using the electrode potential method. A summary of the tests with the known defect blocks described in Chap. 5 showing the sensitivity of this method is given in Table 11-2.

Alloy Constitution. The close connection between the electrical conductivity and the constitution of alloys has been known for many years. The conductivity of a pure metal is always decreased appreciably by the addition of alloying constituents held in solid solution. Conductivity is sensitive to order-disorder phenomena, precipitation, strain, and cold-working. The conductivity can be expected to be affected by any factors that produce distortion of the atomic arrangement in the metal. Astbury and Roper[6] describe a technique for the determination of silicon up to 5 per cent in silicon iron sheets. Rotherham and Morley[7] describe a similar technique for estimating magnesium in aluminum alloys and aluminum in magnesium alloys.

An electrical resistance technique is used for determining corrosion rates. This is done by measuring the loss of metal from a probe installed in the corrosive system under study.[8] The electrical resistance increases as corrosion removes metal from the surface of the specimen. By measuring this change of resistance the corrosion rate can be determined. Using probes having a high ratio of length to cross section, a high degree of sensitivity can be achieved. The instrument is made independent of current and temperature by use of a second probe of the same material. The second probe is placed adjacent to the first probe. The second probe is covered with a corrosion resistant coating to retain its original dimensions. The two probes are each connected in one arm of a Kelvin bridge.

Triboelectric Effect. When two metallurgically dissimilar metals in contact are mechanically agitated, a voltage is generated. This voltage ranges in value from a fraction of a microvolt to several millivolts, depending upon the compositions of the metals or alloys in contact. This phenomenon is known as the "triboelectric effect." This effect is caused by the redistribution of electrons when dissimilar substances are brought into intimate contact and then agitated. The redistributed electrons produce a measurable electric current. The friction necessary to produce triboelectrification does not need to be great enough to rupture the surfaces. Although the specimens are in mechanical and electrical contact with other dissimilar metals, the validity of the test is not affected. Heat-treatment produces chemical changes in most alloys so that the triboelectric effect can be used to distinguish between heat-treated and hot-worked alloys coming from the same heat or melt. In general all metals and alloys may be arranged in a triboelectric series. In such a series a metal which precedes another metal is positive with respect to the succeeding metal. For example, five metals A, B, C, D, and E are arranged in such a series. Metal B will be positive with respect to C but negative with respect to A. Metal C will be less negative when compared to B than when compared to A.

The Triboelectric Series

Magnesium	Nickel
Lead	Tin
Aluminum	Silver
Manganese	Carbon
Zinc	Brass
Cadmium	Platinum
Chromium	Silver
Iron	Gold

The technique of creating a moving contact between two metallic specimens in order to produce triboelectrification is not critical. The output from the mechanical agitation is independent of the speed and length of the stroke. The output is independent of the pressure exerted above the minimum to prevent physical and electrical discontinuities. Experiments have shown that surface roughness conditions have little or no effect on the triboelectric output. The surfaces of the metals being tested should be free of scale or other oxides since scale is triboelectrically very powerful with respect to metal and the surface of the metal being tested is representative of the general metallurgical and chemical constitution of the material. The triboelectric output is not affected by variations in the thickness or size of the specimens.

The circuit diagram for a commercial instrument[9] utilizing the tri-

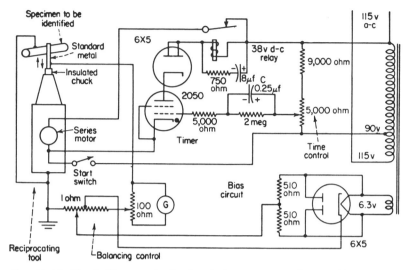

FIG. 11-14. Circuit diagram of triboelectric instrument. (*Courtesy Electronics.*)

boelectric effect is shown in Fig. 11-14. A motor driven reciprocating tool having a ⅜-in. stroke provides the necessary motion between the two metallic specimens. The method of operation consists of mounting the standard in the insulated chuck of the reciprocating tool. The standard is then placed in contact with the test specimen. The necessary electrical connections are made to the standard and the specimen. The scale of the microvolt meter is set at zero to balance out any parasitic or thermoelectric current. The reciprocating action is then started. An electronic circuit is used to control the frequency of the reciprocating motion and the operating time of the instrument. The microvolt meter is read after the reciprocating motion has stopped.

This test is able to detect a major change in the phosphorus, sulfur, and silicon content of carbon steels. It is especially sensitive to the presence of intentional additions of copper. The low-alloy series of steels can be separated from the carbon steels and from each other. The 400 series of magnetic stainless steels are readily sorted as to chromium content. The 300 series of austenitic stainless steels can be sorted from other classes of steels. Nickel-iron and nickel-copper base alloys are readily identified. This technique has been used to sort copper base alloys according to lead content and lead base alloys according to antimony content. Brasses and bronzes are rather difficult to sort. It has been found that 2S aluminum is readily distinguished from the 3S grade by this technique.

Thermoelectric Effect. Figure 11-15 shows a circuit consisting of two dissimilar metallic conductors A and B. If the two junctions of A

and B are at different temperatures, a current will flow in the circuit. This current continues to flow as long as the two junctions are at differing temperatures. The electromotive force producing this current is sometimes called a "thermal electromotive force" (emf). This emf will vary with the temperature difference between the junctions and the materials. Such a combination of dissimilar metals is known as a "thermocouple."

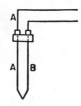

Fig. 11-15. Thermocouple.

The temperature–emf curves for thermocouples are purely empirical, having no precise theoretical derivation. The emf for a particular thermocouple depends only on the temperature difference and increases as the temperature difference increases. If a third metal C is introduced to the circuit the emf in the circuit will be unaffected. A number of couples may be connected in series to form a thermopile. The emf generated by thermocouples in series is equal to the product of the number of thermocouples and the emf for a single thermocouple. The various materials can be put into a thermoelectric series. In such a series each material in the list is thermoelectrically negative with respect to those above it and positive with respect to those below it. Table 11-3 gives a thermoelectric series for selected metals and alloys.

Thermoelectric Metals Comparator. A metals comparator based on the thermoelectric principle has been developed by the General Motors Corporation. This instrument can be used to measure uniformity of metal stock, foil thickness, plating thickness, and thickness of nonconductive films. The instrument can be used to detect variations in metallurgical structure, heat-treatment, and chemical composition. It can find variations in hardness, surface stress, case depth decarburization,

TABLE 11-3. THERMOELECTRIC SERIES FOR SELECTED METALS AND ALLOYS

212°F	932°F	1652°F
Antimony	Chromel	Chromel
Chromel	Nichrome	Nichrome
Iron	Copper	Silver
Nichrome	Silver	Gold
Copper silver	Gold	Iron
Platinum	Iron	Platinum
Palladium	Platinum	Cobalt
Cobalt	Cobalt	Alumel
Alumel	Alumel	Nickel
Nickel	Nickel	Palladium
Constantan	Constantan	Constantan
Bismuth		

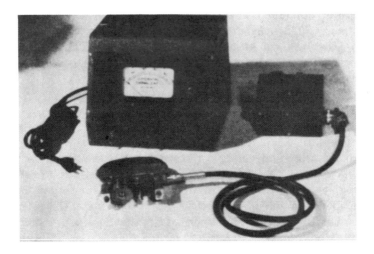

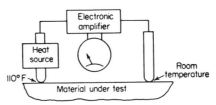

FIG. 11-16. Probe units in thermoelectric metal comparator. (*Courtesy Research Laboratories, General Motors Corporation.*)

cold-working, and adhesion. The instrument works equally well on ferrous and nonferrous metals. The instrument uses a pair of probes of known alloy which make contact with the specimen. One probe is heated to a fixed temperature, while the other probe assumes the temperature of the specimen shown in Fig. 11-16. Since the temperature difference of the probes is kept constant, a change in specimen will result in a different thermal emf. The signal from the probe is fed to an amplifier and a phase sensitive detector. The temperature of the hot junction is held at a certain temperature by the probe heater. When the probe is placed on a specimen the temperature drops, the amount of the drop depending on the thickness and thermal conductivity of the specimen. This fact is used in measuring plating and nonconductive film thicknesses. The difference between the temperature of the junctions varies with the plating thickness, the thermal conductivity of the plating, and the thermal conductivity of the base material. A calibration curve like that shown in Fig. 11-17 can then be made for various combinations of base and coating materials.

Static Electric Field. A static electric field is used to test for defects in dielectric or insulating materials. Such materials include glass, ceramics,

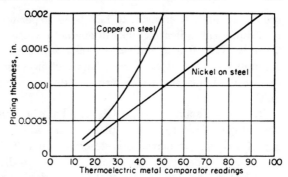

FIG. 11-17. Calibration curve for plating. (*Courtesy Research Laboratories, General Motors Corporation.*)

porcelain, electrical insulating material, and other nonmetallic materials. The electric field patterns may be detected by the use of electrometers, electrified particles, dielectric breakdown, dielectric loss, and change in dielectric properties. The dielectric material may be made part of a condenser, and as the dielectric changes the capacitance changes. A change in capacitance can be detected in various ways, one of the simplest being a change in frequency of an oscillator.

Statiflux. A technique for detecting cracks in porcelain coatings on metal is somewhat similar to the magnetic particle method. This technique is sometimes known by the trade name Statiflux.[10] The principle of this technique is given in Fig. 11-18. A fine grade of calcium carbonate powder is sprayed by means of an air jet on the surface under inspection. Some of the calcium carbonate particles become positively charged when blown through the nozzle. The charged particles are attracted preferentially to the cracks in the enamel coating because of local field conditions. This causes a sharply defined build-up of powder over any cracks in the surface. The reason for the preferential attraction at a crack may be explained briefly as follows. The positively charged powder particles attract free electrons in the metal backing to the underside of the enamel coating. At the crack the potential gradient is greater, and the positive charges are attracted. Leakage of charge also probably

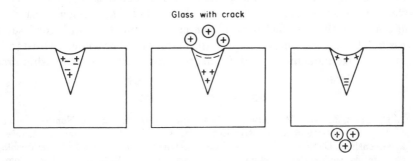

Glass with crack

FIG. 11-18. Principle of statiflux. (*Courtesy Magnaflux Corporation.*)

occurs at such a crack. It has been experimentally found that this technique will indicate cracks in enamel coatings that are narrower than 0.1 μ.[11] It will also indicate cracks which do not extend completely through the enamel coating. This technique is used to locate defects in electrical bushing which are not detectable by other electrical test methods.

Dielectric Testing. Dielectric tests have been used for nondestructively testing glass-fiber-reinforced plastic laminates. An electronic thickness gage for dielectric material is described in U.S. Patent 2,601,649. This instrument is designed to detect variations in dielectric materials. The dielectric properties of the material under test are used to vary the capacitance of a condenser in a critically tuned resonant circuit. As the dielectric between two plates of a condenser is varied, the capacitance of the condenser is also varied. The contact electrode of the instrument forms one side of the condenser, while the grounded shell of the probe forms the other side. When the contact electrode is placed against a laminated test specimen, the capacitance between the electrode and ground is changed. This change in capacitance causes a change in the amount of plate current being drawn by the crystal-controlled oscillator circuit. The plate current is indicated on the panel milliameter on the face of the instrument. The meter readings may be plotted to indicate laminate thickness and, in some cases, to indicate the resin content of the laminate. A gross defect in the laminate will cause a sudden meter deflection as the probe passes from normal laminate to a position over the defective area. The contact electrode consists of a 1-in.-diameter brass roller mounted on a brass shell of the probe by a polystyrene bushing.

Dielectric Strength. Determination of the quality of electrically operated products is of importance. One of the most important techniques offering a solution to this problem is the dielectric strength test. The purpose of this test is generally to evaluate the probable electric strength and endurance of insulation. The dielectric strength test is the deliberate application of a specified overvoltage to a specimen. The overvoltage is maintained for a specified period of time. Such a test will determine if the specified overvoltage can or cannot be withstood successfully for a given time interval, if the material will or will not withstand rated voltage indefinitely, and if defects in material or workmanship are or are not present.

SPECIFIC REFERENCES

1. Diamond, M. J.: "Magnetic and Resistance Methods Used in Nondestructive Testing," Society of Automotive Engineers, Atlantic City, N.J., June 6–11, 1954.
2. Greenslade, G. R., and W. J. Eisenbeis: Electrical Methods, in ASME Handbook, "Metals Engineering—Design," Sec. 1-4, pp. 260–266, McGraw-Hill Book Company, Inc., New York, 1953.

3. Sciaky, D.: Apparatus for Checking the Quality of Welds, U.S. Patent 2,142,619, Jan. 3, 1939.
4. Hirst, G. W. C.: An Apparatus for an Electrical Determination of the Depth of 'Transverse Cracks, *J. Inst. Eng.*, vol. 19, pp. 145–150, 1947.
5. McGonnagle, W. J., J. H. Monaweck, and W. G. Marburger: Methods of Bond Testing, *J. Soc. Nondestructive Testing*, vol. 13, no. 2, pp. 17–22, March–April, 1955.
6. Astbury, N. F., and S. P. Roper: A Direct-Reading Silicon Meter for Electrical Sheet Steels and a Note on Resistivity, *J. Sci. Instr.*, vol. 25, pp. 191–193, 1948.
7. Rotherham, L., and J. I. Morley: A Rapid Method for the Analysis of Light Alloys Based on Electrical Resistivity, *J. Inst. Metals*, vol. 73, p. 213, 1947.
8. Stormont, D. H.: Corrosion Rates Directly Measured by New Resistance Method, *Oil Gas J.*, Jan. 21, 1957.
9. Agnew, N. F.: Sorting Alloys, *Electronics*, vol. 19, pp. 124–125, September, 1946.
10. Staats, Henry N.: The Testing of Ceramics, *J. Soc. Nondestructive Testing*, vol. 10, no. 3, pp. 23–26, Winter, 1952.
11. Orr, Stanley C.: Methods for Testing for Enamel Coating Discontinuities, *J. Soc. Nondestructive Testing*, vol. 10, no. 4, pp. 23–27, Spring, 1952.

CHAPTER 12

EDDY CURRENT METHODS

Eddy current techniques can be used to inspect electrically conducting specimens for defects, irregularities in structure, and variations in composition. Applications of eddy current testing include metal sorting, detection of cracks, voids, and inclusions, measurement of plate or tubing thickness, determination of coating thickness, and measurement of thickness of nonconducting films on electrically conducting base material. Eddy current tests are most effective for locating irregularities near the surface of the specimen. Such testing results in evaluation of characteristics of the test specimen indirectly. A correlation between the measured quantities and the desired structural or serviceable characteristics must be made.

When a coil carrying alternating current is brought near a metal specimen, eddy currents are induced in the metal by electromagnetic induction. The magnitude of the induced eddy currents depends upon the magnitude and frequency of the alternating current; the electrical conductivity, magnetic permeability, and shape of the specimen; the relative position of coil and specimen; and the presence of discontinuities or inhomogeneities in the specimen. Mechanical and thermal treatment of a specimen will affect its electrical conductivity. In the case of austenitic stainless steels, mechanical and thermal treatment will affect the magnetic permeability. The eddy currents induced in the metal set up a magnetic field which opposes the original magnetic field. The impedance of the exciting coil or any pickup coil in close proximity to the specimen is affected by the presence of the induced eddy currents. The path of the eddy currents is distorted by the presence of a defect or other inhomogeneities. The apparent impedance of the coil is changed by the presence of a defect. This change in impedance can be measured and used to give an indication of defects or differences in physical, chemical, and metallurgical structure. Unfortunately, in many cases acceptable variations in quality of a specimen may cause a greater effect on the flow of eddy currents than an unacceptable defect.

The induced eddy currents are concentrated near the surface of the specimen, resulting in the so-called skin effect. In the case of a plane conductor, the current falls off exponentially with depth below the sur-

346

face. The depth of penetration for a plane conductor is given by

$$\delta = \frac{1}{(\pi f \mu \sigma)^{1/2}} \qquad (12\text{-}1)$$

where δ = depth of penetration, meters
f = frequency, cps
μ = magnetic permeability ($4\pi \times 10^{-7}$ henry/meters for non-magnetic materials)
σ = volume electrical conductivity, mhos/meter

The depth of penetration in a plane conductor in a uniform field is the depth at which the current is equal to $1/e$ (37 per cent) times its value at the surface. For nonuniform fields and with other than plane surfaces, the current does not follow Eq. (12-1). The standard depths of penetration for several metals at various frequencies are shown in Fig. 12-1. Such charts serve as a convenient means to determine approximate depths of penetration. The diagonal lines represent the ratio of volume resistivities ρ, in microhm-inches to the relative magnetic permeability μ_R so that the effect of magnetic characteristics of certain metals may be taken into account. For nonmagnetic materials, μ_R equals unity. For example, the depth of penetration in 2S aluminum at a frequency of 5 kc may

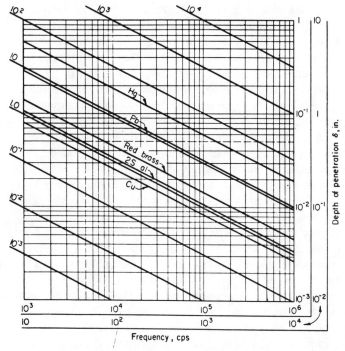

FIG. 12-1. Depth of penetration.

be found as shown by the dotted lines on the chart. The resistivity of 2S aluminum at 70°F is 1.15 microhm-in. The relative permeability μ_R for aluminum, a nonmagnetic metal, is unity; thus $\rho/\mu_R = 1.15$. It should be emphasized that in many cases the reduction in current flow with depth is also governed by the exciting coil geometry. Consequently, Eq. (12-1) should be used only as a guide, and in many cases the depth of penetration can only be and must be determined experimentally.

The magnitude of the inducted eddy currents can be calculated using Faraday's law of induction. The magnitude of the inducted currents as a function of depth in the media is given below. It is assumed here that the magnetic field of the exciting coil varies in a single periodic manner.

$$i_Z = i_0 \exp\left[-(\pi f \mu \sigma)^{\frac{1}{2}}Z\right] \exp j\left[(2\pi ft) - (\pi f \mu \sigma)^{\frac{1}{2}}Z\right] \qquad (12\text{-}2)$$

where i = current at depth Z
i_0 = current just inside surface boundary $(Z = 0)$
f = frequency
μ = magnetic permeability
σ = electrical conductivity
Z = depth in media
$j = -1$
t = time

The first exponential term in Eq. (12-2) represents the decrease in magnitude as the depth increases. The second term describes the phase of the eddy currents at a given depth in relation to the phase at the surface. The magnitude of the induced magnetic flux as a function of depth is given by a similar expression. From Eq. (12-1) it can be seen that the greater the exciting frequency, conductivity, or permeability, the less the depth at which eddy currents can be induced in the metal. For nonferrous metals, whose value of μ is unity when expressed in emu, the frequency can be chosen to achieve the desired penetration. In the case of ferrous metals which have a high value of μ, the penetration, even at very low frequencies, is small. By producing magnetic saturation in the specimen being inspected, the effective value of the permeability can be reduced. Thus sufficient penetration is obtained to make possible eddy current inspection.

Impedance Diagram. There is an apparent change in impedance when a test coil is brought into the presence of a conductor. This apparent change is complicated and is affected by a large number of variables. Changes in the impedance occur both in amplitude and phase. A plot showing the variations in amplitude and phase of the coil impedance is known as an "impedance diagram." Such a graph for a coil encircling a solid cylindrical metal rod is shown in Fig. 12-2. The test coil impedance has been resolved into two components, a reactive component and a resistive component. The dashed curves indicate how the impedance changes when specimens of the same conductivity, but different diameters, are placed in the coil. To eliminate any dependence upon the particular construction or geometry of the encircling coil, the curves of Fig. 12-2 are normalized. This is done by using the ratio of the inductance with specimen to the inductance without specimen (wLo). A number of such diagrams have been prepared by Förster,[1*] Libby,[2] and Oliver.[3] The reactive axis can be related to the energy stored in the coil and specimen during each cycle of the alternating current. The resistive axis can be related to the energy dissipated in the specimen during each cycle. When the specimen has zero electrical conductivity, an insulator, no eddy currents will be induced. If the conductivity of the specimen increases to some finite value, eddy currents will be induced. This will affect the impedance of the test coil in two ways. The

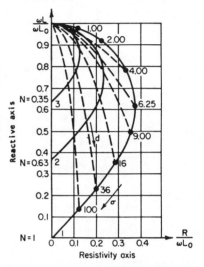

Impedance diagram

$N = \dfrac{\text{Radius of specimen}}{\text{Radius of coil}}$

ωL = Impedance with specimen in coil

ωL_0 = Impedance with empty coil

R = Resistance

FIG. 12-2. Impedance diagram for coil encircling solid cylindrical specimen.

induced eddy currents will create their own electromagnetic field, which will oppose the field of the coil. This results in a decrease of the energy stored in the system and dissipation of energy in the form of heat. Increasing the conductivity causes the reactive component to decrease; however, the resistive component first increases to a maximum and then decreases. Three experimentally determined impedance curves plotted as a function of conductivity with test frequency as a parameter are shown in Fig. 12-3. The three curves are similar in shape, differing only in size and orientation. Similar curves can be plotted showing the impedance of the coil as a function of either frequency or thickness. The effects of variation in permeability and coating materials on the impedance diagram are shown in Figs. 12-4 to 12-7.

One of the most detailed investigations of the use of eddy currents for nondestructive testing is the work of Förster[1] in Germany. Förster prefers to plot his impedance plane diagrams as shown in Fig. 12-8 in terms of a parameter f_g, the value of which is given by the following expression:

$$f_g = \frac{5,060}{\sigma \mu D^2} \tag{12-3}$$

* Superscript numbers refer to Specific References listed at the end of the chapter.

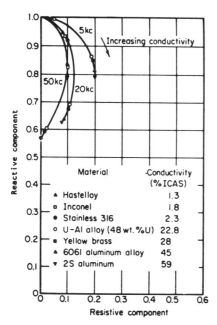

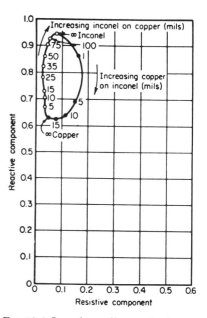

FIG. 12-3. Impedance curves as functions of conductivity and frequency. (*Courtesy Oak Ridge National Laboratory, operated by Union Carbide Corporation for the U.S. Atomic Energy Commission.*)

FIG. 12-4. Impedance diagram for increasing thickness of Inconel on copper. (*Courtesy Oak Ridge National Laboratory, operated by Union Carbide Corporation for the U.S. Atomic Energy Commission.*)

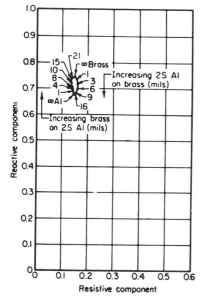

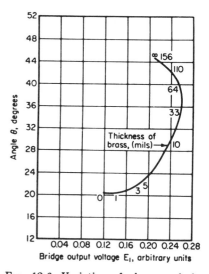

FIG. 12-5. Impedance diagram for increasing thickness of 2S aluminum on brass. (*Courtesy Oak Ridge National Laboratory, operated by Union Carbide Corporation for the U.S. Atomic Energy Commission.*)

FIG. 12-6. Variation of phase angle for brass on zirconium. (*Courtesy Argonne National Laboratory.*)

350

where μ = permeability, henrys/m

σ = conductivity, ohms/m

D = diameter, m

Figure 12-8 shows that the phase separation between the conductivity and the rod diameter axis is essentially constant from $9 \leq f/f_c \leq 100$. By using the ratio f/f_g, Fig. 12-8 can be used for materials of any conductivity, permeability, and diameter. In choosing a test frequency it is only necessary to pick a value of f/fg which lies in the linear region of the impedance plane. Förster refers to the above as the law of similarity. The specimen must behave the same for a specimen of smaller diameter than the coil, and the fill factor N can be defined in terms of the diameter of the specimen and coil

$$N = \frac{d^2 \text{ (specimen)}}{d^2 \text{ (coil)}} \qquad (12\text{-}4)$$

The amplitude of the magnetic flux in the rod is shown in Fig. 12-9. When the rod specimen is replaced with a tube, the thickness of the tube wall becomes another parameter. Figure 12-10 shows an impedance

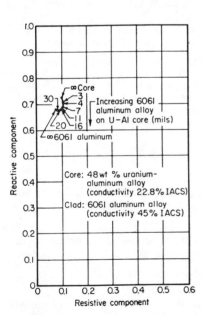

FIG. 12-7. Impedance diagram for increasing thickness of 6061 aluminum on uranium-aluminum alloy. (*Courtesy Oak Ridge National Laboratory, operated by Union Carbide Corporation for the U.S. Atomic Energy Commission.*)

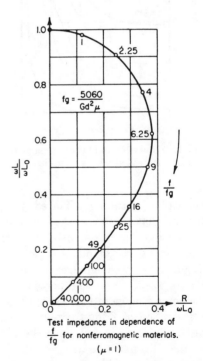

Test impedance in dependence of $\frac{f}{fg}$ for nonferromagnetic materials.

$(\mu = 1)$

FIG. 12-8. Normalized impedance diagram as a function of f/fg. (*Courtesy Magnaflux Corporation.*)

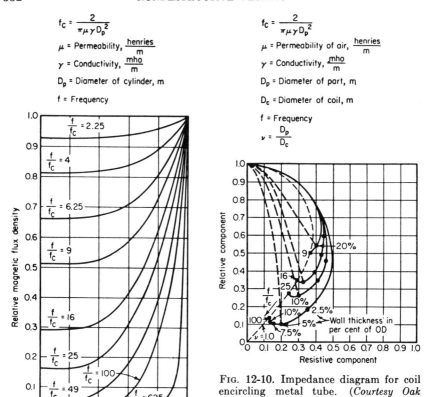

$$f_c = \frac{2}{\pi\mu\gamma D_p^2}$$

μ = Permeability, $\frac{henries}{m}$

γ = Conductivity, $\frac{mho}{m}$

D_p = Diameter of cylinder, m

f = Frequency

FIG. 12-9. Relative amplitude of magnetic flux in a metal cylinder vs. depth.

$$f_c = \frac{2}{\pi\mu\gamma D_p^2}$$

μ = Permeability of air, $\frac{henries}{m}$

γ = Conductivity, $\frac{mho}{m}$

D_p = Diameter of part, m

D_c = Diameter of coil, m

f = Frequency

$\nu = \frac{D_p}{D_c}$

FIG. 12-10. Impedance diagram for coil encircling metal tube. (*Courtesy Oak Ridge National Laboratory, operated by Union Carbide Corporation for the U.S. Atomic Energy Commission.*)

plane plot for tubing. The eddy current distribution in the wall of a tube is somewhat similar to that in a solid rod. The amplitude of the calculated magnetic flux density in a tube is shown in Fig. 12-11. Consequently, by use of phase sensitive circuits it should be possible to distinguish between changes in dimensions and changes in conductivity. Figure 12-12 shows the percentage impedance change for a 1 per cent diameter change in the test specimen.

Detection of Eddy Currents. The detection problem in eddy current testing is complicated since the currents must be observed by inductive means. In addition any variation in the coil to specimen spacing or in the specimen will affect the flow of eddy currents. Fortunately, in many instances such effects may be discriminated against by utilizing the time or phase relations or amplitude of the eddy currents.

The presence of flaws or variations in physical, chemical, or metal-lurgical structure can be detected using the change in the apparent impedance either of the exciting coil or of an independent coil. The exciting coil or probes can be either resonant or driven.

Test Coils and Probes.   Three general types of test coils are commonly used: the concentric coil, the point probe, and the inside or bobbin type coil. The concentric coil completely surrounds the specimen. The concentric coil "interrogates" an annulus of the specimen having a width equal to the effective width of the coil. The point probe consists of a small coil that can be placed near the surface of the specimen. The point probe "interrogates" an area essentially equal to the cross-sectional area of the probe. The inside or coil bobbin is made to be moved through the tube or pipe. Figure 12-13 shows these three types of coils.

Another way of classifying test coils is as "absolute" or "differential." The term absolute is used here to mean that the measurement is made

$$f_c = \frac{2}{\pi \mu \gamma D_p^2}$$

μ = Permeability, $\dfrac{\text{henries}}{\text{m}}$

γ = Conductivity, $\dfrac{\text{mho}}{\text{m}}$

D_p = Outside diameter of tube, m

f = Frequency

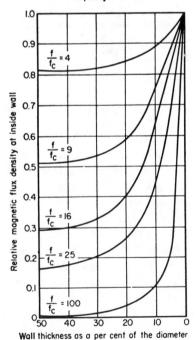

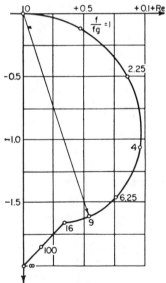

Fig. 12-11. Relative amplitude of magnetic flux in wall of tube vs. depth. (*Courtesy Oak Ridge National Laboratory, operated by Union Carbide Corporation for the U.S. Atomic Energy Commission.*)

Variation of the impedance (amplitude and direction) for 1% thickness variation in dependence of $\dfrac{f}{f_g}$

Fig. 12-12. Change in impedance for a 1 per cent variation in diameter as a function of f/f_g. (*Courtesy Magnaflux Corporation.*)

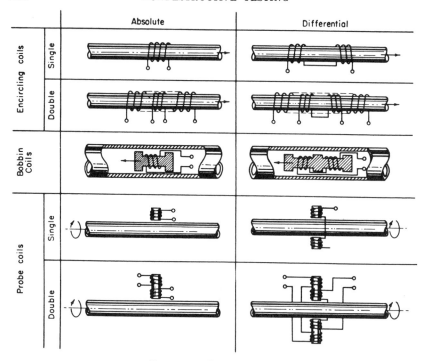

FIG. 12-13. Types of coils.

without a direct reference to or comparison with a standard. In the differential type of test coils two coils connected in series opposition are used. Such an arrangement can be used in either of two ways. In one of the techniques one of the coils surrounds or rests on a standard or specimen known to be good or free of harmful defects; the other coil surrounds or rests on the specimen to be tested. If the specimen being tested is good, the voltage output from the two coils is zero. If the two specimens differ for some reason, there will be an output voltage. In the other technique the two coils are arranged coaxially so that the specimen to be tested goes through both coils, one section of the specimen being compared to an adjacent section of the same specimen. The differential test coil is not very sensitive to gradual changes in diameter or structural properties, but is very sensitive to short cracks or seams. If a defect is long or large enough to extend through both cells, this technique will not reveal the presence of the flaw, except when the flaw enters or leaves the coil system. The configuration and frequency of the exciting and pickup coils must be varied to fit the different test conditions and specimens.

Test coils and probes vary greatly in size, depending upon the specific application. Some consist of one winding and others several windings for special applications. Coils may be of the air core type, or they may have magnetic cores for the purpose of increasing sensitivity or resolution. Test coils may also be tuned to increased sensitivity. Coils also may be shielded with magnetic material or with copper to increase resolution. The initial design of test coils or probes may be carried out using the theoretical properties of coils in relation to the test specimen as previously shown. Then the design must be checked experimentally, making the necessary changes in design as indicated by the test results. Such tests include the use of test specimens with known defects. This experimental work is necessary because the analytical treatments of problems of this sort usually deal with ideal coil configurations, such as a solenoid very long in comparison to its diameter in the cylindrical geometry case. Use of coils having practical dimensions gives results differing from those predicted by calculations based on more or less ideal conditions. The effect of defects on the impedance of a coil is usually too involved to treat analytically with the desired accuracy, and the use of special test specimens or simulated defects results in specific information regarding coil performance. In addition, in many cases, especially in the case of small test coils, there may be more inductance in the leads leading to the coils than in the test coil itself. Also at the higher test frequencies the distributed capacitance and stray lead capacitance affect the results.

The size and type of test coil are governed by the material of the specimen and the kind and size of defect to be detected. If a measurement of the over-all conductivity of a cylindrical specimen is desired, a cylindrical coil could best be used. A frequency low enough to permit reasonable penetration of currents would be selected. Such a coil has considerable end effect, which can be reduced by electrical shielding or by avoiding having the coil extend to the end region of the cylinder. If inspection of the surface or immediate subsurface region of the test specimen is desired, a small flat coil assembly can be used. Such a coil has a limited extent of effective field; thus its useful sensitivity is limited to depths less than about one or two coil diameters, irrespective of the calculated standard depth of eddy current penetration. If the defect is about the same dimension as the coil, a signal will be obtained as the coil is scanned over the defect. However, if the coil is larger than the defect and has a definite "ring" shape, the defect will give two signals, one for each time it passes beneath one edge of the coil. For a coil of given diameter, the sensitivity to a defect decreases rapidly as the defect size becomes less than that of the coil diameter. The characteristics of the various test coil signals depend upon the test specimen properties and geometry in relation to the test coil.

There is no complete theory for the operation of probe-type coils. A considerable amount of empirical information is available, however, and is summarized in Figs. 12-14 and 12-15. These figures represent the

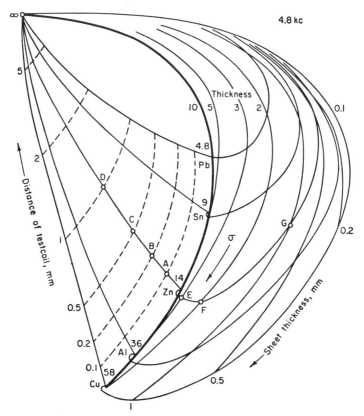

FIG. 12-14. Impedance diagram for probe type test coil at 4.8 kc.   (*Courtesy Magnaflux Corporation.*)

variation in impedance for probe-type coils as a function of the conductivity of the test material.   The heavy curve shown in Fig. 12-14 gives the results obtained when a probe coil is placed on infinite thicknesses of test material.   Infinite thickness means that the field of the probe coil does not penetrate completely through the material.   Thus infinite thickness varies with the operating frequency.   The heavy curve is marked with the chemical symbols of metals corresponding to various conductivity values.   As the conductivity σ is increased from low values to higher values, the impedance variation is indicated by the arrow. If the probe is placed on a surface of the metal, for example, zinc, the inductance will be represented by the point marked Zn on the heavy curve.   If the probe is lifted 0.1 mm away from the surface, the impedance will change to a point marked A in Fig. 12-14.   Lifting the probe farther from the surface causes the impedance to change to points $\bar{B}$, C, D, etc.   The direction of impedance change with distance from the surface is also shown for other materials.   The direction of the impedance change as the probe is lifted from the surface is different from the direction of change of the impedance for conductivity changes. This difference in phase indicates that it is possible to measure conductivity independent of an insulating film on the metal. Conversely, it shows that it is possible to measure the thickness of a nonconducting layer independent of the conductivity of the base metal. If the thickness of the zinc test object is 0.3 mm, the impedance of the test probe will move to point E on Fig. 12-14. For smaller thicknesses, the impedance changes to points F, G, and so forth. As the thickness of the test object is decreased, therefore, the impedance of the test coil will vary along one of the curves indicated in the figure and corresponding to the particular metal under test. It is apparent from the figure that an infinite thickness of metal varies according to the conductivity of the metal. For example, 5.0 mm of tin is an infinite thickness, whereas, for copper, only 2.0 mm represents the same quantity.

After the signal is obtained from the test coil assembly, it is necessary to display it so that it may be read, or analyzed, or used to activate automatic circuits. A wide variety of means to accomplish these ends are used in industry. The signal from the test coil assembly varies in a manner depending upon the test conditions, and it is the function of the detector to indicate the variations of this signal or to indicate when the signal variation exceeds a specified value. Detectors may range from simple ammeters or voltmeters responding to signal amplitude to impedance bridges, vectorscopes, and phase-amplitude detector circuits which make a more complete analysis of the signal. In one type of detector the test coil is included in a portion of an oscillatory vacuum tube circuit, and the effect of the test specimen on the coil is determined by the behavior of the oscillatory circuit. In some detectors the test coil signal can be used directly; in others it is necessary to modify the signal by combining it with a balance signal of such amplitude and phase as to present the detector with the type of signal required for its proper operation.

Probe-to-Specimen Compensation. In the nondestructive testing of

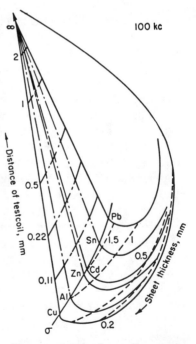

FIG. 12-15. Impedance diagram for probe type test coil at 100 kc. (*Courtesy Magnaflux Corporation.*)

metals by eddy current methods, as mentioned previously, there is the problem of the effect of varying probe-to-sample spacing on the test results. In the testing of specimens of circular cross section by means of circular test coils, changes in coil-to-specimen distance may be caused by diameter variations of the specimen; while in the testing of plane specimens such as a plate with a point type probe, the variation in probe-to-specimen spacing may be caused by irregularities of the surface of the specimen or by vibrations of the probe or specimen. Production testing requires constant movement of the probe parallel to the surface of the specimen, and it is preferred that no contact occur between the probe or probe holder and the specimen. Some previous work on the reduction of the effect of varying probe-to-specimen spacing has been done in Germany[4] and in this country.[5,6] Variation in probe-to-specimen spacing can often be controlled mechanically, such as using a Teflon tape spacer between the probe and specimen or by using a servo-actuated probe carrier. However, in some applications surface contour variations and geometric limitations may cause difficulties so that electronic circuits must be designed to provide additional discrimination against probe movement.

Another method of reducing this effect, using a phase angle meter which is corrected automatically for changes caused by spacing variations, is presented here. Theoretical and experimental work[7] and Fig. 12-16 show how the impedance of a probe in proximity with a plane piece of non-ferromagnetic metal varies as the probe-to-specimen spacing is varied. The ordinate is the fraction of the free space inductance obtained when the metal is in proximity; the abscissa is the incremental change in resistance from the free space value divided by a constant, the free space reactance, the axis of the probe being maintained perpendicular to the plane of the metal. Figure 12-16 is typical of the type of curve obtained when the probe-to-specimen spacing is maintained constant and the conductivity is varied.

Any information to be gained from such a test must be carried by changes in the probe resistance and inductance. Measurements of small changes in probe resistance and inductance are slow and tedious; so the information to be derived from a test should be presented in another, more convenient form. One way to do this is to connect the probe as an element of an a-c bridge. This bridge can then be balanced with the probe in a particular test situation. Any change in the probe's position with respect to the metal, such as a change in probe-to-specimen spacing, or a change in specimen conductivity will disturb the balance of the bridge and cause a sinusoidal voltage to appear at the detector terminals. In the system used by Renken, an Owen bridge was used and balanced with the probe far away from any sample. The impedance looking into

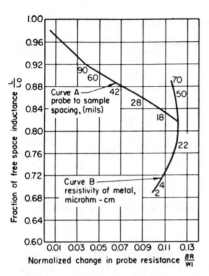

Fig. 12-16. Change in impedance as probe-to-specimen spacing is varied. (*Courtesy Argonne National Laboratory.*)

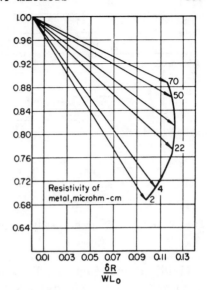

Fig. 12-17. Phasors for various probe-to-specimen spacings. (*Courtesy Argonne National Laboratory.*)

the bridge output terminals can be given by

$$Z = \frac{K}{(1 - L/L_0)(jR/WL_0)} \qquad (12\text{-}5)$$

assuming that the bridge is balanced with the probe at a great distance from a metal and then brought close to it. The quantity K is nearly constant for small changes in the inductance and resistance of the probe. The phase angle of the quantity in the denominator of the expression for Z may be represented by the set of phasors shown in Fig. 12-17, where probe-to-specimen spacing is varied. This phase angle also varies when changes in the other test variables, such as sample conductivity, are made. From Fig. 12-17, it can be seen that the phase angle change is relatively small over a considerable range of probe-to-specimen spacing. This is the reason for balancing the bridge with the probe far away from any specimen and is an advantage inherent in this system. A compensating network such as the one mentioned above will eliminate one of the effects of varying probe-to-specimen spacing. Another effect to be considered here is how the sensitivity of this instrument changes with different probe-to-specimen spacing. Experimentally[7] it has been found that over a useful range of probe-to-specimen spacings the sensitivity of the instrument does not change with different probe-to-specimen spacings.

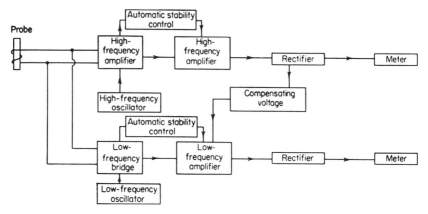

FIG. 12-18. Multifrequency eddy current system. (*Courtesy Argonne National Laboratory.*)

Multifrequency System. Renken and Myers[8] have developed a multi-frequency eddy current testing system. Eddy currents of two separate frequencies are induced to flow in the specimen. One frequency (high frequency channel) is used to compensate for the effect of variations in probe-to-specimen spacing, and the other frequency (low-frequency channel) is used to obtain the desired or required information from the specimen. Figure 12-18 shows the block diagram for the instrument. The probe is a common element of each bridge circuit. The high-frequency channel is not affected by the material of the specimen, but is very sensitive to changes in probe-to-specimen spacing. The output voltage of the high-frequency channel changes the gain of the low-frequency channel just sufficiently to compensate for the effect of variations in the probe-to-specimen spacing. Figure 12-19a shows that the output voltage of the low-frequency channel varies as a function of distance from the specimen. Figure 12-20a shows the change in output voltage as the

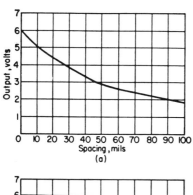

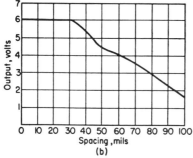

FIG. 12-19. Effect of probe-to-specimen compensating circuit. (*Courtesy Argonne National Laboratory.*)

coating on a specimen varies. Figures 12-19*b* and 12-20*b* show the effect of adding the probe-to-specimen compensation.

A number of eddy current instruments are commercially available. These include the Cyclograph, Probolog, Dermitron, Metals Comparator, Magnatest, the other Förster type instruments, and the Metrol type of instruments. In addition specialized eddy current instruments have been described in the literature. The various instruments will be described and typical results presented.

Cyclograph. A block diagram of the Cyclograph is shown in Fig. 12-21. Basically the Cyclograph consists of an oscillator which drives the test coil. The frequency of oscillation is the resonant frequency of the test coil when shunted jointly by the test coil and by the connecting cable and Cyclograph input capacity. Experiment has established that the axis of operation of the Cyclograph is at an angle of about −2° with respect to the resistance axis of the impedance diagram. Hence, on the impedance diagram, the resistive component, for all practical purposes, is the only component of impedance change which is effective in changing the amplitude of oscillation of the Cyclograph. From the impedance diagram it may be seen that an increase in resistivity and a decrease in diameter

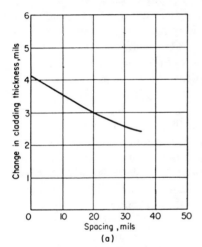

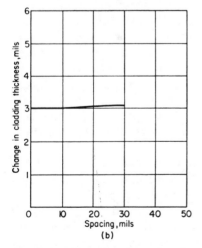

FIG. 12-20. Effect of compensating circuit on coating thickness. (*Courtesy Argonne National Laboratory.*)

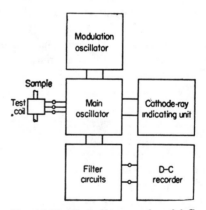

FIG. 12-21. Block diagram of model C Cylograph. (*Courtesy J. W. Dice Company.*)

of the test sample when projected on the axis of operation of the Cyclograph give opposite changes in the resistive impedance of the coil. It is, therefore, possible for a diameter change to cancel a resistivity change, giving a false indication of the magnitude of a defect. This is definitely a disadvantage of the Cyclograph, especially in the case of the concentric coil.

Each particular type of material will have a characteristic resistance per unit length which will determine the required settings of the Cyclograph. For example,[9] it was found when testing uranium clad with aluminum that the point of operation is to the left of the minimum. Bare uranium, on the other hand, has high enough resistance to cause the Cyclograph to operate on the right side of the minimum of the parabola. Thus the effect of a crack in a bare uranium rod is to increase resistance, which produces an increase in amplitude. If the uranium rod is canned, the Cyclograph is now operating to the left of the minimum and the crack causes a decrease in amplitude. Since the change in resistance caused by defects is usually only of the order of a few per cent, the operation is over only a small portion of the parabola. Since this small portion is nearly linear, the Cyclograph has essentially a linear response.

The Cyclograph is most sensitive to changes in resistance when the oscillations are of low amplitude (i.e., a defect encountered when the oscillations are of large amplitude does not produce so large a fractional change in amplitude on the cathode ray tube as when a defect is encountered when the oscillations are of small amplitude). However, the response time of the Cyclograph is a function of the sensitivity of the instrument. When the instrument is set for maximum sensitivity, the response time is increased to such an extent that rapid scanning of a specimen is impossible. This difficulty can be overcome by operating the instrument so that the amplitude of the oscillations (as observed visually on the cathode ray tube) remains near the maximum during scanning. This loss in sensitivity can be compensated for by increasing the sensitivity of the recorder. The small variations in amplitude can be "magnified" by use of a bucking voltage applied to the output signal.[9] This arrangement "magnifies" the changes in voltage due to the flaws in the specimen and at the same time gives a linear response to variations in resistance.

The resolution of the Cyclograph when using concentric coils is determined by the length of the coil; the shorter the coil, the better the resolution. However, if the length of the coil is less than approximately $\frac{1}{8}$ in., no appreciable increase in resolution is obtained, and the sensitivity is reduced because of a reduction in penetration of the flux. The sensitivity of the instrument is dependent to a large extent upon the inside diameter of the coils, because as the clearance between the specimen and coil increases, the sensitivity to defects decreases. It was necessary that the inside diameter of the coil be only 0.010 to 0.015 in. greater than the specimen diameter. The test coils are made by winding simultaneously (bifilar winding) two wires on the proper form. During winding, the two wires should be kept as close together as possible. When the proper number of turns had been completed, the two intimately wound coils were connected "series-aiding" (i.e., the inner terminal of one coil connected to the outer terminal of the other coil). The three resulting terminals are connected by means of a cable to input terminals of the Cyclograph. The frequency of the Cyclograph is determined by the frequency of self-resonance of the search coil when in parallel with the lead and internal stray capacitance.

Tube Inspection Using the Cyclograph. The Cyclograph has been used by Oliver[10] to inspect tubing. By proper selection of operating frequency, optimum sensitivity to OD and ID defects could be obtained, regardless of the diameter, wall thickness, and material of the tubing. Also by using a consistent method of choosing the frequency, the same relative sensitivity can be maintained regardless of the properties of the particular tube. Figure 12-22 is a plot of the Cyclograph reading as a function of tube wall thickness, with the ratio f/f_c as a parameter. Experience has indicated that the optimum OD and ID flaw sensitivity is obtained for an f/f_c which causes the curve to peak near the wall thickness under question, which is the condition of maximum power loss or "core loss" in the tube. In general, discontinuities within the tube wall cause signals which appear as reductions in the wall thickness. Therefore, operating points are selected just to the left of negative peaks in the curves of Fig. 12-22, such that discontinuities, wall thickness reductions, and diameter reductions cause upward deflections in the

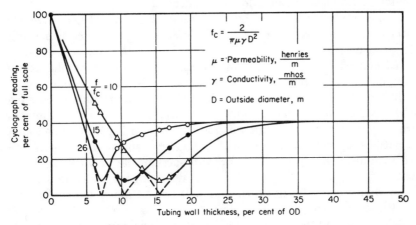

$$f_c = \frac{2}{\pi\mu\gamma D^2}$$

μ = Permeability, $\dfrac{\text{henries}}{\text{m}}$

γ = Conductivity, $\dfrac{\text{mhos}}{\text{m}}$

D = Outside diameter, m

$\dfrac{f}{f_c}$ = 10

Cyclograph reading, per cent of full scale

Tubing wall thickness, per cent of OD

FIG. 12-22. Cyclograph reading vs. tubing wall thickness with f/f_g as parameter. (*Courtesy Oak Ridge National Laboratory, operated by Union Carbide Corporation for the U.S. Atomic Energy Commission.*)

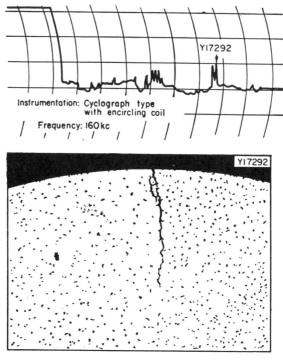

Fig. 12-23. Record and macro defect in ¼-in. diameter by 0.049-in. Hastalloy B tubing. (*Courtesy Oak Ridge National Laboratory, operated by Union Carbide Corporation for the U.S. Atomic Energy Commission.*)

instrument readout, whereas increases in the wall thickness or diameter or foreign metal pick-up cause downward deflections.

Although the Cyclograph signals in themselves are not definitive, the interpretation of these signals can be increased by utilizing a properly designed mechanism to feed the tubing through the coil at a constant speed and recording the resultant signals. Signals from wobble of the tube in the coil, which produce spurious signals, can be eliminated by utilizing a feeding mechanism in which the test coil is allowed to move with freedom on the tube. Signals due to relatively large discontinuities are usually very easily separated from those resulting from dimensional variations in the tube because of their large size and very abrupt nature. This is illustrated by the sharp upward spike in the signal trace of Fig. 12-23 which resulted from a 0.036-in. deep outside radial crack in a ¼-in.-diameter by 0.049-in. Hastalloy B tube. Indicative of the very small defects which may be detected by this method is the 0.001- to 0.002-in.-deep intergranular attack at the inside surface shown in the photomicrograph of Fig. 12-24. This condition was detected in a ¼ in.-diameter by 0.025-in. wall Inconel tube and is indicated on the accompanying trace by the distinct upward hump. Foreign metallic particles having a high magnetic permeability are sometimes found embedded in the inside wall of redrawn tubing, probably having been picked up during the redrawing operation. The presence of such inclusions is indicated by sharp downward spikes, which are in the opposite direction to those caused by cracks or intergranular attack. It should be emphasized that all eddy current techniques are very useful for gaging or inspecting tubing for dimensional tolerances, since a small change in diameter or wall thickness produces a large change in the apparent impedance of the test coil.

It is difficult to make a general statement about the sensitivity of an encircling coil test to small discontinuities. Since the coil is affected by all the metal within its length, it is possible for a long shallow discontinuity to produce a signal as large as or greater than a short deep discontinuity. This condition may be improved by shortening the length of the coil, but there is a limit beyond which the effective length of the coil may not be reduced, and therefore this type of test is not sensitive to pinholes, except as they occur in clusters. The other limitation imposed

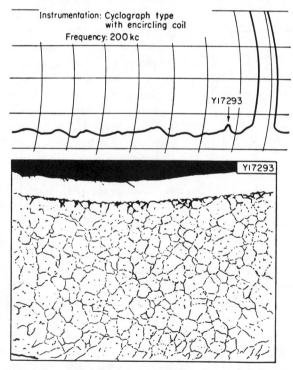

FIG. 12-24. Record and macro defect in ¼ in. diameter by 0.025 in. Inconel tubing. (*Courtesy Oak Ridge National Laboratory, operated by Union Carbide Corporation for the U.S. Atomic Energy Commission.*)

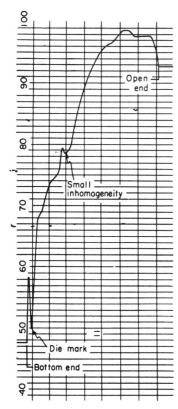

FIG. 12-25. Cyclograph record showing gradient and inhomogeneity in aluminum cans. (*Courtesy Argonne National Laboratory.*)

upon this test in the detection of very small discontinuities is the "signal-to-noise ratio"; that is, a defect cannot be resolved if the signal it causes is small compared with signals from all other variations (including those from dimensional variations) observed as the tube is moved relative to the coil. The maximum in sensitivity which may be achieved with the Cyclograph for practical inspection purposes is the detection of discontinuities which, have lengths comparable to that of the coil (or longer) and depths of 5 per cent or more of the wall thickness. The great advantage of the probe coil for testing tubing is its inherent ability to be definitive. Because it is a surface probe, it is not affected by changes in the over-all height of the oscilloscope pattern and eccentricity is indicated by a pattern having a sinusoidal envelope.

Fuel Element Inspection Using the Cyclograph. The Argonne National Laboratory[9] has used the Cyclograph for inspecting cylindrical nuclear reactor fuel elements and fuel element components. The fuel elements tested were 1-in. OD uranium rods encased in extruded 1100 aluminum cans. The uranium was bonded to the can with an aluminum-silicon (Al-Si) alloy. The nonbonded areas may occur either at the uranium-Al-Si interface or at the Al-Si aluminum interface. The nonbonding may be due to failure of the Al-Si to wet the aluminum or uranium or foreign material on the surface of the aluminum or uranium. Since the process used for jacketing the uranium could cause a reduction in the clad thickness, it was necessary to determine that the cladding meet the required specifications. Measurement of the cladding thickness is important since the service life of a fuel element is predicated on the minimum jacket thickness. The objective of this work was to evaluate the usefulness of the Cyclograph for locating voids and cracks in the bonding layer, nonbonded areas, and thin spots in the cladding elements.

A total of 968 empty 1100 aluminum cans were examined using the Cyclograph. No significant localized inhomogeneity was discovered, but considerable knowledge was gained as to the inherent possibilities of the Cyclograph and some interesting effects were observed. It was found that the empty aluminum cans had a definite "gradient," consisting of a decrease in amplitude of the oscillations (an increase in resistivity) as the coil was moved from the open end to the bottom end of the can. The removal of the bottom (closed end) of the can had no effect on the gradient. Annealing of cans produced no change so that residual stresses from the extrusion were not the cause. A check of the dimensions of cans indicated that the wall thickness might taper from 0.0003 in. to 0.0009 in. from open end to bottom end. It has also been shown that the cold-working of aluminum does produce a change in resistivity (1 per cent for a 40 per cent reduction of area at 75°F), and while it is difficult to talk of reduction of area in the case of impact extrusion, it is quite likely that some dislocation resistance does exist. It is believed that either the taper or cold-working, or perhaps a combination of the two, is responsible for the gradient observed. Additional tests, using a hollow cylinder having 0.0005-in. changes in wall thickness, indicate that the average change in over-all gradient is as indicated in Fig. 12-25. Thirty-six of the cans tested caused the oscillations of the Cyclograph to cease completely at the same sensitivity setting which permitted examination of other cans. Later tests, using a different sensitivity, indicated that these cans also had a gradient. One of these cans was analyzed by metallography and compared with a normal can. It was found that the grain size of the bottom of the abnormal can was two to three times that of a normal can, Fig. 12-26. It is believed that the grain size is responsible for a very small, but detectable, change in resistivity. It was found that the cooling of a can by 40 to 50°F resulted in a full-scale increase in amplitude of the oscillations (i.e., the gradient was of the same general shape but roughly double its original amplitude). Approximately

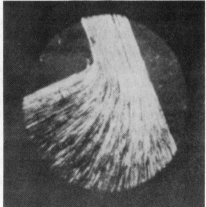

FIG. 12-26. Grain size variations in aluminum. (*Left*) Normal can; (*right*) dead can.

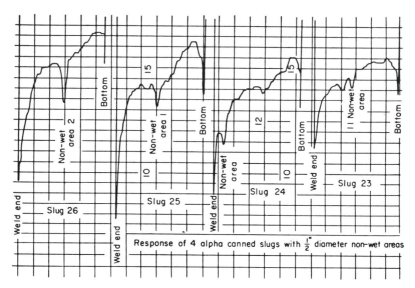

FIG. 12-27. Cyclograph response of four reactor fuel elements.

one-third of the total number of cans gave a definite sharp dip in amplitude at a point 1 in. from the bottom of the can. Visual examination revealed that these cans have a small hair-thin ring mark at this location. These ringlike marks were tentatively identified as die marks. Tests with potassium ferrocyanide showed small dots of iron at random places on the cans, but physical defects of any magnitude were not observed. These rings were visible to the naked eye and offered no deterrent to testing. The indication given by a die mark is shown in Fig. 12-25.

Eighteen canned fuel elements with "manufactured" nonwet areas on the uranium were examined with the Cyclograph. Twelve of the samples gave very definite and sharp indications of the nonwet areas; the other six did not show too sharp an indication of the nonwet areas. The detection of nonwetting is of interest since a nonwet area between bonding alloy and uranium represents an extremely small decrease in cross-sectional area to the flow of the circumferential eddy currents. It should be emphasized that nonwet areas give indications on the Cyclograph which are in the same direction as those given by a thin can wall so that the two effects are not distinguishable for this type of fuel elements. The response of four fuel elements of a group of 12 is shown in Fig. 12-27. The nonwet area on each fuel element was approximately ½ in. in diameter. The location of the nonwet areas as given by the Cyclograph agreed with that given by the fabricator. Mechanical strippings of one specimen showed nonwetting. The response, aside from the nonwet areas, is relatively smooth, with a rather steep gradient or drop-off in response toward the weld end. From results with these 18 fuel elements it may be concluded that nonwet areas give an identifiable indication. The difference in magnitude of indication between the two groups is not readily explained. It is possible, however, that the thickness of the oxide film varied in the two cases. Prolonged heating with the torch will of course lead to grain growth, with an extremely small increase in resistance, but the heating was brief in this case. The effect of nonwet areas, as mentioned previously, on the flow of circumferential eddy currents is small, and therefore indications on the Cyclograph will be small.

Special specimens were made in which small localized depressions were machined into the inner wall of several aluminum cans to determine if thin spots in the aluminum can were detectable. These cans were then put through the regular canning procedure. Metallographic sectioning of these specimens indicated perfect wetting of the depressions. The depressions were machined to various depths, so that fuel elements having can wall thicknesses of approximately 0.005, 0.010, 0.015, and 0.020 in. were available. A recording, Fig. 12-28, shows the deepest "penetration" as a rather sharp deviation in the general overall pattern of the fuel element. These samples were used as "standards" to allow comparison of samples tested on different days at slightly different sensitivities and to give an indication of the relative magnitude of inhomogeneities encountered.

It was found that certain specimens had a large number of inhomogeneities over the length of the fuel element; it was believed that a good percentage of these inhomogeneities were due to very small bubbles throughout the bonding layer. Autoradiographs seem to confirm this interpretation, in that 25 or more of these small bubbles are often observed in a length of 2 to 3 in.

Reproducible results were obtained with both types of probes. Good correlation was obtained between Cyclograph results and destructive tests on this particular fuel element. In addition to the disadvantages previously mentioned there is the problem of distinguishing small bubbles in the bonding layer from thin spots in the aluminum cans. A small bubble is able to cause a deflection comparable

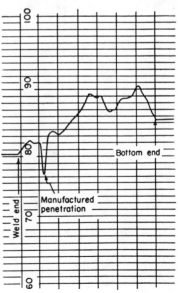

Fig. 12-28. Cyclograph response to manufactured penetration. (*Courtesy Argonne National Laboratory.*)

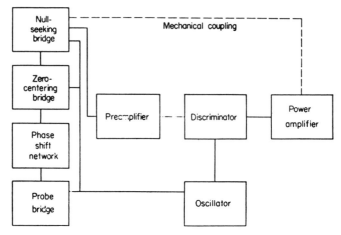

FIG. 12-29. Probe bridge of Probolog. (*Courtesy Shell Development Company.*)

to that of a thin area, because of the high resistivity of the gas in the bubble. Another limitation is in examining the extremities of the fuel element, because at the top and bottom of the element there is an aluminum 'end plug. This sudden discontinuity produces a very rapid change in amplitude over the small end length, and the change is so rapid that any inhomogeneity there has little chance of being recorded. Using the coil type probe the frequency is not critical as the same profile was obtained for a standard within the range of 4 to 8 kc, although localized inhomogeneities do not give as sharp an indication at the lower frequency.

Probolog. The Probolog is a commercial eddy current instrument which is finding application in the field of nondestructive testing: It was designed primarily for detecting corrosion of nonmagnetic tubing and has been used for a number of years for testing heat exchangers in refineries, chemical plants, public utilities, and marine service. The Probolog, developed by the Shell Development Company, is essentially a recording impedance bridge which operates at a fixed frequency. The sensing element is a high-permeability core on which are wound a pair of coils connected as two

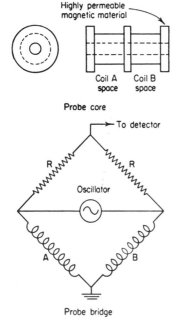

FIG. 12-30. Schematic diagram of Probolog. (*Courtesy Shell Development Company.*)

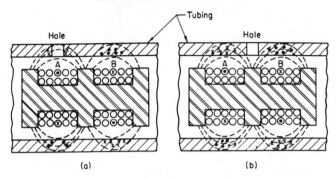

FIG. 12-31. Probolog probe.   (*Courtesy Shell Development Company.*)

legs of the impedance bridge.   Figure 12-29 shows a block diagram of the Probolog.   As long as the material in the neighborhood of both coils has the same electrical and magnetic properties, the bridge will remain in balance.   When some condition arises that alters the eddy current distribution in the field of one coil, altering the apparent impedance of the coil, the bridge will be unbalanced and the pen of the recorder will be deflected.   As can be seen from Fig. 12-29 the Probolog has two main bridges in series, the probe bridge and the null seeking bridge.   The phase shift network and the zero centering bridge are used to correct and balance the electrical signal from the probe bridge into the preamplifier and compensate for the inherent unbalance of the probe.   The model E Probolog operates at 1 kc and has a four-position phase shifter, whereas the adapted Probolog operates at frequencies from 1 to 18 kc and has a continuous phase shifter.

Figure 12-30 shows the probe bridge used in the Probolog.   Potentiometers are provided for making the necessary reactive and resistive balance due to probes and probe cables.   The sensitivity is varied by means of a potentiometer, which determines the percentage of the signal from the phase shift network that will be applied to the preamplifier.

Figure 12-31 shows the core configuration for a typical probe.   Figure 12-32 shows the eddy current pattern within a tube.   The eddy currents will flow predominantly in a circumferential direction within the tube.   In Fig. 12-32 one coil is opposite a hole in the tube; consequently, the impedance of the two coils is different and an unbalance signal appears at the bridge output. When the other coil is opposite the hole, a signal of opposite phase appears at the output terminals.   Figure 12-33 shows the record produced for holes of different

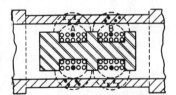

FIG. 12-32. Eddy current distribution in tube. (*Courtesy Shell Development Company.*)

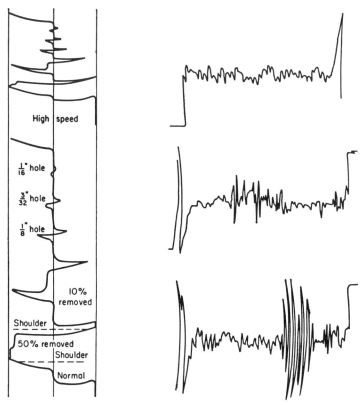

FIG. 12-33. Recording for simulated defects.   FIG. 12-34. Probolog record of welds.
(*Courtesy Shell Development Company.*)   (*Courtesy Shell Development Company.*)

sizes as the probe is drawn through a tube at a constant speed. The magnitude of the recorder deflection due to a given type of defect depends upon a number of factors such as kind of defect, defect shape and orientation, and changes in the volume of material present. If the tubing or a section of tubing is uniformly thin, there will be no signal from the bridge because the impedance of the two coils will be the same. This problem can be eliminated by making one coil axially longer than the other. By proper design of the individual coils, there will be an unbalance which will provide a signal to deflect the recorder pen, as shown in Fig. 12-33.

Applications of Probolog. Warren[11] mentions the use of the Probolog for determining the relative hardness of nonmagnetic material, variations in plating thickness, hidden changes in geometry of structural assemblies, and the quality of welds in aluminum tubing. Figure 12-34 shows typical records made of aluminum welds; the three records are for a good weld, for a weld with minor defects, and for a weld with major defects. Betz[12] has used the Probolog for testing Zircaloy tubing for cracks, seams, laps, pits and gouges, and wall thickness. He reports that the Probolog readily detects longitudinal cracks and seams, particularly those through the wall. Longitudinal cracks or seams 0.005 in. deep by $\frac{1}{2}$ in. long can be detected. However, the instrument is insensitive to circumferential cracks, very short cracks, and very shallow cracks. Whether or not laps can be detected with the Probolog depends on their size and orientation. Likewise, the size of pits or gouges that can be detected depends on the total volume of material removed; about 1 per cent material removed in a $\frac{1}{2}$-in. length of tube is detectable. Changes in wall thickness of approximately 1 per cent in a $\frac{1}{2}$-in. length can be found.

Metals Comparator. The General Electric metals comparator is basically an impedance comparator. The impedance of a test coil varies with the electrical and magnetic properties of the test specimen; thus the instrument will indicate differences in the chemical and physical properties as these are correlated with the electrical and magnetic properties. The instrument consists of an oscillator, balancing network, and test coils. The instrument provides test frequencies of 50, 250, 1,000, 4,000, and 10,000 cps. The range of frequencies provided permits testing both magnetic and nonmagnetic materials. In using the instrument the test coil is placed on a reference specimen, and the balancing resistors are adjusted to give a zero reading on the instrument. When a test specimen of different properties is brought near the test coil, the impedance will change, causing an unbalance in the instrument. This instrument has been used primarily for[13] metal separation, hardness, case depth, and plating thickness. The metals comparator was not intended for defect detection. It has been tried for this latter use; in some applications it has been successful, in others it has not been successful. The instrument is not generally recommended for this application.

Förster Eddy Current Instruments. Other types of commercial eddy current test instrument are those developed by the Institute der Förster in Germany. The principle of these instruments has been discussed by Dr. Förster in a number of papers,[14,15] in addition to those previously mentioned. The Magnatest (Multitest), the most versatile instrument of this group, is briefly described. The Magnatest uses phase discrimination by which electronically it is possible to eliminate variables such as diameter having no relationship to the physical property being measured. Dr. Förster has made a number of specialized modifications of the Magnatest (Multitest) which are sold under different names such as the Sigmatest, Defectometers, Isometer, Sigmaflux, the wire tester, crack detector for tubes and bars, steel ball tester, sheet thickness tester and Argentometer, and Riss-Detektor. The Defectometer is a probe-type instrument for detecting cracks in nonmagnetic material. The Sigmatest measures electrical conductivity of nonferrous metals. The Sigmaflux is used for testing nonferrous wires or rods for alloy variations or gross cracks independently of small-diameter variations. The wire tester was designed primarily for testing tungsten wire for cracks and diameter variations. The steel ball tester is used to inspect finished ball bearings for diameter variations. The sheet thickness tester is used to indicate variations in thickness of nonferrous sheets. The Isometer measures the thickness of nonconducting coatings on nonmagnetic base metal. The Argentometer measures the thickness of nonmagnetic metal coatings on nonmagnetic base metal.

A block diagram of the Magnatest is shown in Fig. 12-35. The fixed frequency oscillator supplies the voltage used in exciting the test coils. The buffer amplifier is used to minimize the loading effects of the solenoids and phase shifter circuit upon the frequency and voltage of the oscillator. The output of the buffer amplifier excites the primaries of two test coils connected in an a-c potentiometer circuit. One of the coils surrounds a standard specimen, while the other coil surrounds the specimen being tested. Controls are provided for balancing this potentiometer circuit. With this arrangement, a test specimen and a standard specimen can be inserted into the coils and the potentiometer circuit balanced until the output terminals have zero voltage across them. If the test specimen is then moved through its coil arrangement, defects will cause an unbalance of the potentiometer circuit which can then be detected by some suitable means. The low-level signal coming from the potentiometer circuit is fed into a cathode follower to minimize current in the secondaries of the coil arrangement and is then amplified with one stage transformer coupled amplifier. The amplifier feeds two phase sensitive detectors the same signal. The reference signal for the phase sensitive detectors is derived from the same signal that originally fed the potentiometer circuit. This signal is split into in-phase and quadrature components which excite the stator windings of a resolver. The rotor voltage of the resolver then is a voltage which

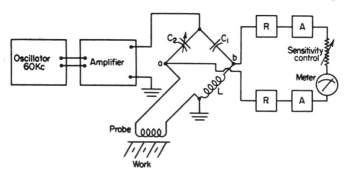

Fig. 12-35. Block diagram of Magnatest. (*Courtesy Magnaflux Corporation.*)

can be continuously phase shifted through 360° with respect to the original output of the buffer amplifier. This resolver voltage is split into in-phase and quadrature components which are connected to the reference voltage inputs of the phase sensitive detectors. The resolver arrangement is used to produce rotation of the axis of the impedance plane. Essentially the voltages presented to the phase sensitive detectors can be shifted through 360° so that the outputs of the phase sensitive detectors can be shifted by the same amount. If the phase shifter were, for example, in the zero position, then the output of the top phase sensitive detector is the demodulated in-phase component of the potentiometer circuit unbalance. The output of the bottom phase sensitive detector is then the quadrature component of the unbalance of the potentiometer circuit. These components are then amplified and used to deflect the beam of a cathode ray tube in accordance with their magnitudes. This instrument presents the impedance plane shown in Fig. 12-2, with the actual operating point indicated by a spot of light. Then as the impedance of the test coil changes because of defects, the position of the spot on the face of the cathode ray tube changes. If, in the case of elements of uniform material, diameter variations were to be eliminated, the phase shifter knob would be rotated until the locus of the diameter changes was, say, horizontal and nearly vertical excursions of the spot would be noted. This is particularly well adapted to recording, since a recorder could be inserted into the circuit such that it recorded only the vertical travel of the beam, hence recording only conductivity changes in the sample.

In addition to the commercial eddy current instruments described above, a number of specialized eddy current instruments have been developed in the United States. A great number of these instruments have been developed in laboratories involved with research and development in the atomic energy field. A number of them have been discussed in the proceedings of the Symposium on Nondestructive Tests in the Field of Nuclear Energy.[16] Some of these instruments and their application will be briefly reviewed.

Inspection of Capillary Tubing. The Knolls Atomic Power Laboratory has developed and used an eddy current instrument for inspecting capillary tubing. In this instrument the reflected change of inductance and resistance of the test coil affects the frequency of the oscillator used to drive the test coil. The frequency of this oscillator is compared to the frequency of another oscillator. Under ordinary conditions, both oscillators have the same frequency and a zero frequency difference; but with a defective specimen the test oscillator frequency changes and the difference frequency is directly proportional to the magnitude of the change in frequency of the test oscillator. If a high frequency is used for the test oscillator, a small percentage change in inductance of the test coil would bring about a large absolute change in frequency. However, at high frequencies the eddy currents are concentrated in the outermost layers of the tube being tested. Consequently, in actual test problems a compromise between sensitivity and penetration depth has to be made. Two identical Hartley oscillators were used, with special attention being paid to the mechanical design of the wiring and chassis. The oscillators were effectively amplitude limited by grid bias resistor and capacitor combinations. The test oscillator was inductively coupled to the tubing sample through a single-turn link lead to the test coil by means of a short length of coaxial low-impedance cable. A frequency sensitive circuit was used to convert small frequency changes to d-c output suitable for recording by a conventional recording system. This is done by use of a frequency discriminator or demodulator featuring a high Q parallel resonant circuit. The best frequency of the two oscillators is adjusted to equal the mid-point of the sensitive range of the frequency discriminator, and the output of the discriminator is rectified to produce a d-c signal proportional to the beat frequency of the test and comparison oscillators. A battery operated zero-suppression circuit completes the circuit design.

In this instrument the resistance changes are minimized and ignored, and interpretation of the test is based solely on the changes in inductance. Therefore one might expect to find that a given effect could arise from more than one cause. For instance, a small dent, hardly visible to the unaided eye, yielded the same kind of indication as that produced by a short radial crack lying longitudinally. The tester was also found to be extremely sensitive to the diameter of the tube. A high order of sensitivity was achieved and maintained. A comprehensive program of destructive examination revealed that the test invariably detected cracks as deep as 0.0025 in. Some of the cracks detected proved, upon metallographic examination, to be as short as 0.070 in. On the other hand, certain unmistakable defects detected visually at high magnification were not detected. These defects proved to be between 0.0005 and 0.001 in. deep and of variable length up to 0.100 in. long. The threshold of sensitivity for this instrument lies somewhere between 0.001 and 0.0025 in. deep, depending on the length of the defect. Testing is impossible within $\frac{1}{4}$ in. of the end of the tube, because of the end effects. The instrument has been used on tubing as small as 0.050 in. OD and as large as 0.226 in. OD with comparative results.

Intergranular Corrosion. In the symposium proceedings mentioned above, R. C. Robinson[17] describes an eddy current instrument which indicates the relative degree of intergranular corrosion present in austenitic stainless steel tubes. Since intergranular corrosion consists of a network of fine cracks along the grain boundaries, the electrical

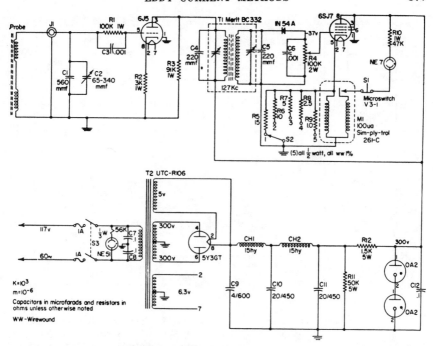

Fig. 12-36. Eddy current tester for intergranular corrosion. (*Courtesy American Society for Testing Materials and R. Robinson.*)

resistance of a corroded tube is different from that of an uncorroded tube. The instrument developed by Robinson is shown in Fig. 12-36. The test coil is the coil of a tuned-plate tuned-grid oscillator. When the coil is placed on a specimen, the amplitude of oscillation is a function of the electrical conductivity of the specimen. The output of the oscillator is rectified by a 1N54A crystal detector operated such as to obtain a negative voltage for biasing the 6SJ7. The higher the resistivity of the specimen, the lower the tube bias and the greater the tube current. The oscillator determines the bias on the 6SJ7 and therefore the tube current. The instrument possessed the sensitivity to distinguish between tubes having a resistivity of 165 and 190 microhm-cm, the former tube being sound or acceptable, whereas the latter was slightly corroded.

Thickness Measurement. An important application of eddy current methods in the field of nondestructive testing involves the measurement of the thickness of one metal plated or clad on another. Both sinusoidal and pulsed eddy currents have been employed by various investigators. Various methods are presented for the purpose of measuring one metal plated or clad on another.

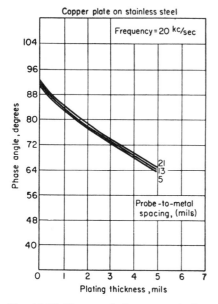

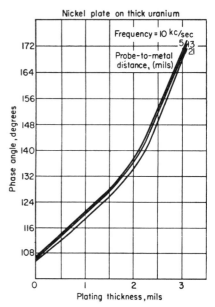

FIG. 12-37. Phase angle for copper plating on stainless steel. (*Courtesy Argonne National Laboratory.*)

FIG. 12-38. Phase angle for nickel plating on uranium. (*Courtesy Argonne National Laboratory.*)

Figures 12-37 and 12-38 show a plot of the phase angle of the output from the instrument developed by Renken[6] for the measurement of plating thickness. Using curves like these, a phase meter could be continuously recorded, thus furnishing a permanent record of the plating thickness of the material. The separate curves of Fig. 12-37 were obtained at three different probe-to-metal spacings of 5, 13, and 21 mils, respectively. Many commercial phase measuring instruments have relative accuracies of 1° at 20 kc/sec, and some are much better than that. Assuming the 1° figure, the accuracy of measurement in Fig. 12-37 is about 0.00025 in. Linsey and Libby[18] describe an eddy current instrument for measuring the thickness of thin layers of nickel plates on a nonmagnetic base metal. The authors report that the instrument can measure nickel thickness over a range of 0.1 to 2.0 mils with an accuracy of 5 per cent. Ross[19] developed an eddy current instrument to measure the thickness of aluminum which was clad on uranium for nuclear reactor fuel plates. Experiments show that the instrument will measure the thickness of aluminum in the range of 0.015 to 0.030 in. to an accuracy of better than 0.001 in.

Hanysz has developed a swept frequency eddy current instrument to measure coating thickness, which is called the Laminogage. It is adaptable to all cases in which the conductivity permeability ratio of the base material differs by at least a factor of 2 from that of the coating.

The instrument uses a reactance tube oscillator, the frequency of operation being determined by the grid and plate circuit parameters. The change in electrical impedance of a coated specimen due to differences in the thickness of the coating cuts off the oscillator at a frequency which is determined by the thickness of the coating. A cathode ray tube or a meter indication can be used as a thickness indicating system. The accuracy claimed for this instrument is not too high. Lead-tin babbitt and lead-indium babbitt overlays on bronze and silver in the range of 0.0003 to 0.0015 in. have been made to an accuracy of better than 10 per cent.

Allen, Nance, and Oliver[20] have used eddy currents to measure the thickness of Mark X MTR fuel plates. These fuel plates had a 48 weight per cent uranium-aluminum alloy core 0.022 in. thick, clad with 0.020 in. 6061 aluminum alloy. In the fabrication process, an area of thin clad may be produced. To reduce the effort of probe-to-specimen

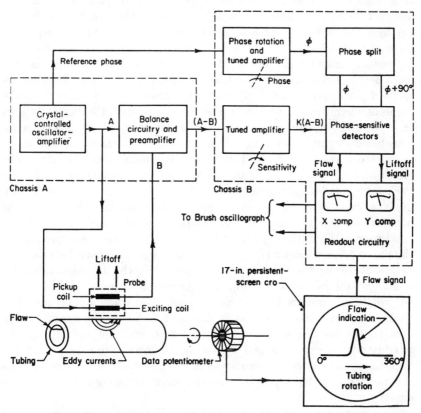

FIG. 12-39. Block diagram of Oak Ridge National Laboratory clad thickness instrument. (*Courtesy Oak Ridge National Laboratory, operated by Union Carbide Corporation for the U.S. Atomic Energy Commission.*)

spacing to a minimum, the test probe was mounted in a heavy brass holder and spring-loaded against the specimen. Figure 12-39 shows a block diagram of this instrument. It consists of a variable frequency oscillator which supplies low-distortion alternating current to a balancing network and to the probe through an isolating network. The isolating network prevents the presence of the probe coil from affecting the voltage supplied to the balancing circuitry and supplies the probe with current of constant amplitude. Thus the voltage at the terminals of the probe is directly proportional to its impedance, which in turn is influenced by the metal specimen. The balancing network alters the input voltage until it is equal to the voltage of the probe, under some initial condition, both in amplitude and phase. This voltage is subtracted from the voltage of the coil and the net difference amplified and indicated on a meter. Any change in the test condition changes the coil voltage in phase or amplitude or both, resulting in a meter deflection. An evaluation of this eddy current measurement was made by destructive examination. It was found that the greatest difference between the metallographic measurements and eddy current measurements was 0.001 in.

Brenner and Garcia-Rivera[21] of the National Bureau of Standards have developed an eddy current instrument primarily for the measurement of plating thickness. A similar instrument is sold under the trade name of Dermitron. This instrument depends upon the characteristics of a parallel resonant circuit for the detection of impedance changes. The basic circuit is shown in Fig. 12-40. An alternating current is impressed across the branch-type circuit from the power supply. The orientation of the germanium diodes provides alternate paths for the positive and negative half cycles of the signal. One path includes a diode, a variable resistor, and a meter. The other path is through the other diode and the tank circuit composed of the probe coil and the condenser. When the resistance is adjusted to equal the impedance of the probe coil condenser $(L\text{-}C)$ combination, equal currents flow through each arm during the conducting half cycle, and the d-c microammeter, unable to follow the high-frequency half-cycle signals, indicates zero current flow. Such a balance may be achieved when a conductor test specimen is at a fixed position in the field of L. If the electrical conductivity of the test specimen changes, the magnitude and distribution of eddy currents are changed, reflecting an impedance change, and the impedance unbalance between areas causes more current to flow through the meter on one half cycle than on the other, resulting in an unbalance signal.

Two measuring techniques are used. For each measurement, the calibration potentiometer R may be adjusted until conditions of no current flow are indicated by the meter and the dial setting of R is

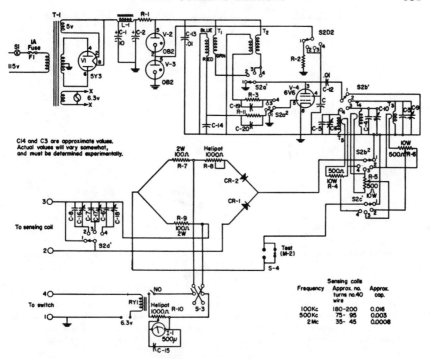

FIG. 12-40. Instrument for measuring coating thickness by eddy currents. (*Reprinted from Plating Magazine, June, 1957, American Electroplaters Society, Inc.*)

recorded. This first method has been found most useful for experiments in which a wide range of conductivities was encountered. In the second method, the meter current may be adjusted to the predetermined value for a reference test specimen by use of R and the various currents for the specimen recorded without change in R. This method is more convenient if all the conductivity readings can be obtained in the meter range. The greatest meter response to changes in the conductivity of the test specimen may be obtained by choosing values of frequency f, capacitance C, and coil inductance L such that the condenser and inductance coil are in parallel, as given by the equation

$$f = \frac{1}{2\pi(LC)^{1/2}} \tag{12-6}$$

where f = resonant frequency, cps
L = inductance of coil at a predetermined position relative to test specimen, henrys
C = capacitance, farads

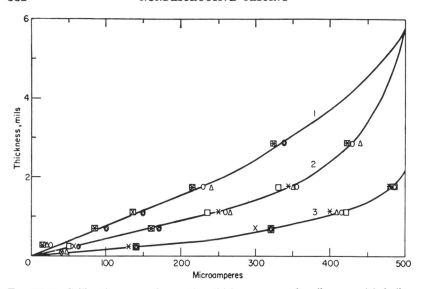

FIG. 12-41. Calibration curve for coating thickness; curve for silver on nickel silver. The gage was set to read zero on nickel silver and to read 500 μa on the thickest-coated standard. Curves 1, 2, and 3 were obtained with frequencies 0.1, 0.5, and 2.0 Mc, respectively. These readings were taken with four different gages (of the same type) to show consistency between gages. (*Reprinted from Plating Magazine, June*, 1957, *American Electroplaters Society, Inc.*)

Under these conditions, the inductive impedance of the coil $2\pi fL$ is equal to the capacitative impedance of the condenser

$$2\pi fL = \frac{1}{2\pi fC} \tag{12-7}$$

the impedance of the parallel circuit is at a maximum, and the current flow through the branch is at a minimum. Thus the choice of operating frequency for such an instrument must be considered in terms not only of the depth of penetration desired but also of the electrical characteristics of the probe coil. The rate at which the resonant condition decays as one of the three controlling parameters f, C, L changes also affects the sensitivity of the method. This fact required proper physical design of the coil to obtain a sharply "tuned" circuit. Figure 12-41 shows the response of the meter under these conditions to test specimens of various conductivities. In the graph the first measuring technique described above (balanced impedance–no current flow) has been used, and the result is a plot of impedance or balancing resistance setting as a function of conductivity. It may be noted that the one ferromagnetic metal included, nickel, does not fall near the curve.

Pulsed Eddy Currents. Pulsed eddy currents have been used for measuring coating thickness.[22] In this technique an electromagnetic field is applied to the surface of the specimen, and the "echoes" from the metallic layers are received. A small single-layer probe coil with its axis perpendicular to the specimen surface serves both as sender and receiver of the electromagnetic waves. These so-called echoes are caused by impedance mismatch between materials of different electrical properties and for ease of explanation are referred to as reflections. As can be seen from Fig. 12-42, the first reflected wave does not

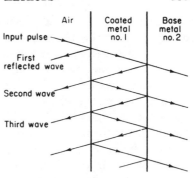

FIG. 12-42. Wave reflection at metallic interface.

contain any information concerning the coating, and consequently must be balanced out by a bridge circuit. One way of doing this is to use (nearly) identical probes. The outputs of the two probes are balanced against each other, and the difference voltage is amplified and presented on an oscilloscope screen. The block diagram of the system used is shown in Fig. 12-43. The pulse circuit produces a pulse which drives both the standard and the test probes through a transformer and bridge circuit. The difference voltage of both probes is passed through an amplifier to the oscilloscope, where the voltage may be observed. It was found experimentally that there were one or more points on the difference voltage curve which were not affected by the probe-to-specimen spacing. The results indicate that the effect of changing the probe-to-metal spacing is to change the slope of the curves, but the height of the crossing points above the axis remains unchanged. To test the above experimentally, metal plates having the same clad metal and the same base metal but with different depths of cladding were inspected. Figure 12-44 shows the correlation between readings on

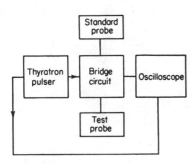

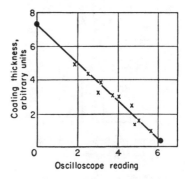

FIG. 12-43. Block diagram of pulsed eddy current system. (*Courtesy Argonne National Laboratory.*)

FIG. 12-44. Correlation of results.

the oscilloscope and clad thickness measured optically. One of the difficulties with this technique is that it is very sensitive to variations in probe-to-specimen spacing, making it difficult to use in the continuous scanning of a moving specimen.

Magnetic Materials. The preceding discussion has concerned only nonmagnetic material. In Fig. 12-45 the normalized impedance for the test coil is given for various values of permeability, conductivity, and diameter for a fill factor of $\frac{1}{2}$. Along the curve corresponding to a permeability of 100 gauss/oersted, values of f/f_g are marked. The narrow curves in this figure join the points representing the impedance of a coil when the only change in the coil and test object system is a change in permeability of the test object. For magnetic materials, a change in diameter causes an impedance change in the same direction as that caused by a change in permeability. For low values of f/f_g, the change in impedance due to a conductivity change is approximately at right angles to the change due to a permeability variation. In this range, it is possible to distinguish between permeability and conductivity effects. The impedance of the test coil when the test object is tubing rather than rod is shown in Fig. 12-46 for an f/f_g ratio of 15. The curve on the extreme right shows the variation in impedance of the test coil as a function of wall thickness. The values of wall thickness, in per cent of outside diam-

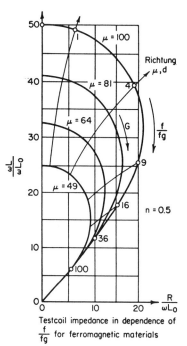

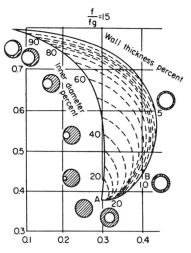

FIG. 12-45. Normalized impedance curves as a function of permeability (fill factor $\frac{1}{2}$). (*Courtesy Magnaflux Corporation.*)

FIG. 12-46. Normalized impedance curves for various kinds of tubes ($f/f_g = 15$). (*Courtesy Magnaflux Corporation.*)

eter, are marked on this curve. Thus, the test coil impedance for solid rod is represented by point A on Fig. 12-46. If, now, a tube having the same outside diameter and a wall thickness of 10 per cent of that diameter is inserted in a coil, the impedance of the coil will be represented by point B.

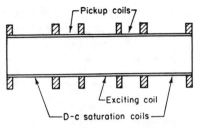

FIG. 12-47. Probe for inspecting ferrous metal.

It may be necessary to decrease the effective permeability of a ferrous metal before eddy currents can be induced in the metal. The d-c coils of the pickup unit, Fig. 12-47, make it possible to apply a d-c magnetic field to saturate the portion of the rod inside the pickup unit and thus lower the permeability. The number of ampere-turns required for saturation depends upon the dimensions of the rod or tubing and its magnetic properties. The end effect for ferrous metals is much greater than in the case of nonferrous metals and is different for metals of different permeabilities. For example, the end effect of the $\frac{1}{2}$-in. steel rod used was approximately 10 in. This large end effect precludes the use of this method for inspection of short rods by present equipment.

Gunn[23] has published a number of papers on eddy current testing. U.S. Patent 2,162,710 issued to Gunn describes the apparatus and method for detecting defects in metallic objects.

Other Applications of Eddy Currents. Eddy current instruments and techniques have been used for a number of other applications. The eddy current techniques can be used for contactless measurement of electrical conductivity at elevated temperatures, determination of extra- and intracrystalline corrosive effects with a sensitivity of 0.1 part per million. Quantitative determination of the reversible and nonreversible portions of the electrical conductivity and specimen diameter during or after elastic and plastic deformation can be made with a sensitivity of 0.1 part per million. Eddy current techniques have been used to detect the presence of voids or gas pockets in the bonding layer of liquid metal bonded fuel elements for nuclear reactors.[24] The particular fuel element considered here consists of an uranium-zirconium alloy contained in a thin-walled (0.008 in.) stainless-steel tube. The fuel element is approximately 17 in. long and 0.175 in. in diameter. Normally there is a 0.006-in. annulus between the uranium rod and the wall of the tubing filled with sodium. Both circular-type coils and point probe-type coils were investigated.

Some results using the point probe are shown in Fig. 12-48. This figure shows that the point probe was able to detect a void on one side of the fuel element from positions not only directly above the void, but

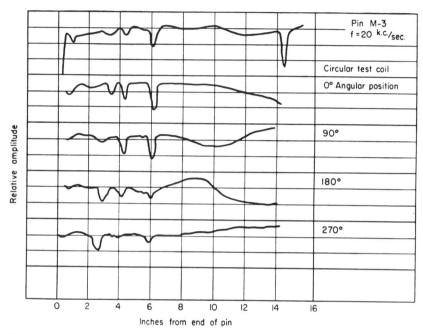

FIG. 12-48. Point probe results for locating bubbles. (*Courtesy Argonne National Laboratory.*)

also with the probe at angles as much as 180° away, measuring around the fuel element circumference. The top trace was obtained using the circular coil, and the lower four traces were obtained using the point probe, which had a diameter of ¼ in. Each successive test was made with the fuel element rotated 90° from its previous position, starting at an arbitrary zero-degree point. The voids in these traces are the short peaks and valleys. The long rises and falls are caused by variations in the thickness of the sodium layer under the probe because of the uranium rod being warped or off center. This effect is partially canceled when the circular test coil is used because of its symmetry. At frequencies in the neighborhood of 25 kc/sec, the eddy currents flowing in the stainless steel have a very minor effect compared to those flowing in the sodium because of the thinness of the can and the much greater conductivity of the sodium. The test coil effectively "sees" through the can. Destructive tests made by "stripping" the stainless steel from the fuel element showed that there was a definite correlation between the amplitude of the void indication and the length of the void around the circumference and, naturally, a correlation between the length of the void indication on the recorder trace and the actual void.

In addition it is possible to use measurements of conductivity as a measure of other properties. For example, during age-hardening the hardness and conductivity of the material alter simultaneously, so that the degree of age-hardening attained can be checked by measuring the electrical conductivity.   Figure 12-49 shows that for copper-chromium alloys there exists a certain relationship between conductivity tensile strength and elongation.   Figure 12-50 shows the effect of phosphorus upon the electrical conductivity of copper.   Other factors which affect conductivity and can be measured by eddy current techniques and methods include porosity, grain size, alloy composition, internal stresses, depth of casehardening, and thickness determinations.

Cavanagh[25] has used the Cyclograph to study internal stresses and to predict fatigue failure of materials.   In a ferromagnetic material, changes in structure are closely related to changes in magnetic properties.   This is because the magnetic domain structure and the facility with which this may undergo changes depend on the physical state of the metal.   Generally, the annealed state is characterized by mobile domain walls, low permeability, and high core losses, the condition being reversed by cold-working or fatigue.   Using a Cyclograph, it is possible to measure relative core losses and to examine the changes which occur while fatigue is in progress.   Various criteria are possible, but at the present time it is felt that the most useful one is the change in stress response of core loss.   For ferromagnetic metals, the Cyclograph coil is placed around the specimen, the latter being subjected to axial tensile fatigue.   In order for any type of failure to occur in metal, some permanent distortion of the crystal lattice must take place.   This distortion will change the magnetic and

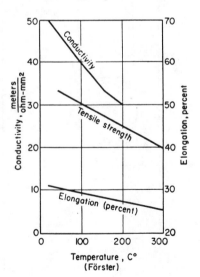

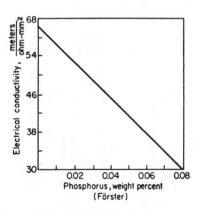

Fig. 12-49. Relationship between con-luctivity, tensile strength, and elongation or copper-chromium alloys.

Fig. 12-50. Effect of phosphorous on electrical conductivity of copper.

electrical properties of the material and is detectable with sufficiently
sensitive instruments. Figure 12-51 shows typical curves obtained by
Cavanagh in the course of his experimental investigations.

Myers and Renken[26] describe an eddy current instrument for testing
the quality of 0.055-in.-diameter zirconium wire. An encircling coil
having an inside diameter 0.005 in. greater than the normal wire diameter
was used. A block diagram of the instrument is shown in Fig. 12-52.

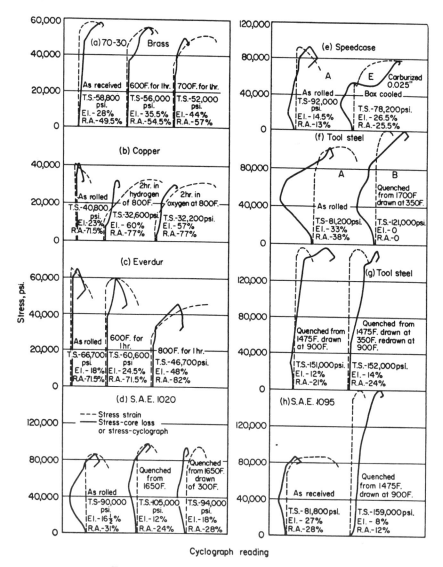

FIG. 12-51. Stress-strain and core-loss curves.

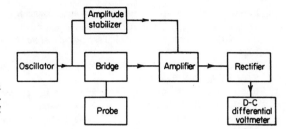

FIG. 12-52. Block diagram of wire checker. (*Courtesy Argonne National Laboratory.*)

Cracks and voids less than 0.001 in. in depth were reliably detected by this instrument.

The use of eddy currents for testing and inspection is in its infancy, but it has a great future. Today the technique has many limitations, most of which are due to a lack of understanding of the basic principles and inadequacy of the instrumentation.

SPECIFIC REFERENCES

1. Förster, F., and K. Stambke: A Method of Nondestructive Testing Employing a Slip-over Coil, *Z. Metallk.*, vol. 45, no. 4, pp. 171–179, 1954.
2. Libby, H. L.: Basic Principles and Techniques of Eddy Current Testing, *J. Soc. Nondestructive Testing*, vol. 14, no. 6, pp. 12–18, November–December, 1956.
3. Oliver, R. B., and J. W. Allen: Inspection of Small Diameter Tubing by Eddy Current Methods, *J. Soc. Nondestructive Testing*, vol. 15, no. 2, pp. 104–109, March–April, 1957.
4. Förster, F.: Die Zerstorungsfiere Messing der Dicke von Nichtmetallischen and Metallischen Oberflachenschicten, *Metall*, vol. 7, p. 320, May, 1953.
5. Yates, W. A., and J. L. Queen: Sheet and Plated Metal Measurements with a Phase Single Type Probe, *AIEE Trans.*, vol. 73, part I, p. 138, 1954.
6. Renken, C. J., and D. L. Waidelich: Minimizing the Effect of the Probe-to-Metal Spacing in Eddy Current Testing, *ASTM, Spec. Tech. Pub.* 223, 1958.
7. Waidelich, D. L., and C. J. Renkin: The Impedance of a Coil Near a Conductor, *Proc. Natl. Electronics Conf.*, vol. 12, p. 188, 1956.
8. Renken, C. J., and R. G. Myers: A Multi-frequency Eddy Current Testing System, Argonne National Laboratory Report ANL-5861.
9. McGonnagle, W. J.: Testing of Cylindrical Fuel Elements Using the Cyclograph, *ASTM, Spec. Tech. Pub.* 223, 1958.
10. Oliver, R. B., and J. W. Allen: Inspection of Small Diameter Tubing by Eddy Current Methods, *ASTM, Spec. Tech. Pub.* 223, 1958.
11. Warren, William J.: Probolog—An Application of Eddy Current Techniques to Nondestructive Testing, *Corrosion*, vol. 10, pp. 318–323, October, 1954.
12. Betz, R. A.: Two Applications of Eddy Current Instruments to Testing of Zircaloy Core Components, *ASTM, Spec. Tech. Pub.* 223, 1958.
13. Smith, B. M.: A Metals Comparator for the Inspection and Classification of Metals, *J. Soc. Nondestructive Testing*, vol. 11, no. 2, pp. 41–46, Fall, 1952.
14. Förster, F., and H. Breitfeld: Theoretical and Experimental Basis for the Nondestructive Material Testing with Eddy Current Methods. V. Quantitative Crack Testing of Metallic Materials with Feed-through Coil, *Z. Metallk.*, vol. 45, no. 4, pp. 188–193, 1954.

15. Förster, F.: Theoretical and Experimental Basis for the Nondestructive Material Testing with Eddy Current Instrument with Feed-through Coils for the Quantitative Nondestructive Material Testing, *Z. Metallk.*, vol. 45, no. 4, pp. 180–187, 1954.

16. Symposium on Nondestructive Tests in the Field of Nuclear Energy, *ASTM, Spec. Tech. Pub.* 223, 1958.

17. Robinson, R. C.: A Nondestructive Test for Intergranular Corrosion in Stainless Steel, *ASTM, Spec. Tech. Pub.* 223, 1958.

18. Linsey, G. D., and H. L. Libby: Nickel Depth Meter, *ASTM, Spec. Tech. Pub.* 223, 1958.

19. Ross, J. D.: A Cladding Thickness Tester for Flat Fuel Elements, E. I. du Pont de Nemours & Company, Savannah River Report DP-117, August, 1955.

20. Allen, J. W., R. A. Nance, and R. B. Oliver: Eddy Current Measurement of Clad Thickness, *ASTM, Spec. Tech. Pub.* 223, 1958.

21. Brenner, Abner, and Jean Garcia-Rivera: An Electronic Thickness Gage, *Plating*, vol. 40, no. 11, pp. 1238–1244, November, 1953.

22. Waidelich, D. L.: Pulsed Eddy Currents Gage Plating Thickness, *Electronics*, vol. 28, no. 11, pp. 146–147, November, 1955.

23. Gunn, Ross: Apparatus and Method for Detecting Defects in Metallic Objects, U.S. Patent 2,162,710, June 20, 1939.

24. Renken, C. J., and W. J. McGonnagle: Eddy Current Techniques for Testing Liquid Metal Bonding, *ASTM, Spec. Tech. Pub.* 223, 1958.

25. Cavanagh, P. E.: A Method for Predicting Failure of Metals, *ASTM Bull.* 143, pp. 30–33, December, 1946.

26. Myers, R. G., and C. J. Renken: Detecting Invisible Flaws in Wire, *Electronics*, vol. 31, no. 39, pp. 72–73, Sept. 26, 1958.

CHAPTER 13

OTHER USEFUL TESTING TECHNIQUES

In this chapter a number of useful nondestructive testing techniques are briefly described to inform the reader of their existence. Such tests include spot tests, spectrochemical analysis, spark tests, surface analysis, the measurement of stress and hardness.

Spot Tests. Rapid identification of common metals and alloys is often necessary for the segregation of materials and for salvage of scrap. Chemical spot-testing procedures are often used for this purpose. The procedures are only qualitative. Feigel[1,*] lists over 200 spot tests for over 40 metals. The term "spot test analysis" is used to refer to sensitive and selective tests based on chemical reactions. The tests are microanalytical or semimicroanalytical and are applicable to both inorganic and organic compounds. The basis of the method depends on the nature of the reagents used and the advantageous use of chemical reactions. Thus, the utmost of sensitivity and selectivity is obtained with a minimum of physical and chemical operations. The tests are ordinarily run by using one of the following four techniques. (1) One drop each of the test solution and reagent are brought together on a porous or nonporous surface such as paper, glass, or porcelain. (2) A drop of the test solution is placed on a medium impregnated with appropriate reagents. (3) A drop of reagent solution is placed on a small quantity of the solid specimen. (4) A drop of reagent or strip of reagent paper is subjected to the action of liberated gases from a drop of the test solution or from a minute quantity of the solid specimen.

Electrographic Printing. In electrographic printing a paper soaked in an electrotype is placed between the specimen and a second electrode, forming an electrolytic cell. When a voltage is applied to the cell, with the specimen as anode, ions from the specimen pass into the paper. The distribution and concentration of the ions can be revealed by a reagent with which they will form an insoluble colored compound. Jirkovsky[2] has used electrographic printing to identify iron in pyrite, marcasite, and other iron-bearing minerals. Electrographic sampling has been applied recently by various workers[3-5] for the identification of both metallic and nonmetallic constituents in alloys. Hilborn and Pugh[6] describe

---

* Superscript numbers refer to Specific References listed at the end of the chapter.

a procedure which is a combination of electrographic printing and colorimetric spot tests. Their procedures were adapted for studying contaminants on aluminum for certain nuclear reactor components. Procedures were adapted to detect and identify minute quantities of the following metals: iron, lead, magnesium, mercury, molybdenum, nickel, tin, and copper, which may be present as inclusions on or alloying constituents in aluminum.

Sulfur Printing. Sulfur printing is used for detecting segregation in alloys. In the Baremann technique a sheet of bromide photographic paper is moistened in a 3 to 4 per cent solution of sulfuric acid. After the excess liquid has drained off, the sheet is placed in contact with the roughly polished specimen for 1 to 4 min. Hydrogen sulfide reacts with the bromide, producing silver sulfide which forms a brown pattern on the photographic paper. The intensity of the discoloration depends on the concentration of the sulfides. Thus the distribution and relative concentration of the sulfide are determined. In the Heyn and Bauer technique a silk cloth soaked in an acid solution of mercuric chloride is pressed against the polished specimen for 4 to 5 min. Sulfur-rich areas produce a black coloration, and phosphorus-rich areas cause a yellow coloration.

Spectrochemical Analysis. Excited atoms or molecules emit light of a definite wavelength. Dispersion of this light by a prism or grating produces a spectrum which is characteristic of the atom or molecule. The means of excitation are an arc, spark, flame, or discharge tube. The discovery by Kirchhoff and Bunsen[7] that each chemical element produces a characteristic set of spectrum lines is the basis for quantitative spectrochemical analysis. Details of the methods and techniques can be found in textbooks.[8-11] The sensitivity obtainable is in parts per million.

Spark Testing. Steels can be classified by visual examination of the sparks produced when the specimen is held against a high-speed abrasive wheel. When a specimen is held in contact with the moving abrasive wheel, small particles of the metal are removed. Because the tearing process is so rapid, the temperature of the particles is raised to incandescence. Each type of steel produces a characteristic spark picture and color. For details of spark testing the reader is referred to the ASM "Metals Handbook." *Iron Age*[12] contains 28 elaborate sketches of sparks from important SAE steels.

Surface Analysis by Nuclear Scattering. Nuclear scattering and nuclear reactions induced by protons and deuterons have been used for the analysis of solid surfaces. Oxygen, aluminum, silicon, sulfur, calcium, iron, copper, silver, barium, and lead have been determined by the scattering method. Carbon, nitrogen, oxygen, fluorine, and sodium have been determined by nuclear reactions. In the analysis of the surfaces and the effects of surface treatments, extremely high sensitivity is required because of the small quantities per unit area involved. Furthermore, detection methods must be selective for only the surface layers. The methods described here are applicable to the detection of all elements to a depth of several microns. The sensitivities obtainable are in the range of 10^{-8} to 10^{-6} g/cm^2. Nuclear properties can be used to distinguish between specific elements or isotopes. The interaction energies are too large to be influenced significantly by chemical or crystalline binding energies. Thus no information is obtained on compounds or crystal structure by these methods.

The large particle energies permit individual particles to be detected, so that very high sensitivities are achievable. Film thickness and variation in concentration with depth can be measured with a depth resolution of about 10^{-2} μ. Nuclear reactions can be produced by bombardment with energetic protons, deuterons, or alpha particles. Such bombardment produces the emission of other energetic particles or radiation, or radioactive isotopes. The elements in the surface can be identified by the energies of particles or the radiation emitted and/or by the half-lives of the isotopes produced. These reaction techniques are best applied to the light elements since adequate yields are obtainable for incident particle energies less than 2 Mev.

The elastic scattering of high-velocity ions provides another means of analyzing surface layers. This technique may be used for analysis of very minute quantities of unknown substances, to detect compositional changes in surfaces, and to study the nature of surface layers. It is complementary to other techniques, such as electron diffraction, X-ray fluorescence, and mass spectroscopy. The scattering nucleus can be identified by the fractional loss in momentum of the scattered particle. This technique can be used for all elements but is generally less rapid than the nuclear reaction methods. For more details of these techniques the reader is referred to a paper by Rubin, Passell, and Bailey.[13] Papers by Laubenstein[14] and by Endt and Kluyver[15] provide extensive data on the nuclear interactions of all the light nuclides. An accelerator is required to provide a proton beam up to 0.1 μa with a momentum spread of about 0.04 per cent. The scattered protons are analyzed by a high-resolution point-focusing magnetic spectrometer with an aperture of 0.0025 steradian. The scattering angle can be varied from 0 to 150°. Data are usually obtained by varying the magnetic field in the spectrometer, while holding the other parameters constant, and counting protons through the exit slit of the spectrometer. From this momentum distribution curve the concentration of elements on the specimen surface can be calculated.

The sensitivity of nuclear reactions decreases drastically with increase in atomic number. The sensitivity of the scattering method increases with increase in atomic number.

Electron Probe. An electron probe microanalyzer has been designed and built at the Battelle Memorial Institute.[16] It is possible to isolate for analysis a region less than 0.0001 in. in diameter on the specimen being studied. The equipment can be used in the analysis of all elements above magnesium in the periodic table. The Battelle analyzer uses a heated tungsten filament to provide electrons. The electrons after being accelerated through a 50,000-volt potential gradient bombard the specimen. The electron beam excites X radiation of the material being studied. The X radiation then passes through a curved crystal vacuum spectrometer. A Geiger counter detects the wavelength passing through the crystal.

Activation Analysis. The irradiation of most atomic nuclei by thermal neutrons causes these nuclei to become radioactive. The subsequent emission of beta or gamma radiations from these nuclei is characteristic of the particular isotope. If the energy and rate of emission of the radiations are known, the isotope can be identified. The magnitude of the radioactivity is directly related to the amount of isotope present in the specimen. The detection of radioactivity is an extremely sensitive technique. Thus neutron activation analysis permits the accurate appraisal of very small amounts of an element. Neutrons are able to pass through a considerable thickness of most substances. It is possible, therefore, to analyze materials in solid, liquid, or gaseous phases. The induced radioactivity is usually of short duration and low intensity, necessitating only reasonable care in handling specimens.

Resistance Strain Gage. The measurement of stress and strain is a problem that frequently confronts the nondestructive testing engineer. Methods of measuring stress and strain are briefly summarized and literature references provided.

The resistance strain gage is the most simple, inexpensive, and most widely used of all transducer devices. The simple construction of the unit makes it expendable. The effect of its mass and stiffness upon the specimen may be neglected in the majority of cases. The principle of the resistance gage is extremely simple. Essentially, the gage consists of a length of very fine wire arranged in various patterns. This wire is bonded to a paper base or into a thin matrix of paper impregnated with Bakelite. In use, the gage is cemented firmly to the specimen and is strained uniformly with the specimen in either tension or compression. An electric current is passed through the wire. The change in the resistance of the wire varies with the applied strain. The measuring instrument may be a potentiometer, a Wheatstone bridge, a graphic recorder, or an oscillograph.

If a wire is subjected to a mechanical strain along its longitudinal axis, a change in length results. The change in unit length, or strain, depends on the load applied, the modulus of elasticity of the metal, and the cross-sectional area. From the relationship

$$R = \frac{\rho L}{A} \tag{13-1}$$

where R = resistance of wire
  ρ = specific resistance of wire material
  A = cross-sectional area of wire
  L = length of wire

it is seen that a change in L will be accompanied by a change in R. Taking the logarithms to the base e of both sides and differentiating gives

$$\frac{\Delta R}{R} = \frac{\Delta \rho}{\rho} + \frac{\Delta L}{L} - \frac{\Delta A}{A} \tag{13-2}$$

If Poisson's ratio is set equal to $1/m$ and if the material is homogeneous and operating below the elastic limit, $\Delta A/A$ is related to $\Delta L/L$ as follows:

$$\frac{\Delta A}{A} = -\frac{2}{m}\frac{\Delta L}{L} \tag{13-3}$$

$$\frac{\Delta R}{R} = \frac{\Delta L}{L}\left(1 + \frac{2}{m}\right) + \frac{\Delta \rho}{\rho} \tag{13-4}$$

and

$$\frac{\Delta R/R}{\Delta L/L} = \left(1 + \frac{2}{m}\right) + \frac{\Delta \rho/\rho}{\Delta L/L} \tag{13-5}$$

$(\Delta R/R)/(\Delta L/L)$ is a property of the wire forming the gage and is defined as the gage factor. This gage factor is not usually less than unity; on the other hand it is often greater than $(1 + 2/m)$. Therefore there must be a change in the specific resistance of the material or in the mechanical constants. The actual gage factor will depend upon the forming of the wire into a completed strain gage. The gage factor of the materials used in gage construction is usually constant up to a mechanical strain of about 0.5 per cent. Important as the gage factor of a material is, there are two other factors of equal importance. One of these is the resistance versus temperature change relationship. The other is the practicability of joining the gage wire to a connecting wire without forming a thermocouple. A suitable compromise has been found in the supro-nickel alloy gages. In these gages a very low thermal coefficient of resistivity is combined with a gage factor of approximately 2.2.

From the above description of the resistance strain gage it is obvious that the change in resistance caused by stresses in the specimen will be small. It is therefore of the utmost importance to use a sensitive circuitry and reliable measuring equipment. A commonly used method of measuring small resistance changes is the Wheatstone bridge in its various forms. Because of the smallness of the changes, it is necessary to remove any initial out of balance from the bridge circuit before measure-

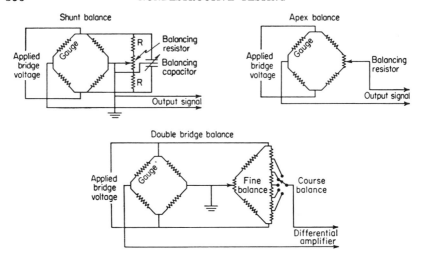

Fig. 13-1. Different types of bridge circuits for strain gage measurements.

ments are taken, and this is usually achieved by one of the three circuits shown in Fig. 13-1. The first and second circuits are quite simple and are called apex balance and shunt balance, for obvious reasons. The third system involves a double bridge, the particular advantage being that it is not necessary to have balancing elements within the measuring bridge and a large degree of initial unbalance can be accommodated without upsetting the sensitivity of the circuitry. This system is most useful when zero balance is achieved with a calibrated dial. It is not necessary to use a complete bridge of strain gages; in fact in many cases only two gages are used, one as a strain detecting element and the other to provide temperature compensations. The other two arms of the bridge are comprised of two fixed resistors. The strain gage bridge may be energized with either alternating or direct current, each system offering its own particular advantage. For the d-c system may be claimed the simplicity and large capacity of the battery power supply, and the simplicity of instrumentation when only a visual indication of static deformations is required. There are, however, disadvantages in that it is necessary to use d-c amplifier or chopper systems if the static signal is to be amplified. However, if purely dynamic signals of above 10 cps are being produced, then the amplifier may be a-c coupled and the scheme is relatively straightforward. Another disadvantage of the d-c system is that in long-term tests the effects of thermal emfs can be large compared with the signal to be measured. On the credit side is the fact that cable capacities do not affect the gage readings; the system is, however, susceptible to hum.

Residual Stress. One way in which residual stress can be determined is by X-ray diffraction techniques. When monochromatic X rays impinge on the atoms in a metallic crystal, the electromagnetic waves are reflected by the electrons, producing a diffraction pattern. Bragg's law

$$N\lambda = 2d \sin \Theta \qquad (13\text{-}6)$$

where N = order of interference (an integral number)

     λ = wavelength of X rays

     d = spacing between atomic planes

     Θ = angle of incidence and reflection

gives the condition for constructive interference of the reflected waves. The strain is determined by measuring the change in the interatomic distance between the planes of the atoms. The Bragg angle must be determined for two conditions: when the X-ray beam is normal to the specimen surface and when the X-ray beam makes an angle of 45° to the specimen surface. The theory and application of X-ray diffraction for residual stress measurements are rather complex. Details have been worked out by a number of authors.[17-19] Norton[20,21] and Kinelski[22] have applied this technique for measuring residual stress in weldments.

For measuring induced stresses during operation, brittle lacquer (Stresscoat) and transparent photoelastic plastic (PhotoStress) are used. Stresscoat is a brittle lacquer made by Magnaflux Corporation. Stresscoat is an air-drying material that is sprayed on the test specimen and allowed to dry. At the same time a simple cantilever beam is sprayed and permitted to dry. This beam is used during testing to calibrate the coating in order that quantitative strain measurements may be made. After the Stresscoat has dried 4 to 6 hr, the specimen is loaded and the coating observed for cracks. The crack pattern provides a visual presentation of the regions of highest stresses. The crack pattern indicates the relative stress concentrations and shows the direction of the principal stresses. Figure 13-2 shows the impact stresses that occur during the firing of a gun.

FIG. 13-2. Impact stresses during firing of gun. (*Courtesy Magnaflux Corporation.*)

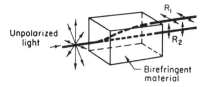

FIG. 13-3. Splitting of light beam by Birefringent material.

Certain transparent materials such as crystals of calcite and mica exhibit the phenomenon of double refraction. They divide an incident ray of light into two beams having different velocities and consequently different indices of refraction. The two beams transmitted through the doubly refractive materials are polarized at right angles, as shown in Fig. 13-3. In photoelectric stress analysis, polarized light is used to reveal the presence of strains in special transparent materials such as glass, celluloid, and Bakelite. These substances are not normally doubly refractive, but become so when strained. In this technique it is necessary to make a transparent model of the structure using glass, celluloid, or Bakelite. Polarized light is transmitted through the model. If the model is in a stress-free condition, it will leave the field of view dark placed between crossed Nicols or polaroids. If stress is applied to the model, the emerging light is in general elliptically polarized and is partly transmitted by the analyzer. The stress distribution is made evident by colored bands of light.

In the PhotoStress[23] technique the specimen is coated with a thin layer of transparent photoelastic plastic. This plastic becomes doubly refractive or birefringent when stressed. The birefringence produced is directly proportional to the intensity of the stress. Birefringence can be measured using polarized light. Under polarized light the strains appear as black and color fringes in the plastic. If there is a reflective surface provided on the specimen or on the face of the plastic bonded to it, the birefringence can be observed using a reflection polariscope. Using such a polariscope the direction of the principal strains can be visually observed, recorded on a photographic film, or observed on an oscillograph. The paper by Zandman and Wood[22] describes the Photostress technique in detail and the instrumentation available for making the necessary measurements.

SPECIFIC REFERENCES

1. Feigel, Fritz: "Spot Tests," vol. I—Inorganic Applications, vol. II— Organic Applications, Elsevier Publishing Company, Amsterdam, 1954.
2. Jirkovsky, R.: Chemical Abstracts 25, 5640, 1931.
3. Hunter, M. S., J. R. Churchill, and R. B. Mears: Electrographic Methods of Surface Analysis, *Metal Progr.*, vol. 42, pp. 1070–1076, December, 1942.
4. Caley, E. R.: *Museum News*, vol. 15, no. 5, pp. 9–11, 1937.
5. ASTM Symposium on Rapid Methods for the Identification of Metals, *ASTM, Spec. Tech. Pub.* 98, 1950.
6. Hilborn, H. S., and R. C. Pugh: Spot Tests for Contaminants on Aluminum, E. I. du Pont de Nemours & Company, Report DP-88, 1954.

7. Kirchhoff, G. R., and R. Bunsen: Chemische Analyse Durch Spectral Beobachtungen, *Ann. d. Physik,* vol. 110, pp. 161–189, 1860.
8. Sawyer, Ralph A.: "Experimental Spectroscopy," Prentice-Hall, Inc., Englewood Cliffs, N.J., 1944.
9. Brode, Wallace R.: "Chemical Spectroscopy," John Wiley & Sons, Inc., New York, 1939.
10. Harrison, G. R., R. C. Lord, and J. R. Lofbourow: "Practical Spectroscopy," Prentice-Hall, Inc., Englewood Cliffs, N.J., 1949.
11. Turyman, F.: "Metal Spectroscopy," Charles Griffin & Co., Ltd., London, 1951.
12. Spark Testing, *Iron Age,* Sept. 26, 1935, and Oct. 3, 1935.
13. Rubin, Sylvan, Thomas O. Passell, and L. Evan Bailey: Chemical Analysis of Surfaces by Nuclear Methods, *Analyt. Chem.,* vol. 29, no. 5, pp. 736–743, May, 1957.
14. Laubenstein, R. A., M. J. W. Laubenstein, L. J. Koester, and R. C. Mobley: The Elastic Scattering and Capture of Protons by Oxygen, *Phys. Rev.,* vol. 84, no. 1, pp. 12–18, October, 1951.
15. Endt, P. M., and J. C. Kluyver: Energy Levels of Light Nuclei, *Rev. Mod. Phys.,* vol. 26, no. 1, pp. 95–166, January, 1954.
16. Battelle Memorial Institute: Electron Microanalyzer, *Battelle Tech. Rev.,* vol. 6, no. 12, p. 14, 1957.
17. Barrett, C. S.: "Structure of Metals," 2d ed., McGraw-Hill Book Company, Inc., New York, 1952.
18. Klug, H. P., and L. E. Alexander: "X-ray Diffraction Procedures," John Wiley & Sons, Inc., New York, 1954.
19. Mihalism, J. R.: A Study of X-ray Stress Measurement Techniques, submitted for partial fulfillment of Doctor of Science degree, Massachusetts Institute of Technology, 1953.
20. Norton, J. T., and B. M. Loring: Stress Measurements in Weldments by X-rays, *Welding J.,* vol. 20, pp. 284s–287s, 1941.
21. Norton, J. T., and D. Rosenthal: An Investigation of the Behavior of Residual Stresses under External Load and Their Effect on Safety, *Welding J.,* vol. 22, pp. 63s–78s, 1943.
22. Kinelski, E. H., and J. A. Berger: X-ray Diffraction for Residual Stress Measurements of Restrained Weldment, *Welding J.,* vol. 36, pp. 5130–5175, December, 1957.
23. Zandman, F., and Marc R. Wood: PhotoStress, *Product Eng.,* pp. 167–178, September, 1956.

CHAPTER 14

THICKNESS MEASUREMENTS

A great variety of industrial materials such as metals, plastic, and paper are produced in sheet form. It is often important that close control be kept of the thickness of the material during the manufacturing or fabricating process since the lack of uniformity in thickness may appreciably affect the ultimate usefulness of the material. The economic factors involved are also often of considerable importance. In this chapter two general types of gages will be considered, contact and noncontact. Among the contact type of gages are ultrasonic resonance, electrical conduction, magnetic, electromagnetic, and mechanical. Such contact gages can be used accurately to measure the thickness of sheet materials. However, for some applications contact gages cannot be used and noncontacting gages are required. Noncontacting gages can be used for continuously measuring material traveling at high speed and at elevated temperatures, with an accuracy of 1 to 2 per cent. The surface of the material is not scratched or deformed in any way. Among the noncontact gages developed in recent years, radiation gages play an important part. A radiation gage consists of a source of radiation, a detector, and indicating or recording equipment. The material to be gaged in some cases passes between the source and detector, and in other such gages the detector and source are on the same side of the material.

In this chapter methods and techniques for measuring the thickness of materials and coatings will be described. The methods and techniques to be discussed for measuring the thickness of material can be applied to either metallic or nonmetallic materials. The methods and techniques to be discussed for measuring coating thickness can be used for measuring metallic as well as nonmetallic coatings on various kinds of base materials.

Thickness Gaging by Radiation Absorption Measurements. The absorption of X rays, gamma rays, beta particles, and alpha particles as they pass through matter is used as the basis for measuring the thickness of material. The type and energy of the radiation used depend on the nature and thickness of the material to be measured. In these applications the material to be gaged is inserted between the source of radiation and a suitable radiation detector. The amount of radiation transmitted through the specimen is determined by the thickness of the material.

The output of the detector is calibrated in terms of the thickness of the material.

X-ray Gage. Figure 14-1 shows absorption curves for X rays in steel. A study of these curves indicates two important factors which affect the design of an X-ray thickness gage. At 78 kV and 0.100 in. of steel a 1 per cent change in voltage will produce the same change in detector current as a 2.8 per cent change in thickness. Likewise a 1 per cent change

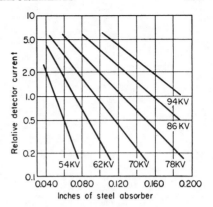

FIG. 14-1. X-ray absorption in steel.

in either sensitivity or intensity will have the same effect as a ⅓ per cent change in thickness. To avoid the necessity for close regulation of the tube voltage and beam current, most X-ray gaging systems employ two X-ray beams in a balanced-bridge, or null, system. One of the beams passes through a standard thickness of material, while the second beam passes through the material to be gaged. The transmitted intensity of the two beams is then compared by a suitable detector, or detectors. The output from the detectors is a measure of the thickness difference between the standard and the specimen.

In one system[1]* two separate X-ray sources are operated from a common electrical power supply, the two transmitted beams impinging on a single detector, Fig. 14-2. Another system,[2] Fig. 14-3, makes use of two beams from a single X-ray source and employs two separate detectors. In both systems, the detectors consist of phosphor phototube combinations used in a comparison circuit. The relationship between the thickness of absorbing material and the intensity of the transmitted beam is not linear. Therefore most gaging systems require a scale-factor adjustment, which must be set at the proper value for each thickness to be gaged. This additional adjustment is eliminated in one gage[1] by inserting a uniformly tapered wedge of material in one of the two X-ray beams. The wedge is

* Superscript numbers refer to Specific References listed at the end of the chapter.

FIG. 14-2. Westinghouse thickness gage. (*Courtesy Westinghouse Electric Corporation, X-Ray Division.*)

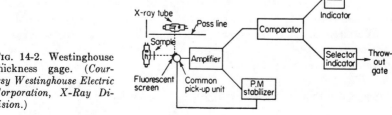

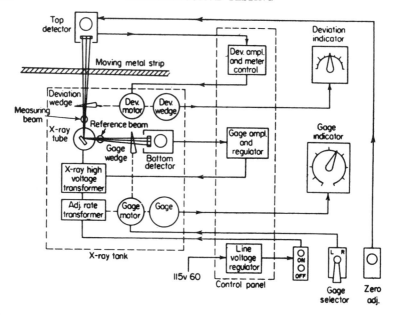

Fig. 14-3. General Electric X-ray thickness gage. (*Courtesy General Electric, X-Ray Division.*)

positioned automatically by a servo system to maintain the transmitted intensity of the two beams in balance. The thickness indication of the gage is therefore derived directly from the position of the balancing wedge. X rays have the advantage that they can be adjusted within broad limits by varying the anode voltage. For example, 0.0004-in. aluminum foil can be measured with 5-kV and 2-in. aluminum plates with 250-kV X rays. An accuracy of ±1 per cent can be achieved over the whole range of thickness.

Ettinger[3] describes a differential X-ray gage in which potassium iodide scintillation crystals are used. A block diagram of the system is shown in Fig. 14-4. The X-ray generator used was of the self-rectifying type which produces short pulses of X rays at a repetition rate of 60 cps. It is therefore possible to use a-c amplifiers so that drifts due to warm-up and aging are very much reduced. To eliminate output variations caused by line voltage changes, the whole system is operated as a self-balancing bridge. When equal absorption takes place along the two paths, no voltage is induced in the secondary of the transformer shown. A difference signal appears across the transformer secondary when the absorption in the two paths is different. This difference signal provides an input to a vacuum tube amplifier which drives a phase sensitive detector. The output of the phase sensitive detector changes if the unbalance between the two X-ray absorption paths reverses.

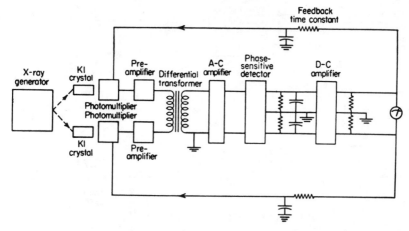

FIG. 14-4. Differential X-ray gage. (*Courtesy Electronics.*)

Figure 14-5 shows a projectile, an X-ray generator, and two photo-multiplier tubes. To keep the two potassium iodide crystals in close proximity, and yet allow for the diameter of the photomultiplier tubes, curved light guides of clear Lucite were employed. The crystals were cemented to one end of the guides whose other ends were, in turn, cemented to the photomultiplier windows.

Howell[4] developed an X-ray thickness gage for measuring the thickness of aircraft propeller blades. Cadmium sulfide crystals were used as detectors. The properties of these detectors were discussed in the chapter on radiography. This was also a differential-type gage consisting of two identical detectors and amplifier channels. The output from the

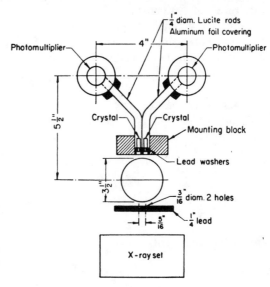

FIG. 14-5. X-ray detector system. (*Courtesy Electronics.*)

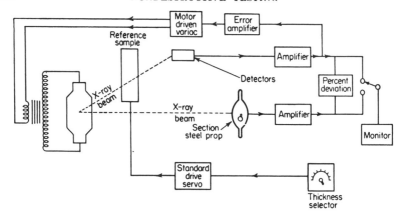

FIG. 14-6. Block diagram of gaging unit.

two channels was metered so as to give a differential reading. When the test specimen and the reference specimen were of equal thickness, the meter reading would be zero. If the thickness of the test specimen varied, the meter reading would indicate the thickness in per cent plus or minus. Figure 14-6 shows a block diagram of the gaging unit.

The noncontact type of equipment is useful for thickness measurement during hot rolling of metal strip. The accuracy of measurement is affected by the temperature of the hot strip. The X-ray absorption of the hot steel is reduced as the temperature increases. Since the standards themselves are cold, compensation has to be made for this difference in absorption. Under normal hot rolling conditions at temperatures around 1400°F, this difference is approximately 2 per cent over a fairly wide range of temperature.

Gamma-ray Sources. In the chapter on radiography it was shown that, for homogeneous narrow beam of X rays or gamma rays, the absorption could be described by the equation

$$I = I_0 \exp(-\mu X) \tag{14-1}$$

This expression can be used to determine the source of X-ray or gamma radiation which will give the maximum sensitivity. Differentiation of Eq. (14-1) both with respect to the thickness X and absorption coefficient μ gives

$$\frac{dI}{dX} = -\mu I_0 \exp(-\mu X) \tag{14-2}$$

$$\frac{d}{d\mu}\left(\frac{dI}{dX}\right) = -I_0 \exp(-\mu X) + \mu I_0 X \exp(-\mu X) = 0 \tag{14-3}$$

For maximum sensitivity Eq. (14-3) should be set equal to zero, giving

$$\mu X = 1 \tag{14-4}$$

Because of statistical fluctuations in actual measurements the above derivation must be revised to maximize the ratio caused by the addition of more absorber and statistical fluctuations. The statistical fluctuations vary as the square root of the counting rate I. The expression to be maximized is now

$$\frac{\partial}{\partial \mu}\left(\frac{\partial I/\partial x}{I^{1/2}}\right) = 0 \tag{14-5}$$

from the equation

$$\frac{\partial I}{\partial x}\frac{1}{I^{1/2}} = \frac{-\mu I_0 \exp(-\mu x)}{[I_0 \exp(-\mu x)]^{1/2}} \tag{14-6}$$

$$\frac{\partial I}{\partial x}\left(\frac{1}{I^{1/2}}\right) = [I_0 \exp(-\mu x)]^{1/2}(-\mu) \tag{14-7}$$

$$\frac{\partial}{\partial \mu}\left(\frac{\partial I}{\partial x}\frac{1}{I^{1/2}}\right) = -[I_0 \exp(-\mu x)]^{1/2} + \left(\frac{x\mu}{2}\right)\left[I_0 \exp(-\mu x)\right]^{1/2} = 0 \tag{14-8}$$

from which

$$\mu x = 2 \tag{14-9}$$

Nickerson[5] refers to Eq. (14-9) as the expression for maximum readability.

TABLE 14-1. HARD GAMMA EMITTERS

Radioisotope	Half-life	Energy,* Mev
Na^{24}	15.06 h	2.754
Ga^{72}	14.3 h	2.51
La^{140}	40 h	2.50
Ir^{194}	19 h	2.1
Sb^{124}	60 d	2.04
As^{76}	26.8 h	1.7
Pr^{142}	19.2 h	1.59
Ag^{110m}	270 d	1.516
K^{42}	12.44 h	1.51
$Eu^{152,154}$	13 y; 16 y	1.40
Co^{60}	5.27 y	1.33
Br^{82}	35.87 h	1.312
Fe^{59}	45.1 d	1.289
Ta^{182}	115 d	1.223
Sc^{46}	85 d	1.12
Zn^{65}	250 d	1.12
Rb^{86}	19.5 d	1.08
Rh^{106} (Ru^{106})	30 s	1.045
Cs^{134}	2.3 y	0.794
W^{187}	24.1 h	0.78
Zr^{95}	65 d	0.754
Nb^{95}	35 d	0.745
Ba^{137m} (Cs^{137})	2.6 m	0.662

* The energy given in each case is the maximum gamma energy which occurs to an extent of 5 per cent or more.

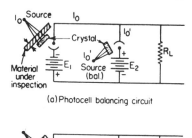

(a) Photocell balancing circuit

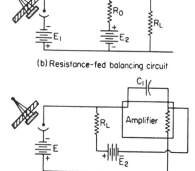

(b) Resistance-fed balancing circuit

(c) Feedback circuit

FIG. 14-7. Compensating circuit. (*Courtesy Los Alamos Scientific Laboratory.*)

The above derivation may be summarized as follows for a source of fixed intensity. The maximum sensitivity is obtained when the product of the linear absorption coefficient and thickness is equal to 2. In making such measurements a source having the highest possible intensity should be employed. Table 14-1 lists the gamma-ray source available from the United States Atomic Energy Commission. Because of the limited selection of gamma-ray sources it is not always possible to satisfy equation (14-9). Consequently, when using gamma-ray sources, one must be content with some approximation to the optimum condition of $\mu x = 2$.

In making thickness measurements with radioactive sources, corrections must be applied to compensate for the radioactive decay. Berman and Harris[6] describe a scheme for continuously compensating for source decay. The transmitted radiation is electrically balanced against the direct radiation from a source which decays at the same rate. This system is shown in Fig. 14-7. The circuit consists of a duplicate assembly of source and detector. Suppose that the two sources have strengths S_1 and S_2, respectively. The detector currents are $i_1 = K_1 S_1$, $i_2 = K_2 S_2$, where K_1 and K_2 are constants depending on the geometry of the system and the efficiency of the detector. The difference of the detector currents can be made equal to zero to establish the balance condition. This condition is maintained at all times, since the two sources decay at the same rate. However, the sensitivity to changes in thickness x does vary with time, since

$$\frac{dI}{dx} = \frac{d}{dx}\left[I_0 \exp\left(-\mu x\right)\right] \qquad (14\text{-}10)$$

Gamma-ray Gages. Berman and Harris[6] describe a precision gage for measuring the uniformity of materials by gamma-ray transmission. In order to achieve high precision, it is necessary to collimate the radiation to define the area of measurement. It is also necessary to collimate the radiation emerging from the specimen being measured to discriminate against radiation scattered in the specimen. Berman and Harris give a detailed analysis of the collimation problem for a gamma-ray source. Some of the problems encountered in the design of such a gage are listed below. Photomultiplier tubes are quite voltage sensitive. Sodium iodide crystals, photomultiplier tubes, and batteries are temperature sensitive. Therefore to achieve high sensitivity, it is necessary to control

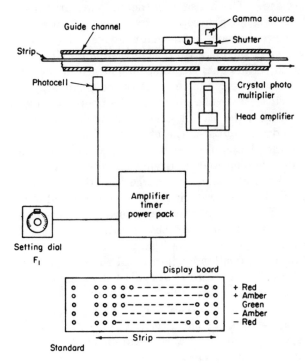

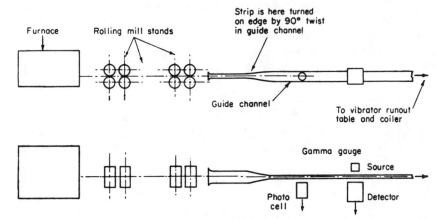

FIG. 14-8. Gamma gage for hot steel strip. (*Courtesy The British Institute of Radio Engineers.*)

the temperature of the crystal, tube, and any batteries used. For measuring small direct currents from a photomultiplier an electrometer tube and d-c amplifiers can be used. However d-c amplifiers tend to drift. The vibrating-reed electrometer[7,8] can be used to measure such currents without appreciable drift.

Hot Strip Gage. There are two rather unique features in a gamma-ray gage for hot steel strip described by Syke.[9] Figure 14-8 shows the diagrammatic arrangement of the gamma gage. The red radiation from the leading edge of the hot strip falls on the photocell and causes a relay to be energized. This puts the electronic measuring circuit in readiness and opens a shutter placed in the path of the gamma-ray beam. As the tail end of a strip leaves the photocell, the photorelay closes the source shutter. Just after the leading edge of the strip reaches the measuring head a timer device initiates the charging of a capacitor. The pulses derived from the photomultiplier pass through a preamplifier, discriminator, and pulse shaping and rectifying circuits. Each registered pulse thus puts a definite amount of charge into the integrating capacitor. The voltage across the integrating capacitor at the end of a time interval is measured. This voltage gives a measure of the mean strip thickness passed through the gamma-ray beam during the interval. The other feature of this equipment is the "lamp board." The lamp board consists of 23 or 48 vertical and 5 horizontal rows of lamps to display the longitudinal profile of the strip. A typical strip may be 600 ft long. At a rolling speed of 1,200 fpm, this will pass through the mill in 30 sec. At a timer drum setting of 1.6 sec cycle, 18 lamps light up on the board, each lamp representing the mean thickness of a 32-ft-long section.

Tube Gage. Another typical application of the gamma gage is the measurement of wall thickness of red hot steel tubes.[10] Figure 14-9 shows the general arrangement for tubes of about 2- to 6-in. bore and 0.1- to 0.6-in. wall thickness. The source is mounted in the end of a steel rod, and the tube, which is rotating in a cradle, is threaded over this. The detector unit is outside the tube and is water-cooled. By rotating the tube and pushing forward at the same time, the wall thickness is measured along a spiral line. Source, detector, and electronics used in the tube wall thickness gage are similar to those used in the strip thickness gage. The wall thickness is recorded on a strip chart recorder. Eccentricity of bore relative to outside diameter of the tube is indicated as a wave on the recorder chart.

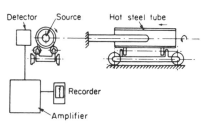

FIG. 14-9. Gamma gage for tubing. (*Courtesy The British Institute of Radio Engineers.*)

Nickerson's[5] paper summarizes the mathematical treatment of the

TABLE 14-2. RANGE OF ALPHA PARTICLES FROM SOME RADIOACTIVE SOURCES

	Range, cm
Source	*(air, 760 mm Hg at 60°F)*
Uranium I	2.73
Radium	3.39
Radium C	6.97
Polonium	3.93
Thorium	2.90
Thorium C'	8.62

absorption process and the factors which affect the sensitivity of radiation gages. Summarizing the range of usefulness of X ray and gamma rays for gaging, the following general conclusions may be made. In the range from 0.001 to 0.050 in. of steel, soft X rays must be used; in the range from 0.050 to 2 in. of steel, X rays or soft gamma rays must be used; and in the range above 2 in. of steel, hard gamma rays must be used.

Alpha Gage. For measuring thin sheet foil such as condenser paper or very thin sheet materials, an alpha gage can be used. Using an ionization chamber as the detector, 1 per cent changes in weight can be detected. Alpha particles are emitted during the radioactive decay of some nuclei. An alpha particle is a helium atom stripped of its orbital electrons. Alpha particles lose their energy as they pass through matter by ionization. But unlike gamma or X rays the absorption of alpha particles does not follow an exponential law. Alpha particles emitted by a given element have a very definite range. This range is usually expressed in centimeters of air at 60°F and 76 mm Hg. The range of the alpha particles from some alpha sources is given in Table 14-2. In discussing the range of alpha particles in material other than air, it is convenient to introduce the term "stopping power." This is defined as the reciprocal of that thickness of substance equivalent to 1 cm of air in its ability to stop alpha radiation. If the stopping power of an element is divided by the number of atoms per cubic centimeter, the atomic stopping power of the substance is obtained, as shown in Table 14-2. The relationship between the range of an alpha particle and its velocity cannot be expressed by any one simple formula; alpha particles of medium range are found to follow Geiger's empirical formula,

$$R = av^3 \tag{14-11}$$

where R = range, cm

$a = 9.6 \times 10^{-28} \text{ sec}^3/\text{cm}^2$

v = velocity, cm/sec

A relationship also exists between the range of an alpha particle and the half-life of the emitter known as the Geiger-Nuttall law

$$\log \frac{0.693}{T_{\frac{1}{2}}} = B \log R + A \tag{14-12}$$

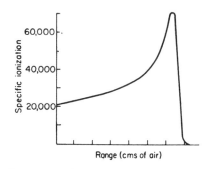

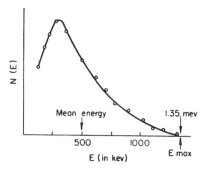

FIG. 14-10. Specific ionization of RaC'   FIG. 14-11. Typical beta-ray spectrum.
alpha particles.

where A and B are constants and R is the range in air. If the range is
given in centimeters, $A = -41.6$ and $B = 60.4$ for the uranium family,
the half-life being in seconds.

In the alpha gage described by Shaw[11] an ionization chamber was used.
The decrease in energy of alpha particles after passing through the specimen was measured to determine specimen thickness. Figure 14-10
shows the ionizing power as a function of range for alpha particles. By
measuring the energy of the alpha particle passing through the specimen,
the thickness of the specimen can be determined. Using a source of
150 μc of radium and an absorber of 0.001 gm/cm², weight changes of
1 per cent could easily be detected above the statistical variations.

TABLE 14-3. BETA EMITTERS

Radio-isotope	Half-life	Maximum energy, Mev
Y^{90}	2.54 d	2.18
P^{32}	14.3 d	1.701
Y^{91}	61 d	1.537
Sr^{89}	53 d	1.463
Bi^{210}	5.02 d	1.17
Ag^{111}	7.6 d	1.04
Pd^{109}	13.6 h	0.961
Pr^{143}	13.7 d	0.932
Tl^{204}	4.0 y	0.765
Cl^{36}	3.08×10^5 y	0.714
Tc^{99}	2.12×10^5 y	0.29
Ca^{45}	163 d	0.254
Pm^{147}	2.6 y	0.223
S^{35}	87.1 d	0.167
C^{14}	5568 y	0.155
H^3	12.46 y	0.01795

Beta Rays. Beta rays are high-speed electrons which are emitted during the disintegration process of certain radioactive materials. The beta rays emitted from a radioactive source are not monoenergetic, but have a continuous energy distribution, as shown in Fig. 14-11. Each beta emitter has a definite maximum energy. The energy given for beta rays in the literature is the maximum energy. Table 14-3 shows the beta emitters available from the United States Atomic Energy Commission. The wide range of beta-ray energies available makes possible thickness measurements from 0.000015 in. of aluminum to approximately 0.050 in. of steel. The energy in million electron volts and the half-life are also included in Table 14-3. Figure 14-12 gives an indication of the weight per unit area of absorber required to completely stop the beta radiation. Figure 14-13 shows proper radiation for various thicknesses of four materials. The ideal source is one which provides the steepest and most linear section of its absorption characteristic over the range of weight per area to be covered.

As beta particles pass through matter they are slowed down by collisions with the electrons of the absorbing material. The thickness of material which will completely stop a beta particle depends on two factors: the initial energy of the radiation and the electron density in the material. The electron density varies from one substance to another. The amount absorbed can be expressed as the product of the density (grams per cubic centimeter) and thickness (centimeters), i.e., grams per square centimeter. In this case the thickness is nearly independent of the nature of absorber since there are about the same number of electrons in a given mass of any material. In the literature the range of beta particles is often given in grams per square centimeter of alumi-

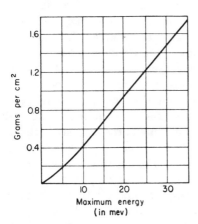

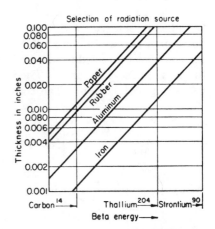

Fig. 14-12. Weight per unit area vs. beta energy.   Fig. 14-13. Radiation source vs. thickness.

num. The range of beta particles can be computed from the empirical
relationship

$$\rho t = 0.54 E_m - 0.16 \qquad (14\text{-}13)$$

where t = thickness of absorbing material, cm

    ρ = density of the material (g/cm^3)

    E_m = maximum energy of the beta particle, Mev

Because of their small mass, beta particles do not travel in straight lines
through an absorber but pursue a more or less random path. The absorption curve for a pure beta emitter looks somewhat like an exponential
curve. The intensity falls off sharply for small thicknesses of absorber,
since the low-energy beta particles are rapidly absorbed.

Beta Gage. Figure 14-14 shows a beta gage. The source of beta
rays is placed on one side of the material and a suitable detector on the
other side. The detector can be an ionization chamber or beta sensitive
scintillation crystal such as anthracene. The components of a complete
beta-ray thickness gage[12] are shown in block form in Fig. 14-15. Between
the source and specimen is a motor-driven chopper which interrupts the
beam 90 times per second. The signal from the ionization chamber is
therefore a 90-cycle signal which can be amplified by a-c amplifiers.
To provide an expanded, easily read thickness scale, a potentiometer-
type measuring circuit is employed. The ionization chamber voltage is
compared with a second 90-cycle voltage of opposite phase which is
adjusted in magnitude until the difference between the two voltages is
zero. The comparison voltage is derived from a small permanent-
magnet alternator driven from the chopper shaft. The attenuator used
to adjust the magnitude of the comparison voltage may be calibrated in
terms of thickness.

A beta gage has been used to measure wall thickness and roundness of
tubing. In this gage a Sr^{90} source is mounted inside a long rod. Two
detector units are mounted 180° apart near the source. In operation,
the source is held fixed inside the tubing while the tubing is passed for its
full length past the source. The tubing is then rotated 90° and returned
to its original position. The beta radiation penetrating the walls varies
with wall thickness and is detected
by the measuring heads. The wall
thickness measurements are recorded
on a two-channel recorder. Since the
entire wall of the tube is examined, it
is possible to pick up any deviation
in wall thickness. If the tube is
eccentric, this will be indicated by the
separation of the two pens of the
recorder. The latter effect occurs

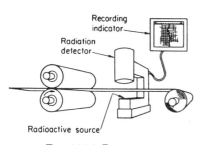

FIG. 14-14. Beta gage.

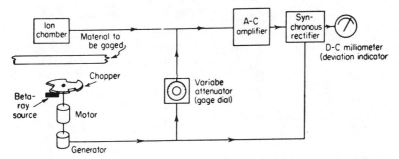

FIG. 14-15. Schematic diagram of gage. (*Courtesy General Electric Company, X-Ray Division.*)

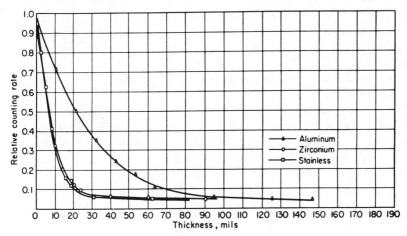

FIG. 14-16. Absorption curve of beta radiation, using anthracene scintillator. (*Courtesy Argonne National Laboratory.*)

because one detector is closer to the source than the other. This causes a deviation in the amount of radiation at the two detectors even though the absorber thickness is the same.

A typical curve showing the absorption of 2.3-Mev beta particles by aluminum, stainless steel, and zirconium is shown in Fig. 14-16. The shape of the curves is nearly exponential. The absorption curves can be represented by the exponential equation

$$I = I_0 \exp{(-\mu_L x)} = I_0 \exp{(-\mu_m \rho x)} \qquad (14\text{-}14)$$

where I = intensity of transmitted radiation
I_0 = intensity of incident radiation
μ_L = linear absorption coefficient
μ_m = mass absorption coefficient
ρ = density of material
x = thickness of absorber

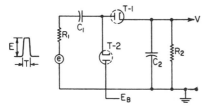

FIG. 14-17. Schematic diagram of count rate meter.

The mass absorption coefficient μ_m is nearly independent of the atomic number of the absorbing material, increasing slightly as the atomic number increases. The value of μ_m can be determined from the relation[13]

$$\mu_m = \frac{\mu_L}{\rho} = \frac{22}{E_m^{1.33}} \quad (14\text{-}15)$$

An instrument which integrates or averages the counts from a radiation detector over a definite interval of time is known as a count rate meter. This instrument can be used for making a permanent record of the intensity from a source. A schematic diagram of a count rate meter is shown in Fig. 14-17. Each pulse to be counted causes a definite amount of charge to be placed on a condenser. The charge on the condenser leaks off through a high resistor; thus the average potential across the condenser is proportional to the counting rate. The value of the capacitance determines the statistical error in the observed counting rate. The value of the resistance determines the sensitivity of the meter.

If each beta ray detected puts a charge q on the condenser and the average counting rate is N, the voltage E across the resistor (R ohms) is

$$E = qNR \quad (14\text{-}16)$$

The current I across the indicating meter will be given by

$$I = GE \quad (14\text{-}17)$$

where G is a constant. The indicating meter may be arbitrarily set equal to zero for any given thickness of absorber x'. If E' is the corresponding voltage, then Eq. (14-17) becomes

$$I = G(E' - E) \quad (14\text{-}18)$$

From Eqs. (14-16) and (14-18)

$$I = GRqN_0 \exp(-\mu x')\{1 - \exp[-\mu(x - x')]\} \quad (14\text{-}19)$$

Defining the sensitivity S of the gage as dI/dx since $N = N_0 \exp(-\mu x)$

$$S = \frac{dI}{dx} = \mu GRqN_0 \exp(-\mu x) = GE\mu \quad (14\text{-}20)$$

Since statistical fluctuations tend to mask the effects of varying thickness, the sensitivity is given by

$$S = \frac{dI}{\sqrt{I}\, dx} \quad (14\text{-}21)$$

Statistical fluctuations set the ultimate limitation to the accuracy obtainable with a beta-ray thickness gage. Fluctuations may occur both in the number of beta particles reaching the detector and in the response to each particle. For the moment consider only fluctuations in N. Other sources of fluctuations, such as noise in the large resistor used with an ionization chamber, will be neglected.

Schiff and Evans[14] have shown that the fractional standard deviation σ_Q in the charge Q on the condenser C_2 shown in Fig. 14-17 is

$$\frac{\sigma_Q}{Q} = \frac{1}{(2NRC)^{1/2}} = \frac{\sigma_E}{E} \tag{14-22}$$

thus the deviation σ_I in meter current is

$$\sigma_I = G\sigma_E = Gq\left(\frac{NR}{2C}\right)^{1/2} = \frac{S}{\mu(2NRC)^{1/2}} \tag{14-23}$$

For a given sensitivity S, the fluctuation in meter current is reduced by using a stronger source or a longer time constant RC. If the value of σ_I from Eq. (14-23) is inserted in the equation

$$dx = \frac{dI}{S} \tag{14-24}$$

where dI is the smallest change in meter reading that can be detected, the relative accuracy is

$$\frac{dx}{x} = \frac{1}{\mu x (2NRC)^{1/2}} = \frac{\exp(1/2\mu x)}{\mu x (2N_0 RC)^{1/2}} \tag{14-25}$$

The accuracy in the measurement is actually determined by the larger of the following two quantities: the smallest change in meter current which can be detected and the fluctuation in meter current due to statistical fluctuations. If the statistical fluctuation is the determining quantity, no improvement can be obtained by increasing the conductance G. In general, there must be a compromise between the fluctuation σ_I and the response time which can be tolerated. If the equilibrium time t is defined as the time required for the difference between the equilibrium value and the actual value of the reading to be less than the probable error due to statistical fluctuations, Schiff and Evans have shown that

$$t = RC(\tfrac{1}{2}\ln 2NRC + 0.394) \tag{14-26}$$

This analysis is applied to a count rate meter no matter what type of radiation is being counted.

Parry[15] describes a combined beta and dielectric gage used in the cigarette industry. Beta gages do not respond to fast changes in mass. The dielectric gage has a very fast response time and is very sensitive to

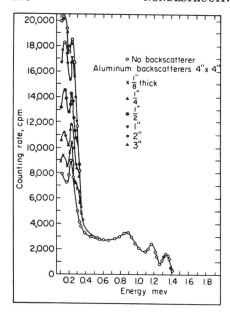

FIG. 14-18. Backscatter of gamma rays by aluminum plates. (*Courtesy Nucleonics.*)

humidity and temperature changes, which under most industrial conditions are relatively slow changes. By using the beta gage with the dielectric gage, an instrument is obtained with most of the advantages of both instruments. Such equipment is now in use in the cigarette industry. The equipment in use on cigarette making machines incorporates a pair of balanced ionization chambers and radioactive sources. The cigarette to be measured is passed between one chamber and source. An absorber having an equivalent mass to the cigarette is placed between the other chamber and source. The polarizing voltages across the ionization chamber are of opposite polarity. The currents from the two chambers pass in parallel through the same load resistor. When a cigarette having the average value of mass passes through the instrument, the voltage across the load resistor is zero. If the mass of a cigarette changes, a voltage is generated across the load resistor.

The beta and gamma transmission gages previously described require the radioactive source and the detector to be placed on opposite sides of the material to be measured. Sometimes this may be a serious disadvantage if only one side of the material is accessible. Also the measurement of coatings is in general not possible when using the transmission technique. For such applications a backscattering or reflection of radiation technique is often used.

Backscatter of X Rays and Gamma Rays. When X rays or gamma rays are absorbed by material, scattered X rays and gamma rays are emitted. Scattered radiation that leaves the absorber through the same sur-

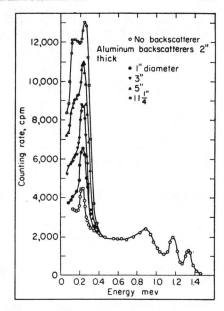

FIG. 14-19. Backscatter of gamma rays by aluminum cylinders. (*Courtesy Nucleonics.*)

face as the primary radiation entered is called "backscattered radiation." The intensity and energy of the backscattered radiation are dependent upon the scattering angle. An expression for calculating the energy of the scattered radiation is given in the chapter on X-ray radiography. The energy and intensity of the backscattered radiation both vary with the angle between the primary beam and the scattered radiation. Figures 14-18 and 14-19 show the backscattering of Co^{60} by gamma rays by aluminum plates and cylinders.[16] Figure 14-20 shows the backscatter of

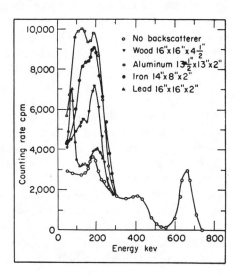

FIG. 14-20. Backscatter of 661-kev gamma rays. (*Courtesy Nucleonics.*)

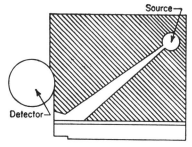

FIG. 14-21. Hare backscatter gage.

the 661-kev gamma from Cs^{137} for various materials. This curve also indicates that materials having a high atomic number such as lead are poor scatterers for gamma rays. These curves also show that there are some backscattered gamma rays even when no material was placed under the source. This was due to the scattering from the frame supporting the source, the surroundings, and the detector. The energy of the gamma ray after scattering to an angle ϕ is given by

$$\lambda' - \lambda = 0.024A(1 - \cos \phi) \qquad (14\text{-}27)$$

$$E_\phi = \frac{E_0}{1 + 1.9576E_0(1 - \cos \phi)} \qquad (14\text{-}28)$$

where E_0 is the incident gamma ray energy, Mev

Gamma-ray Backscatter Gage. A gamma-ray backscatter gage is described by Hare.[17] Figure 14-21 shows such an instrument. The gamma rays from a radium or cobalt source enter the metal and a certain portion are backscattered to the detector. The detector is shielded from the direct radiation by an absorber. The source and detector are kept in the same relative position to the surface of the metal. The amount of scattered radiation received by the detector depends on the density and thickness of the metal. The gamma rays must be of fairly high energy if steel specimens having thicknesses in the order of $\frac{1}{2}$ to 1 in. are to be measured. Consequently, the absorber between source and detector must be rather thick, and relatively large sources are needed. Such an instrument produces a poor localization of the area of measurement and may not detect local variations in thickness.

If Co^{60} gamma rays are scattered through an angle of 180° they will have energies of 0.214 and 0.209 Mev, respectively. It is important to note that there is a large energy difference between the direct and scattered radiation. This effect is used in one backscatter gage.[18] In this instrument the detector is a scintillator crystal such as sodium iodide. The pulses produced by a scintillation detector are proportional to the energy of the radiation impinging on the scintillator. By using a pulse height analyzer, it is possible to count only the pulses which are due to the backscattered radiation. Consequently, even though the primary radiation from the cobalt source is present, only the lower energy scattered radiation is counted. The pulses can be fed to a count rate meter to measure the amount of backscattered radiation. The instrument

Fig. 14-22. Putman and Jefferson backscatter gage. (*United Kingdom Atomic Energy Research Establishment photographs.*)

developed by Putman and Jefferson is shown in Fig. 14-22. The source used in this instrument was a 3-mm length of ½-mm-diameter cobalt wire with an activity of 21 μc. There is a contribution to the counting rate even in the absence of any specimen. There is also a contribution from the source holder. Both of these contributions are proportional to the activity of the source. By applying a preset potential, these two contributions can be suppressed, giving a zero meter reading in the absence of a specimen.

In order to obtain the maximum possible backscattering from a given source and also the best ratio of the backscattered to direct radiation, the source should be mounted as close as possible to the surface to be measured. The optimum position for the detector is less well defined.

Figure 14-23 shows the relation between counting rate and the distance of the crystal face from the surface. Figure 14-24 shows calibration curves for this instrument. An accuracy of approximately 4 per cent is obtained for thicknesses from 0.10 to 0.75 in. A special feature of this instrument is that the measurement of thickness is confined to a small area.

Beta-ray Backscatter. Whenever beta rays impinge on material a fraction of the electrons may be reflected or backscattered. The intensity of the backscattered radiation increases with increasing thickness

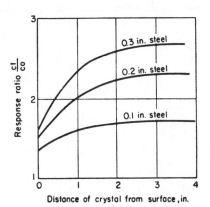

Fig. 14-23. Counting rate vs. distance. (*United Kingdom Atomic Energy Research Establishment photographs.*)

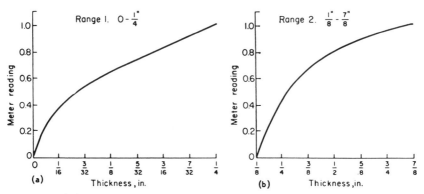

Fig. 14-24. Calibration curve for (a) range 0 to ¼ in. steel, (b) ⅛ to ⅞ in. steel. (*United Kingdom Atomic Energy Research Establishment photographs.*)

of material until the thickness equals approximately one-fifth the range of the electrons. A further increase in thickness does not increase the intensity of the backscattered radiation. The ability of a material to backscatter beta rays is a function of the atomic number Z of the material (Fig. 14-25). The backscattering is directly proportional to Z^n, where n is an empirical constant having a value of 0.7 to 0.8.[19] Figure 14-26 shows variation of counting rate for various thicknesses of material or mass per unit area. As the thickness of the material is increased a value of thickness is reached which produces no further increase in counting rate. This thickness is known as infinite thickness or saturation backscattering and depends on the maximum energy of the beta particle. The saturation or infinite thickness is reached when the most energetic beta particles do not possess sufficient energy to penetrate the thickness of the material and return to the detector. The value of infinite thickness is determined by the material and the beta particle energy. The

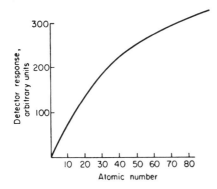

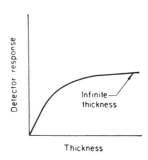

Fig. 14-25. Backscatter of beta rays vs. atomic number material. (*Courtesy Tracerlab, Inc.*)

Fig. 14-26. Backscatter of beta rays vs. thickness. (*Courtesy Tracerlab, Inc.*)

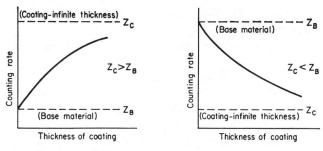

FIG. 14-27. Effect of coating material on count rate.

magnitude of the detector response at infinite thickness is determined by the atomic number of the scatter. If a second layer of material having a different atomic number is placed upon the first material, the counting rate will change. The counting rate will increase if the second material has a higher Z and will decrease if the Z is lower, as shown in Fig. 14-27. It is necessary that the base material be of infinite thickness. The greater the difference in atomic number between the two materials, the

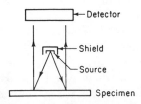

FIG. 14-28. Beta backscatter gage.

greater the difference in the counting rate and, consequently, the sensitivity in the measurement.

The curves in Fig. 14-26 can be represented by the empirical equation

$$I = I_{\bullet}[1 - \exp(-kx)] \qquad (14\text{-}29)$$

where I = detector response at any thickness x, less than saturation or infinite thickness

$I_{\bullet}$ = detector response for infinite thickness

k = backscattering coefficient

x = thickness of material

Beta Backscatter Gage. Figure 14-28 shows a beta backscatter gage. The detector is positioned in such a way that direct radiation from the source cannot enter the detector. Only the radiation reflected or backscattered from the specimen under the source will be detected.

One problem in using beta gages is the sensitivity of the gage to variations in the distance between the source–detector unit and the specimen. Figure 14-29[20] shows the variations in detector response with changes in distance for the gage shown in Fig. 14-28. There is a certain optimum value of distance which makes the instrument insensitive to variations in distance. This optimum value is independent of the nature of the material being measured and almost independent of the radioisotope. It is a function primarily of source detector and scatter geometry.

Some types of materials which can be measured by backscattering methods but which cannot be measured by conventional radiation absorption methods are tin or zinc coating on steel; paint or lacquer on metallic surfaces; rubber and plastic sheets or film on calendering rolls; selenium on aluminum or other backing materials; barium coating on photographic paper; chromium or brass on steel; fillers in paper and plastics; porcelain coatings; metal platings, such as nickel or chromium, superimposed on other metallic surfaces; and plastic coatings on wire. The sensitivity of this gage is extremely high for thin films. Beta gages have been found to be capable of measuring extremely thin coatings of tinplate to an accuracy of a millionth of an inch. Empirically, it has been found that the error of reading that can be achieved in a practical instrument can be expressed as ± 0.3 mg/cm^2 divided by atomic number units or ± 1 per cent of full-scale meter reading, whichever is greater. For example, if the atomic number difference between the two surfaces is 30 units (tin or steel), then measurements can be made to ± 0.01 mg/cm^2.

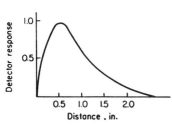

FIG. 14-29. Effect of distance on detector response. (*Courtesy Tracerlab, Inc.*)

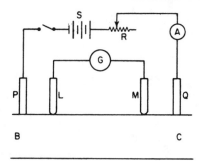

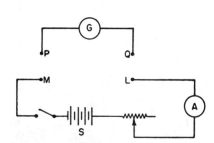

FIG. 14-30. Resistance method for measuring thickness.

FIG. 14-31. Resistance method for measuring thickness.

Ultrasonic Gaging. The ultrasonic resonance technique can be used to measure the thickness of metal. The principle of the resonance technique was described in Chap. 8. Details of this technique and instruments can be found in the chapter on ultrasonics.

Electrical Resistance Gaging. The resistance method can be used for measuring the wall thickness of plates, sheets, and tubes from one side only. In thin sheet material the resistance between two points on the surface is inversely proportional to the thickness of the specimen. For a given voltage the current flow is proportional to the specimen thickness. This relationship does not necessarily hold true for thicker specimens. It is necessary experimentally to determine the relationship between current flow and thickness. The electric current is introduced in the test specimen by means of a pair of current electrodes. The potential drop is measured by means of an independent set of potential electrodes.

Thornton and Thornton[21] and Buchanan, Marsh, and Thurston[22] describe a resistance method for measuring thickness. Figure 14-30 shows the principle of the method. The current enters and leaves the specimen by electrodes P and Q, and the potential drop is measured between electrodes L and M. A current of approximately 10 amp is supplied to the current electrodes. The potential between the electrodes L and M is measured by a suitable instrument capable of accurately measuring potentials in the millivolt range. Thornton states that with currents of 10 amp mild steel up to $1\frac{1}{4}$ in. thick and 3 in. thick cast iron can be measured to an accuracy of about 2.5 to 5 per cent If the instrument is calibrated at one temperature and then used to measure specimens at a different temperature, a correction must be made to compensate for the change in resistivity. One disadvantage of the system is that measurements cannot be made near the edge of the sheet.

Warren[23] has modified the above method so that it is independent of the conductivity of the specimens. Warren's method is shown in Fig. 14-31. Two potential and two current electrodes are located at the four

FIG. 14-32. Magnegage.
(*Courtesy American Instrument Company.*)

corners of a square. The potential difference is measured with a known current flowing. This potential difference is compared with the potential difference measured with the same current in a square of different size. The current value normally employed is 4 amp. With contact square sizes of 4, 2, 1, ½, and ¼ in., respectively, the ratio between the potential differences of the two squares determines the proportionality factor between the specimen thickness and the size of one of the squares. The accuracy of this method has been found to be greater than 3 per cent.

Magnetic Attraction Gages. Magnetic techniques can be used for measuring coating thickness. Such techniques are limited to combinations where either the base material or the coating is ferromagnetic. The general technique is to measure the force necessary to pull a small magnet off the specimen. A small permanent magnet is suspended freely from a horizontal lever which is actuated by a spiral spring. The spring is wound by rotating a graduated dial until the magnet is freed from the specimen. Sideways motion of the magnet is prevented by moving the magnet vertically through a glass guide tube. The dial reading used in conjunction with a calibration curve gives a measure of the thickness. Brenner[24-26] found that for a magnetic coating on a magnetic base material the attractive force is proportional to the thickness of the coating. For nonmagnetic coatings on iron or steel the attractive force decreases with increasing coating thickness. Figure 14-32 shows a commercial form of this instrument known as the Magnegage. The accuracy of the instrument is better than 10 per cent. The instrument can measure nonmagnetic material on iron or steel in the thickness range of 0 to 0.080 in., nickel on iron or steel 0 to 0.002 in., and nickel on monmagnetic material such as copper, brass, or zinc in the range of 0 to 0.001 in. The sensitivity of the instrument is affected by the surface finish, surface curvature, base metal thickness, and the magnetic properties of the base metal and the coating material. For best results the following precautions should be observed in using the instrument. The radius of curvature of cylindrical specimens should not be less than ¾ in. Readings should not be taken close to the edges of sheet specimens. Readings cannot be taken on areas less than ¾ in. without a correction factor.

The Tinsley Thickness Gage consists of a special magnet attached to a spring and contained within a pencil-like tube. To make a measurement the magnet is placed on the specimen and the body of the gage drawn away extending the spring. The spring extension which can be observed on the scale is proportional to the force required to detach the magnet from the surface. The Tinsley gage can be used to measure coatings in the range from 0.0002 to 0.015 in. with an accuracy ±15 per cent.

Figure 14-33 shows a permanent magnet thickness gage developed by the General Electric Co. This gage has a range of 0 to 0.060 in., with an accuracy of ±5 to 10 per cent within 50 per cent of the calibrated value or 0.2 mil, whichever is greater. An internal Alnico magnet provides flux to the feet. The variable air gap in the magnetic circuit, due to the thickness of the film on the backing material, causes flux changes in the circuit. For thick films, more flux will leak between two iron legs (located between the magnet and the feet) than for thin films. This internal leakage flux is measured by a gaussmeter-type movement used in a null balanced manner. The leakage flux is a measure of the film thickness.

Fig. 14-33. General Electric permanent-magnet thickness gage. (*Courtesy General Electric Company.*)

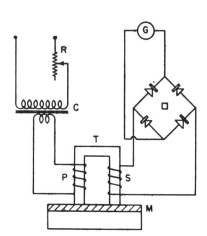

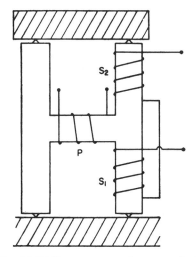

FIG. 14-34. Tait transformer method for measuring tin on tinplate.

FIG. 14-35. Stevenson transformer method for measuring thickness.

Magnetic Induction Gages. Berthold has developed a magnetic method for measuring thickness.[27] In this method a magnetic yoke is placed on the specimen and is excited by a current of about 6 amp from a storage battery. The steel underneath the yoke becomes magnetically saturated. On switching off the current the magnetic field breaks down, thus giving a pulse. This pulse is fed to a ballistic galvanometer. Since the magnetic flux depends on the thickness of the specimen, the thickness is indicated by the deflection of the galvanometer.

An instrument for measuring the coating of tin on tinplate has been developed by Tait.[28] The instrument shown diagrammatically in Fig. 14-34 consists of a small transformer with a U-shaped core T. A primary winding on one leg is connected through a stepdown transformer and a variable resistance to an a-c supply. A secondary winding on the other leg is connected through a rectifier to a galvanometer G. The open magnetic circuit forms an inefficient transformer, and a relatively small reading is obtained on the galvanometer. When the instrument is placed on a mild steel or iron surface, the efficiency of the magnetic circuit is improved and the reading on the galvanometer increased. Separation of the poles from the metal will give readings intermediate between those for the open and closed magnetic circuit. The thickness of tinplate can be gaged with an accuracy of approximately 0.0004 in. The sensitivity of the instrument is greater for thinner coatings. The instrument can be used to measure the thickness of nickel coatings on steel and nickel coatings on brass. The thickness of sheets of non-magnetic material such as tinfoil, aluminum foil, or paper can be measured by placing them between the instrument and a magnetic base.

Using the transformer principle, Stevenson[29] developed the differential type of gage shown in Fig. 14-35. The primary winding is put on the crossbar of the H-shaped transformer. The two secondary windings are connected in series opposition and wound on opposite ends of one limb of the transformer. The specimen being examined forms one magnetic circuit. The other magnetic circuit is formed by a standard of the same composition and thickness as the specimen. If the two specimens are of equal thickness, the resultant of the two induced voltages will be zero. For unequal thickness, the resultant induced voltage will be proportional to the difference in thickness.

Magnetic Reluctance Gage. Cook[30] describes two instruments, utilizing magnetic reluctance for measuring the thickness of electrodeposited nickel in the range of 0.0001 to 0.001 in. on a nonmagnetic base. The voltage induced in the secondary of a transformer is dependent upon the flux linking the primary and secondary windings. The flux, in turn, is dependent upon the reluctance of the magnetic circuit linking the primary to the secondary. The nickel coating forms part of the magnetic circuit. The secondary voltage is dependent upon the reluctance of the nickel. The reluctance is proportional to the path length and inversely proportional to the cross-sectional area. The width and length of the path are constant; so the reluctance is inversely proportional to the plating thickness. The secondary voltage is then a function of the plating thickness that bridges the transformer gap. The secondary voltage is rectified and compared directly with a reference voltage.

X-ray Techniques. X rays are limited in their ability to measure thin materials accurately because of the penetrating nature of the radiation. Friedman and Birks[31] describe a technique for measuring the thickness of thin coatings on crystalline bases. They use an X-ray source and a detector positioned on the same side of the coating. The X rays pass through the coating and are reflected by the base back to the detector. The intensity of the X rays is reduced by the double absorption path through the coating. Figure 14-36 shows the technique. Crystalline materials reflect monochromatic X rays at angles given by the Bragg formula

$$n\lambda = 2d \sin \Theta \qquad (14\text{-}30)$$

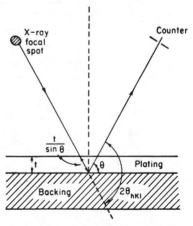

FIG. 14-36. Reflection of X rays from base material. (*Courtesy U.S. Naval Research Laboratory.*)

TABLE 14-4. MEASUREMENT OF COATING BY X RAY

Coating	X-ray reflection measurement, cm	Comparison measurement, cm
Al	0.00254	0.0025*
Zn	0.00097	0.0010*
Al leaf	0.00016	0.00014†
Au leaf	0.00010	0.00009‡

* Micrometer.   † Spherometer.   ‡ Weighing.

where n = order of diffraction

λ = X-ray wavelength

d = crystalline spacing giving rise to a reflection at angle Θ

In Fig. 14-36 a collimated beam of monochromatic X rays strikes a coated specimen at a diffraction angle Θ. The incident radiation is absorbed in penetrating the coating before and after reflection from the backing. If t is the coating thickness and μ_m the mass absorption coefficient (the ratio of the diffracted intensity I from the coated specimen to I_0) the intensity from the uncoated base material is

$$\frac{I}{I_0} = \exp\left[-2\mu_m\rho\left(\frac{t}{\sin\Theta}\right)\right] \tag{14-31}$$

If I/I_0 is measured, the thickness t may be computed from the above equation.

The variation in the intensity ratio I/I_0 with wavelength for silver plating on a copper base is illustrated in Fig. 14-37. Soft radiation reflected at small diffraction angles should be chosen for thin coatings and hard radiation at large angles for thick coatings. To test the accuracy of the technique, Friedman and Birks prepared a number of metallic foils and determined their thickness by micrometering, by weighing a given area, with a spherometer, and by X-ray reflection. Table 14-4 lists the results obtained by the various methods. This technique is useful for measuring the thickness of electroplated metal films, evaporated electrode coatings, and thin pigment layers in the range from 10^{-5} to 10^{-2} cm. This technique is also applicable to metallic or nonmetallic base material and to multilayer coatings of different materials.

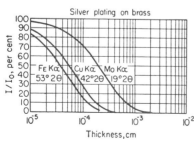

FIG. 14-37. Variation of ratio I/I_0 for silver plated brass. (*Courtesy U.S. Naval Research Laboratory.*)

Fluorescence. When material is irradiated with photons of sufficient energy, the energy is absorbed and partially reradiated. This phenomenon is known as "fluorescence." The fluorescent radiation is monochromatic, having a wavelength which is characteristic of the fluorescing material. Fluorescent X-ray spectroscopy is used in the measurement of coating and plating thickness. When a specimen such as silver-plated copper is excited, the fluorescent copper spectrum will be partly absorbed by the silver plating. The reduction in intensity of the $CuK\alpha$ line is a direct measure of the silver thickness. It is easy to set up a calibration curve from standards, or calibration curves may be calculated from known absorption coefficients.

A simpler method of determining plating thickness is feasible when the plating and the base material differ appreciably in the energy required to excite their X-ray spectra. For instance, in cadmium plating on iron, iron is excited at about 7 kV, whereas cadmium requires about 26 kV. Thus when cadmium plated steel is irradiated, the cadmium will not fluoresce appreciably but will absorb the fluorescent iron spectrum strongly. For example, the total scattering from an unplated steel specimen was found to be 17,400 counts per second, whereas a cadmium-plated specimen gave only 600 counts per second under the same conditions. This difference can be easily calibrated in terms of cadmium thickness.

If only fluorescent radiation from the coating is excited, then the intensity of the fluorescent radiation is proportional to the number of coating atoms that are excited. With an incident X-ray beam of constant intensity, the number of coating atoms excited will be proportional to the total number of coating atoms present. Thus, if a given surface area is observed, the intensity of the fluorescent radiation is proportional to the thickness of the coating. It has been found experimentally that more than 90 per cent of the total radiation received at the detector is fluorescent radiation due to the coating. Thus, with reasonable accuracy, it may be said that the radiation received by the detector is proportional to the thickness of the coating on the base material. In using this technique the following conditions must be fulfilled: the thickness of the coating is small; the fluorescence of the base material is small; and a relatively large area of the specimen is irradiated, but only a small portion of this area is observed. Figure 14-38 shows an experimental setup to measure the fluorescent radiation from a coating. A Geiger counter and count rate meter can be used to measure the quantity of the radiation. It should be reemphasized that it is necessary to adjust the equipment so that the radiation being measured is only the fluorescent radiation from the coating material. Figure 14-39 shows a correlation curve for nickel-plated specimens. For two plating layers the procedure is slightly more involved but practical. Under certain conditions, three plating layers may be considered and their thicknesses determined.

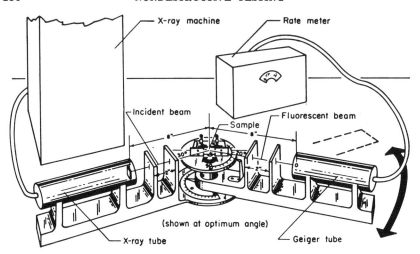

FIG. 14-38. Experimental arrangement for making fluorescent radiation measurements. (*Courtesy Battelle Memorial Institute.*)

The North American Phillips Company has designed and developed a fluorescent X-ray gage for measuring tin coating thickness on an iron base. Fortunately, it takes a relatively strong primary X-ray beam to cause tin fluorescence, compared to that required for iron. More specifically, iron fluorescence requires 1.74 A or less, whereas tin requires 0.424 A. By selecting X-ray energies between these figures the desired fluorescence could be obtained. The 1.54 A wavelength from a copper target Machlett OEG-50A X-ray tube was used as the primary beam because the radiation lies between 1.74 and 0.424 A; the use of a copper target tube with a nickel filter provides a simple source for this wavelength. Tin absorption of the large wavelength is appreciable and therefore gives large changes in detected radiation with changes in tin thickness, and by employing a monochromatic radiation the calibration curve of radiation versus tin thickness can be readily calculated. The fluorescent iron radiation had a wavelength of 1.93 A. The X-ray tube and detector were kept at right angles, the angle Θ being 70°. The mass absorption coefficients in tin for the 1.54 A primary beam and the 1.93 A fluorescent radiation were 247 cm/g² and 470 cm/g², respectively.

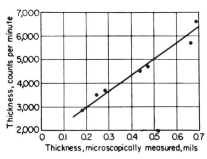

FIG. 14-39. Correlation curve for nickel-plated specimens. (*Courtesy Battelle Memorial Institute.*)

Figure 14-40 shows the principle of the method. The primary beam from the X-ray source penetrates the coating at an angle Θ. The fluorescent radiation from the base material is emitted in all directions. Some of the fluorescent radiation emerges through the coating at an angle of ϕ and arrives at the detector. If I_0 is the in-

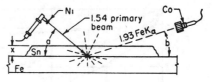

FIG. 14-40. Principle of the fluorescent X-ray gage.

tensity of the X-ray beam and I' the intensity of the beam impinging on the base material, then

$$\frac{I'}{I_0} = \exp\left(-\frac{\mu_1 x}{\sin \Theta}\right) \tag{14-32}$$

where μ_1 is the linear absorption coefficient and x the thickness of the coating material. The intensity I of the fluorescent X-ray beam arriving at the detector is

$$\frac{I}{I'} = \exp\left(-\frac{\mu_2 x}{\sin \phi}\right) \tag{14-33}$$

where μ_2 is the linear absorption coefficient for the fluorescent radiation. The intensity of the fluorescent radiation depends on the intensity of the radiation impinging on the base material. Substituting for I' in Eq. 14-33

$$\frac{I}{I_0} = \exp\left[-\left(\frac{\mu_2}{\sin \phi} + \frac{\mu_1}{\sin \Theta}\right) x\right] \tag{14-34}$$

By selecting ϕ small the effective thickness can be increased, resulting in a greater absorption of the fluorescent radiations.

Autoradiography. In the case of nuclear reactor fuel elements, another technique is available for measuring coating thickness. Two of the radiations emitted by the radioactive uranium in the fuel element have been used to measure the cladding thickness of fuel elements: the 2.3-Mev beta particle emitted by a daughter of U^{238} and the 0.18-Mev gamma ray emitted by U^{235}. The intensity of the radiation transmitted through the cladding can be measured by an X-ray film placed in contact with the fuel element. This requires several hours' exposure time as well as development and measurement of film density. A scintillation counter can also be used to measure the intensity of the transmitted radiation.

Figures 14-41 and 14-16 show absorption curves obtained by Bradley, McGonnagle, and Gonzales[32] for aluminum, zirconium, and stainless steel. These curves were obtained by placing sheets of varying thickness over a flat uranium source; the relative number of counts per unit time is plotted against the sheet thickness. For the measurement of cladding in the range of 0.020 in. in thickness, the beta sensitive scintillator gives a

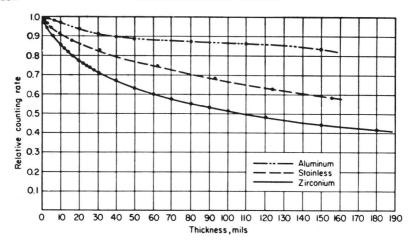

FIG. 14-41. Absorption curve of gamma radiation using sodium iodide scintillator. (*Courtesy Argonne National Laboratory.*)

change of 10 per cent in the counting rates per mil change in thickness. In the range of 0.020 to 0.030 in. the counting rate changes only approximately 5 per cent per 0.001 in. Also, the use of beta radiation from the uranium to measure the clad thickness makes the shielding problem an easy one. A ⅛-in. sheet of lead provides sufficient shielding to stop the 2.3-Mev beta radiation.

A block diagram of the scintillator counter and associated circuits is shown in Fig. 14-42. A zero set adjustment was provided on the count rate meter so that small differences in rates could be more easily detected. This modification made the count rate meter a differential type of thickness gage. Each nuclear disintegration is a completely random and independent process and obeys the laws of statistics. These laws predict that the number of disintegrations actually counted in a given time will show deviations from this average. Consequently, when any type of radiation measuring instrument is used, the readings will show fluctuations. Figure 14-43 shows a typical record from a reactor fuel element.

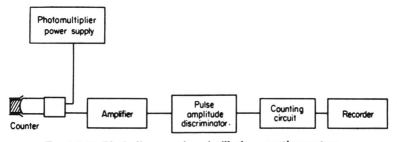

FIG. 14-42. Block diagram of a scintillation counting system.

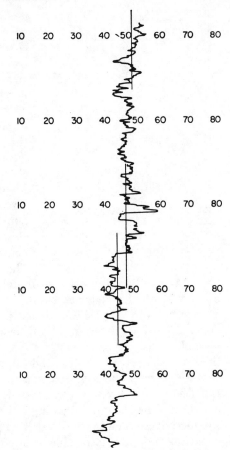

FIG. 14-43. Radiation record from reactor fuel element. (*Courtesy Argonne National Laboratory.*)

Eddy Current Techniques. Eddy current techniques provide a way of measuring the thickness of platings or coatings of one metal on another. Some of these techniques have already been discussed in Chap. 12. Brenner and Garcia-Rivera[33] describe an eddy current thickness gage for measuring coatings on metals. The conductivities of the coating and base metal must differ sufficiently. Measurements can be made on combinations of materials regardless of whether they are magnetic or nonmagnetic. This instrument measures the change in current through the test coil as different types or thicknesses of material are brought into contact with or near the probe. In calibrating such an instrument the calibrating curves have two limits. One limit is the reading of the meter with the probe on the bare base metal. The other limit is the reading with the probe on a thick layer of metal composing the coating.

Two types of measuring circuits have been used in this instrument. The bridge type of circuit is shown in Fig. 14-44. The power supply S

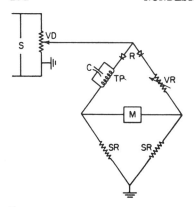

FIG. 14-44. Bridge-type circuit.
(*Reprinted from Plating Magazine, November, 1953, American Electroplaters Society, Inc.*)

is a commercial type of oscillator with an output of 6 watts. The amount of current through the circuit is controlled by means of the voltage divider. The two germanium diodes permit current flow in only one direction; so only half of the current cycle is used. The variable resistance is used to balance the circuit. The resistances which form the two arms of the bridge have about the same resistance as the probe. Figure 14-45 shows the branch type of circuit that is also used. The power supply is a commercial oscillator, and the current through the circuit is controlled by the voltage divider. The diodes are in opposite phase, and the current alternately passes through each leg of the circuit. The test probe with the capacitor across it is balanced by the variable resistance. During half of the cycle the current flows through the test probe and the meter. During the other half of the cycle the current flows through the variable resistance and meter. The meter registers the difference of these two currents. With either circuit the meter is adjusted to read zero when the probe is placed on a standard specimen. When the probe is placed on the specimen to be tested, the resulting change of inductance is indicated by the meter reading. The meter reading depends upon the conductivity of the layer through which the induced current flows. A capacitor of appropriate value must be placed across the probe; otherwise the sensitivity is poor. It is essential to have a probe of the proper inductance.

Renken[34] has developed a method of measuring metal plating using sinusoidal waves. The probe is maintained some distance away from the plated material. In this instrument variations in probe-to-metal spacing are automatically compensated for by means of an electronic circuit.

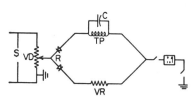

FIG. 14-45. Branch circuit. (*Reprinted from Plating Magazine, November, 1953, American Electroplaters Society, Inc.*)

The resistance and inductance variations of the probe are changed into a sinusoidal voltage of varying amplitude and phase angle. A change in probe-to-metal spacing causes a change in both phase and amplitude. The electronic compensating circuit makes the change in phase angle independent of the probe-to-metal spacing. Figure 14-46 shows a block diagram of the system. Figures 12-37 and 12-38 show a plot of the phase angle of the output from the instrument versus plating thickness. The separate curves of Fig. 12-37 were obtained at three different probe-to-metal spacings of 0.005, 0.013, and 0.021 in., respectively. This illustrates the use of the compensator in keeping the measurements nearly independent of probe-to-metal spacing. Assuming a 1° relative accuracy for the phase meter, the accuracy of measurement in Fig. 12-37 is about 0.00025 in. Figure 12-38 shows the measurement of nickel plating on uranium. Nickel is slightly ferromagnetic,

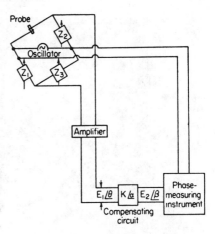

FIG. 14-46. Block diagram of eddy current thickness-measuring equipment. (*Courtesy Argonne National Laboratory.*)

which makes the measurement of such platings easier. Again three separate curves are given for the 0.005, 0.013, and 0.021 in. probe-to-metal spacings. For this case a relative accuracy of 1° in phase measurement should give a sensitivity of at least 0.0001 in.

Renken[35] has developed a second system to compensate for variations in probe-to-specimen spacing, Fig. 14-47. Two separate channels are used. The frequency of the one channel is chosen so that this channel is very insensitive to cladding thickness variations but still sensitive to probe-to-metal variations. This signal is applied to the other channel where it electrically compensates for the effect of probe-to-metal spacing

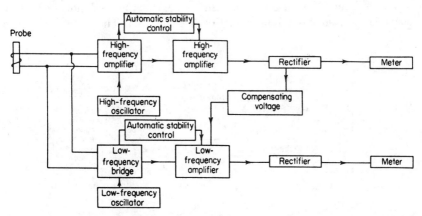

FIG. 14-47. Two channel eddy current thickness gage. (*Courtesy Argonne National Laboratory.*)

FIG. 14-48. Boonton metal film gage. (*Courtesy Boonton Radio Corporation.*)

variations on this second channel while not impairing the cladding thickness signal. This system has been applied successfully to a number of different types of plating and cladding. It will measure a 0.002-in. change in 0.020 in. of Zircaloy cladding on uranium alloy fuel.

Figure 14-48 shows the Boonton Metal Film Gage. The instrument consists of an oscillator, driver, probe, and reference phase circuits. The probe circuit is made resonant at the oscillator frequency, with the probe placed on the base material. The probe signal and the reference signal after amplification are impressed on the grids of a gated-beam phase detector tube. The plate current of the phase detector varies with the phase difference between the probe and reference signals. A more thorough understanding of the operation and design of the instrument can be obtained by referring to a fundamental block diagram, Fig. 14-49.

The Filmeter[36] is used to measure the thickness of nonconducting coatings deposited on nonmagnetic base metals. The inductance of a

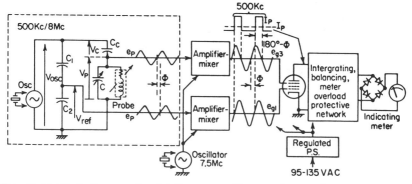

FIG. 14-49. Block diagram of metal film gage. (*Courtesy Boonton Radio Corporation.*)

coil is affected by the coated base metal. The current flowing in the coil induces eddy currents in the base metal. The magnitude of the induced eddy currents varies with the distance between the coil and the base metal. These eddy currents produce a change in the inductance of the coil. This in turn causes a change in frequency of the oscillator to which the coil is connected. Thickness from zero to 0.005 in. can be measured with an accuracy of 3 per cent of full scale and thicknesses from zero to 0.001 in. with an accuracy of 5 per cent of full scale.

Colten[37] describes a gage for measuring the thickness of nonferrous metals, based on the principle of electromagnetic absorption. Two coils are used, an exciting coil and a pick-up coil. The sheet to be measured passes between the two coils. The electronic circuit consists essentially of a stable oscillator, amplifier, and vacuum tube voltmeter. For gaging copper 0.003 to 0.006 in. thick, the exciting coil was operated at a frequency of 6,000 cps. For gaging aluminum $\frac{1}{8}$ to $\frac{1}{2}$ in. thick, a frequency of 18 cps was used. Copper strip between 0.0003 and 0.006 in. thick has been gaged at a rate of about 300 fpm with an accuracy of ± 2 per cent.

Förster[38] has also developed an eddy current gage in which the flat conductor is placed between an exciting coil and a pick-up coil. The presence of the metal between the coils alters the potential of the pick-up coil. The change is determined by the frequency of the eddy currents, the kind of material, and the thickness of the material. The field of the exciting coil induces an alternating current in the pick-up coil. When no specimen is interposed between the two coils, the induced current at the terminals of the pick-up coil is designated E_0. When a specimen is introduced between the two coils, eddy currents are induced in the metal. The field of the induced eddy currents is superimposed on that of the exciting coil and attenuates it. The field in the pick-up coil will be altered by the presence of the specimen. The potential at the terminals of the pick-up coil can now be designated as E_M. The ratio E_M/E_0 is a function of conductance, permeability, thickness of the metal, frequency, and the distance between exciting and pick-up coils.

The use of pulsed eddy currents for measuring coating thickness has been discussed by Waidelich.[39] An electromagnetic field is applied at the surface of the clad metal by means of a probe coil. Echoes are produced whenever the wave encounters a sudden discontinuity such as an interface, Fig. 14-50. The same probe coil is

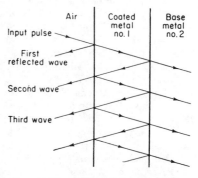

FIG. 14-50. "Echoes" from interface.

used for sending the electromagnetic radiation and for receiving the echoes. The echo received from the front surface of the sample is balanced out by means of a bridge circuit. The echo from the coating–base metal interface contains information concerning the thickness of the coating. This echo has a positive peak followed by a small negative peak. The amplitude or crossing point between the positive and negative peak can be used to determine the thickness of clad. The magnitude of the reflection between the clad and base metal is determined by the permeability and conductivity of the two metals.

Optical Techniques. The use of interferometry for measuring the thickness of films was discussed in Chap. 4. This subject is more fully discussed in the book by Tolansky.[40] Interferometry has added appreciably to our knowledge of thin films by providing an accurate method for measuring film thickness down to about 25 A.

SPECIFIC REFERENCES

1. Lundahl, N. N.: X-ray Thickness Gauge for Cold-Rolled Steel, *AIEE Trans.*, vol. 67, pp. 83–90, 1948.
2. Clapp, C. W., and R. V. Pohl: An X-ray Thickness Gauge for Hot-strip Rolling Mills, *Elect. Eng.*, vol. 67, no. 5, pp. 441–444, May, 1948.
3. Ettinger, George M.: A Differential X-ray Absorption Gauge of High Sensitivity, *Proc. Nat. Electronics Conf.*, vol. 8, pp. 113–120, 1952.
4. Howell, John F.: Automatic Metal Gauging Using X-rays, *Proc. Natl. Electronics Conf.*, vol. 8, pp. 121–126, 1952.
5. Nickerson, Richard A.: The Fundamentals of Differential Radiation Measurements, *J. Soc. Nondestructive Testing*, vol. 16, no. 2, March–April, 1958.
6. Berman, Arthur I., and John N. Harris: Precision Measurement of Uniformity of Materials by Gamma Ray Transmission, *Rev. Sci. Instr.*, vol. 25, no. 1, pp. 21–29, January, 1954.
7. Palevsky, H., R. Swank, and K. Grenchek: Design of Dynamic Condenser Electrometers, *Rev. Sci. Instr.*, vol. 18, no. 5, pp. 298–314, May, 1947.
8. Harris, John N., and Lawrence R. Megiel: Techniques Used in Measuring Uniformity of Materials with Gamma Radiation, *J. Soc. Nondestructive Testing*, vol. 11, pp. 9–13, July, 1953.
9. Syke, G.: A Gamma Ray Thickness Gauge for Hot Steel Strips and Tubes, *J. Brit. Inst. Radio Engrs.*, vol. 14, no. 9, pp. 419–425, September, 1954.
10. Skye, G.: A Gamma Ray Thickness Gauge for Hot Steel Strips and Tubes, *J. Brit. Inst. Radio Engrs.*, vol. 14, no. 9, pp. 424–425, September, 1954.
11. Shaw, E. N.: The Alpha Gauge, *J. Brit. Inst. Radio Engrs.*, vol. 14, no. 9, pp. 414–418, September, 1954.
12. Clapp, C. W., and S. Bernstein: Noncontacting Beta-ray Thickness Gage, *Gen. Elec. Rev.*, vol. 53, no. 3, pp. 31–34, March, 1950.
13. Goodman, Clark: "Introduction to Pile Theory," Addison-Wesley Publishing Company, Reading, Mass., 1952.
14. Schiff, L. I., and R. D. Evans: Statistical Analysis of the Counting Rate Meter, *Rev. Sci. Instr.*, vol. 7, pp. 456–462, December, 1936.
15. Parry, R. Y.: A Combined Beta and Dielectric Gage, *J. Brit. Inst. Radio Engrs.*, vol. 14, no. 9, pp. 427–432, September, 1954.

16. Hire, Gerald J., and Richard C. McCall: Gamma Ray Backscattering, *Nucleonics*, vol. 12, no. 4, pp. 27–30, April, 1954.

17. Hare, D. G. C.: Method and Apparatus for Measuring Thickness, U.S. Patent 2,227,756, Mar. 31, 1942.

18. Putman, J. L., and S. Jefferson: Tube Wall Thickness Gage with Selection of Back Scattered Gamma Radiations, A.E.R.E. Report 1/R 1369, Dept. of Atomic Energy, Harwell Berks, England, 1954.

19. Zumwalt, L. R.: The Best Performance from Beta Gages, *Nucleonics*, vol. 12, no. 1, pp. 55–58, January, 1954.

20. Clarke, Eric, J. R. Carlin, and W. E. Barbour: Measuring the Thickness of Thin Coatings with Radiation Backscattering, *Elec. Eng.*, vol. 70, no. 1, pp. 35–37, January, 1951.

21. Thornton, B. M., and W. M. Thornton: The Measurement of the Thickness of Metal Walls from One Surface Only by an Electrical Method, *Proc. Inst. Mech. Engrs.*, vol. 140, p. 349, 1938.

22. Buchanan, J. G., F. W. Marsh, and R. C. A. Thurston: The Measurement of Wall Thickness of Metal from One Side Only, by the Direct Current Conduction Method, *J. Soc. Nondestructive Testing*, vol. 16, no. 1, pp. 31–35, January–February, 1958.

23. Warren, A. G.: Measurement of the Thickness of Metal Plates from One Side, *J. Inst. Elect. Engrs.*, vol. 84, pp. 91–95, 1939.

24. Brenner, A.: Magnetic Method for Measuring the Thickness of Nickel Coatings on Nonmagnetic Base Metals, *Bur. Standards, J. Research*, vol. 18, pp. 565–583, May, 1937.

25. Brenner, A.: Magnetic Method of Measuring the Thickness of Nonmagnetic Coatings on Iron and Steel, *Bur. Standards, J. Research*, vol. 20, pp. 357–368, March, 1938.

26. Brenner, A., and E. Kellogg: A New Method for Magnetic Measurement of the Thickness of Composite Copper and Nickel Coatings on Steel, *Bur. Standards, J. Research*, vol. 40, pp. 295–299, April, 1948.

27. Berthold, K.: A New Wall Thickness Measuring Instrument for Ferromagnetic Materials, *Stahl u. Eisen*, vol. 70, pp. 233–234, 1950.

28. Tait, W. H.: An Instrument for Measuring the Thickness of Coatings on Metals, *J. Sci. Instr.*, vol. 14, no. 9, pp. 341–343, 1937.

29. Stevenson, A. B.: A Magnetic Comparator Metal Thickness Tester, *J. Sci. Instr.*, vol. 15, pp. 156–158, 1938.

30. Cook, L. H.: Nickel Thickness Gage, *ASTM, Spec. Tech. Pub.* 223 1958.

31. Friedman, H., and L. S. Birks: Thickness Measurements of Thin Coatings by X-ray Absorption, *Rev. Sci. Instr.*, vol. 17, no. 3, pp. 99–101, March, 1946.

32. Bradley, G. E., W. J. McGonnagle, and P. R. Gonzales: Measurement of the Cladding Thickness on Uranium Fuel by Autoradiation, *ASTM, Spec. Tech. Pub.* 223, 1958.

33. Brenner, Abner, and Jean Garcia-Rivera: An Electronic Thickness Gage, *Plating*, vol. 40, no. 11, pp. 1238–1244, November, 1953.

34. Renken, C. J., and D. L. Waidelich: Minimizing the Effect of Probe-to-Metal Spacing in Eddy Current Testing, *ASTM, Spec. Tech. Pub.* 223, 1958.

35. Renken, C. J., R. G. Myers, and W. J. McGonnagle: Status Report in Eddy Current Theory and Application, *Argonne National Laboratory, Report* 5861, pp. 37–48, November, 1958.

36. Dunn, E. J.: Film Thickness Measurements, *ASTM Bull.*, no. 172, pp. 35–39, February, 1951.

37. Colten, Robert B.: Noncontacting Gages for Nonferrous Metals, *Electronics*, vol. 29, no. 3, pp. 171–173, March, 1956.

38. Förster, F.: Quick and Precise Routine Determination of the Thickness of Steel Sheet from One Side, Report 26, Institute Doctor Förster, Reuthngen, Germany, 1953. Brutcher Translation 3387.

39. Waidelich, D. L.: Reduction of Probe-Spacing Effect in Pulsed Eddy Current Testing, *ASTM*, *Spec. Tech. Pub.* 223, 1958.

40. Tolansky, S.: "Multiple-Beam Interferometry of Surfaces and Films," Oxford University Press, New York, 1948.

CHAPTER 15

CONCLUSION

All nondestructive testing methods and techniques have their limitations whatever they may be. No nondestructive testing program can be designed effectively without the engineer fully comprehending the advantages, disadvantages, and limitations of each method or technique. This is the basis for determining the choice of methods and techniques. Recognizing those limitations a testing program should be designed so that the selection of the methods and techniques is compatible with the demand upon the product. There must be a need for the test, and the testing cost must be justified in reduced costs, improved quality, or increased production of the particular product. A well-designed, logically developed system will usually meet these three important criteria.

Those using nondestructive testing must carefully select the proper methods and techniques to be used for the tests and place them at the proper point in the production effort. Bearing in mind that most tests are for the purpose of seeking known harmful defects, the method must be capable of determining their presence. Harmful defects can be traced to their point of origin or related to cause and effect. The nondestructive testing process should be placed at this point whenever practical, either to eliminate the defects immediately after occurrence or to control the process to prevent their occurrence. This means that some causes may be traced to the raw materials, some to processing, and others to the manufacturing sequence, but in any event the harmful defect must be eliminated at the earliest practical point. It is obvious that further expenditure of time and effort on such defective materials is not warranted.

Success has been achieved by nondestructive testing, and new methods and techniques are being used through judicious and practical efforts by imaginative, yet realistic, engineers. It is clearly evident that the development of nondestructive testing has only begun. Its growth will be supported by management where results show their value in meeting competitive consumer demands for increased production, reduced costs, and improved quality.

Because of the very nature of most nondestructive testing techniques, they involve the interpretation of defect indications. Herein lies a

441

tremendous work area for those in this field. Standards must be established by which to measure and classify test findings into harmful or harmless categories. The lack of standards and specifications has caused endless and even needless testing, discussion, and argument. Furthermore, where standards are available, more consideration must be given to the selection of the proper levels to be used, again, consistent with the needs of the product. This is especially true in a relatively new and untried field such as atomic energy. For example, great interest is expressed in the subject of nondestructive testing of pressure piping for nuclear piping installations. However, many of the specifications presented to manufacturers of pressure piping and its component parts are unrealistic. Part of the exorbitant cost of atomic energy comes from materials and processes specifications that attempt to provide insurance against any and all contingencies. Considerable experience with pressure piping and component parts has shown engineers that many precautions as to nondestructive testing written into specifications were unnecessary and failed in their intent. Quite frequently the desired effect obtained by one was canceled by another. There is no doubt that many unusual hazards exist from radioactive containment and the thought that these hazards require extra precaution is a valid one. Therefore, it is necessary to make a realistic engineering evaluation of the real worth of these special requirements. The rigid enforcement of existing specifications with proved standards of acceptance or rejection is a starting point; but, at present, specifications for nuclear power plants often contain significant differences from each other and from those written for the same materials used in nonnuclear service. All specifications or requirements specified in addition to those existing in the codes must be carefully evaluated before they are included as a mandatory requirement in any specification.

Generalizations about the selection and application of particular nondestructive testing procedures are not of much value but are suggestive only of the test methods that may be considered. Each case is a particular problem, and usually there will be particular features that make it unique and different from previous applications. The advantages and disadvantages as well as the possibilities of each type of test must be known before the proper test procedure can be used. As an example, a simple leak test uses halogenated gases with a mass spectrometer. This is an excellent device, but it must never be used to check a system containing stainless steel piping because the chloride residue from the test is sufficient to cause accelerated corrosion under certain conditions.

Every physical method of measuring "something" can be considered as a potential nondestructive test, since the "result" or answer from such a measurement is connected by some law of physics to a property which one may wish to measure. The development of new and better methods and techniques requires: (1) improvements in making the measurements, (2) new ways of making the measurements, and (3) correlations of the relationships between the physical properties measured and the characteristics which may be of interest in materials.

To meet the needs of the world's advancing technology, improvements must be made in existing methods, new methods developed, and adequate instrumentation provided.   In addition, the inspection processes must be speeded up and made less operator dependent.   This requires that the inspection process be automated.

A number of factors must be considered when searching for or selecting the best test method and technique for a testing or inspection problem.

1. Why nondestructive tests are being made
2. Materials or material to be tested
   a. Magnetic or nonmagnetic
   b. "Light or heavy" material
   c. Coated material
   d. Composition
   e. Service history or conditions
3. Methods of fabrication
   a. Cast
   b. Wrought
   c. Powder metallurgy
   d. Welded
   e. Coating process
   f. Metallurgical treatment
   g. Physical treatment
   h. Chemical treatment
4. Geometry
   a. Shape
   b. Thickness
   c. Surface condition
   d. Dimensions
5. Defects possible or expected
   a. Surface or subsurface
   b. Kinds
   c. Location
   d. Size
   e. Which are considered harmful
   f. Variations in chemical, physical, and metallurgical properties
6. How and when and where defects occur
   a. During processing of raw material
   b. During fabrication
   c. During service

d. Where in the forming or fabrication the test is to be applied

e. Whether components as well as finished product can be tested

f. Select test specimens that are truly representative of production

7. Sensitivity and resolution desired or required

a. Whether "synthetic defects" can be made

b. Whether standards are realistic

c. Destructive tests

8. What information concerning material and processing the engineers, metallurgists, shop personnel, and others can provide

There is no universal nondestructive test applicable to all and every kind of situation. Only with a thorough understanding of the basic principles of each type of nondestructive test can one specify the correct test. Consideration and understanding of the possibilities and limitations of the methods are to be appreciated if one is to choose the proper relationship between the many variable factors. One must get away from the idea that the "instrument" is the thing. The idea is too prevelant that "all one has to do is to get an instrument and one's troubles are over." Those unfamiliar with the basic nature of nondestructive tests expect them to work miracles and to provide magic solutions to production problems. This is far from the truth. Nondestructive tests usually reveal only the specific kinds of defects they were designed to reveal.

The purpose of this book has been to discuss:

1. The basic principles of the various methods of nondestructive testing

2. Correct application of the various methods

3. Techniques which have been used

4. The kind of results to be expected

Provided with this type of information it is hoped that the test engineer:

1. Knows what factors should be considered in applying a particular test

2. Will be able to select the best test for a particular problem

3. Adjusts the variables to give the best results

4. Knows how to select the best equipment for a particular problem

5. Applies methods and techniques skillfully and intelligently

The following points should be kept in mind by the reader:

1. In general, nondestructive testing results are indirect—nearly all nondestructive tests measure material properties differing from those which control serviceability.

2. Each testing problem is unique or has some unique features.

3. Nondestructive tests are qualitative and require correlation with destructive tests.

4. Experienced operators are needed to interpret results correctly.

5. A great deal of skill, experience, and interpretation are needed to select the best method and to establish the best technique for a given testing problem.

6. Only by experience and experimentation does one gain confidence in methods and techniques.

7. Nondestructive tests are not a substitute for ignorance, and they will not work miracles.

8. There is no universal nondestructive test, and each new problem requires new methods or techniques.

9. Each test has its advantages, disadvantages, and limitations.

10. In addition to the basic knowledge concerning testing, related information in other fields is needed.

11. Although most of the tests are simple in principle, the choice of test method, the technique used, and the interpretation of the test results require skill and experience.

INDEX

447